Structural Engineering Documents
20

Sustainability in Structural Concrete Design

Jorge de Brito
Rui Vasco Silva
(Editors)

International Association for Bridge and Structural Engineering (IABSE)

ISBN: 978-3-85748-201-4 (print)
978-3-85748-202-1 (ePub)
978-3-85748-203-8 (eBook)

DOI: https://doi.org/10.2749/sed020

Publisher

IABSE
Jungholzstrasse 28
8050 Zürich
Switzerland

Phone: Int. +41-43-443 9765
E-mail: secretariat@iabse.org
Web: www.iabse.org

This book was produced in cooperation with Structurae, Dresdener Str. 110, Berlin, Germany (https://structurae.net).

Copyediting: Hilka Rogers
Layout & typesetting: Florian Hawemann

Preface

This Structural Engineering Document has the purpose of providing a clear insight into the most recent developments in academia and industry concerning the practices leading to the sustainability of structural concrete. It presents a wide range of information from low-carbon and recycled aggregate-containing structures to up-to-date developments on alkali-activated concrete, all intending to enlighten the reader on how to improve the sustainability indicators of these systems by using technologies with a high enough readiness level and thus more likely to be used in practice.

The three opening chapters offer a broad perception of what can be achieved by implementing a sustainable design for structural concrete from different perspectives. In the first chapter, Vanderley, Quattrone, Abrão, and Pillegi provide notions on a multiscale approach for low-carbon structures. This is followed by the chapter of Silvestre, which offers considerable insight into the life cycle assessment of sustainable concrete containing different waste materials. Kurda, de Brito, Ahmed, and Hafez deliver information on the use of various types of waste materials in concrete.

The large scale of waste coming from construction and demolition activities is absolutely undeniable. It corresponds to the largest percentage of waste coming from all economic sectors and, within the scope of circularity promotion and decreased carbon emissions-related goals of the United Nations 2030 Sustainable Development Goals, it should be dealt with accordingly. The tremendous worldwide research effort put forward towards the reincorporation of construction and demolition waste has demonstrated the high technical and economic feasibility of using recycled aggregates to produce structural concrete. For these reasons, the current Structural Engineering Document focuses extensively on this topic in chapters four to eleven. González-Fonteboa, Seara-Paz, and Martínez-Abella provide a crucial review of the short- and long-term performance of structural recycled aggregate concrete. Tošić, Kurama, and Torrenti then present interesting notions on the effect of recycled aggregates on the long-term deformation of concrete. Furthermore, Etxeberria goes into detail on several durability-related aspects affected by the use of recycled aggregates from construction and demolition waste. Sainz-Aja and Thomas deal with the understudied, albeit fundamental, fatigue phenomenon occurring on recycled aggregate concrete subjected to cyclic loading. After that, Silva and de Brito offer empirical practical rules on how to design recycled aggregate concrete structures, whereas Pacheco and de Brito tackle the matter of design based on reliability analysis. Poon, Shen, Jiang, and Xuan discuss how total recycling of concrete can be achieved by resorting to accelerated carbonation. Xiao, Cheng, Pan, and Fang describe an interesting new case study on large-scale structural applications of recycled aggregate concrete at the Shandong Shanghe Industrial Base.

Finally, given the considerable advancement of other low-impact technologies, the three last chapters focus on the latest findings on: the use of alternative aggregates for concrete production, namely those from electric arc furnace slag; alkali-activated concrete, which normally uses reactive industrial byproducts; and the potentially low impacts associated with concrete made with fibre-reinforced elements. Faleschini, Zanini, and Ortega-Lopez highlight the use of electric arc furnace slag in concrete production. Puertas and Mejía de Gutiérrez then provide recent in-depth insights on the development of alkali-activated materials. Firmo, Rosa, Correia, D'Antino, and Matos subsequently show how the corrosion of reinforcement can become an outdated aspect by using glass-fibre-reinforced elements in structural concrete.

The total amount of time and effort by all parties in putting this document together is tremendous. The Editors would like to thank all authors for their contributions and for allowing this project to come to fruition. Finally, the Editors also deeply acknowledge the remarkable work of the external reviewers, Professor Mark Richardson and Professor Yury Villagrán-Zaccardi, as well as the IABSE's Chief Reviewer from the Editorial Board, Ann Schumacher.

Jorge de Brito and Rui Vasco Silva
September 2023

Table of Contents

Chapter 1
Notes for a Multiscale Approach for Low Carbon Concrete Structures 1

1.1 Introduction 1

    1.1.1 The Implication of Net Zero CO_2 Target of the Cement Industry
           to Concrete Producers 2

    1.1.2 Carbon Sources in Concrete and CO_2 Footprint Measuring 3

    1.1.3 Objective 4

1.2 Fundamentals of Designing Low-Carbon Concretes 5

    1.2.1 Low Carbon Concrete and Clinker Replacement 5

    1.2.2 Defining Low Carbon Concrete 5

1.3 Engineering the Cement Paste for a Low Carbon Concrete 8

    1.3.1 Procurement and Characterization of Raw Materials 9

    1.3.2 Dispersant Selection and Dosage 9

1.4 Engineering Particle Size Distribution 12

    1.4.1 Modelling Viscosity 12

1.5 Measuring Mixing Water Demand for Desirable Rheological Behaviour 13

1.6 Defining the Optimum Particle Size Distribution 14

1.7 Modelling Compressive Strength of Paste Mixtures 15

1.8 Selecting a Paste with Minimum Carbon Footprint for Each Application 15

1.9 Developing a Low Carbon Concrete Formulation 16

1.10 Results From Practical Applications 20

1.11 Final Remarks 22

1.12 Acknowledgements 24

1.13 References 25

Chapter 2
**Life Cycle Assessment in Support of the Sustainability of Current and
Alternative Types of Structural Concrete** 29

2.1 Introduction 29

2.2 Environmental Life Cycle Assessment of Concrete 30

    2.2.1 Life Cycle Inventory (LCI) 31

           LCA Databases for Raw Materials and Construction Products 32

2.2.2 Life Cycle Impact Assessment and Interpretation.........................33
2.2.3 Life Cycle Assessment of Traditional and Alternative Raw Materials
     for Concrete 34
     Cement 35
     Aggregates..41
     Transportation of Raw Materials.......................................43
2.2.4 Life Cycle Assessment of the Use Stage 43
2.2.5 Life Cycle Assessment of the End-of-Life Stage.........................43
2.3 Economic and Environmental Life Cycle Assessment of Concrete..............46
2.4 Life Cycle Assessment-Driven Structural Design of Concrete 48
2.5 Conclusion...52
2.6 References 52

Chapter 3
Optimizing the Performance of Green Concrete Using Nonconventional
Recycled Materials 57
3.1 Introduction 57
3.2 Nonconventional Aggregates and Standards.................................58
3.3 Agricultural Wastes and Aquaculture Farming as Aggregates 59
3.4 Industrial Wastes Used as Aggregates 62
3.5 Municipal Wastes as Aggregates...64
    3.5.1 Municipal Solid Waste Incinerator Bottom Ash.........................64
    3.5.2 Sewage Sludge Ash 66
3.6 Insulating Aggregates 67
    3.6.1 Tire-Rubber Aggregates...67
    3.6.2 Plastic Waste Aggregate 68
    3.6.3 Glass Waste Aggregates 70
3.7 Other Types of Aggregates...71
3.8 Conclusion...71
3.9 Acknowledgements 72
3.10 References 72

Chapter 4
Short- and Long-Term Performance of Structural Recycled Aggregate Concrete ...87
4.1 Introduction and Objectives 87
4.2 Short-Term Behaviour 88
    4.2.1 Flexural Performance..88
    4.2.2 Shear Performance 91
4.3 Long-Term Performance 93

    4.3.1 Performance Under Sustained Load.....................................94

    4.3.2 Performance After Sustained Loading: Recovery and Failure Behaviour ...95

4.4 Conclusions...98

4.5 Acknowledgements 99

4.6 References 99

Chapter 5
Long-Term Deformation of Structural Recycled Aggregate Concrete 103

5.1 Introduction 103

5.2 Shrinkage of Recycled Aggregate Concrete.................................104

    5.2.1 Experimental Results on the Shrinkage of Recycled Aggregate Concrete ..105

    5.2.2 Modelling of Recycled Aggregate Concrete Shrinkage....................109

5.3 Creep of Recycled Aggregate Concrete.....................................111

    5.3.1 Experimental Results on RAC Creep....................................112

    5.3.2 Modelling of the Creep of Recycled Aggregate Concrete 115

5.4 Long-Term Deflections of Recycled Aggregate Concrete Structures..........117

    5.4.1 Experimental Results on Deflections of Recycled Aggregate
           Concrete Structures 118

    5.4.2 Modelling of Deflection Behaviour of Recycled Aggregate
           Concrete Structures 123

5.5 Conclusions..126

5.6 Acknowledgements 127

5.7 References 128

Chapter 6
Durability of Recycled Aggregate Concrete.................................. 135

6.1 Introduction 135

6.2 Carbonation Resistance 135

6.3 Chloride Ion Penetration 137

6.4 Permeability...138

6.5 Alkali-Silica Reaction 140

6.6 Freeze-Thaw Resistance 142

6.7 Sulphate Attack..143

6.8 Conclusions..144

6.9 References 146

Chapter 7
Fatigue of Recycled Aggregate Concrete 155

7.1 Introduction 155

7.2 Fatigue Test Methodologies 157

    7.2.1 Fatigue Limit and Fatigue Life 157

    7.2.2 Strength Range 158

    7.2.3 Wöhler Curve or S-N Curve 158

    7.2.4 Wave Type and Frequency 159

    7.2.5 Test Specimens 159

7.3 Fatigue Behaviour of Recycled Aggregate Concrete Under Compression
and Bending Cyclic Loadings 160

    7.3.1 Strain Versus Cycles 161

    7.3.2 Fatigue Limit Versus Compressive Strength 163

7.4 Fatigue Failure Micro-Mechanisms by Micro-CT Analysis 164

7.5 Resonance Fatigue Behaviour of Recycled Concrete 165

7.6 Effect of Temperature on Recycled Concrete 167

7.7 Fatigue for Railway Superstructure Applications 168

7.8 Conclusions 169

7.9 References 169

Chapter 8
Structural Design of Recycled Aggregate Concrete 175

8.1 Introduction 175

8.2 Methodology 176

    8.2.1 Certified Recycled Aggregates 176

    8.2.2 Structural Design Methodology 178

    8.2.3 Building Design 180

8.3 Results and Discussion 182

8.4 Concluding Remarks 186

8.5 References 187

Chapter 9
**Design Guidelines for Recycled Aggregate Concrete Based on
Reliability Analysis** 191

9.1 Introduction: Partial Factor Formats and Code Calibration 191

9.2 Recycled Aggregate Concrete 193

    9.2.1 Coarse Recycled Aggregates Produced From Concrete Waste 193

    9.2.2 Properties of Concrete Made With Coarse Recycled
Concrete Aggregates 194

    9.2.3 Uncertainty in Resistance Models and Recycled Aggregate Concrete 197

9.3 Methodology for Calibration of Design Equations 199

    9.3.1 Code Calibration 199

9.3.2 Design Equations for Ultimate Limit State Recycled Aggregate
        Concrete Design 201

9.3.3 Deemed-to-Satisfy Provisions for Concrete Cover Design of
        Recycled Aggregate Concrete. 209

9.4 Conclusion. 213

9.5 Acknowledgements 214

9.6 References 214

Chapter 10
Total Recycling of Concrete Waste Using Accelerated Carbonation 219

10.1 Introduction 219

10.2 Carbonation and Application of Coarse Recycled Concrete Aggregate 221

        10.2.1 Physical and Microstructural Properties of Carbonated Coarse
                 Recycled Concrete Aggregate. 221

        10.2.2 Microstructural Properties of Carbonated Coarse Recycled
                 Concrete Aggregate. 222

10.3 Application of Carbonated Coarse Recycled Concrete Aggregate in
        Recycled Aggregate Concrete. 224

        10.3.1 Compressive Strength of Recycled Aggregate Concrete. 224

        10.3.2 Microhardness of Recycled Aggregate Concrete .224

        10.3.3 Elastic Modulus of Recycled Aggregate Concrete 226

        10.3.4 Bulk Electrical Conductivity of Recycled Aggregate Concrete. 226

        10.3.5 Water Absorption of R Recycled Aggregate Concrete AC 226

        10.3.6 Chloride Penetration Resistance of Recycled Aggregate Concrete 227

        10.3.7 Gas Permeability of Recycled Aggregate Concrete 228

        10.3.8 Carbonation Resistance of Recycled Aggregate Concrete. 229

        10.2.9 Relation Between Water Absorption and Permeability of RAC 230

10.4 Carbonation and Recycling of Fine Recycled Concrete Aggregate 230

        10.4.1 Physical and Microstructural Compositions of Carbonated
                 Fine Recycled Concrete Aggregate 230

        10.4.2 Mineralogical Compositions and Chemical Reactivity of the
                 Carbonated Fine Recycled Concrete Aggregate. 232

        10.4.3 Influence of Carbonated Fine Recycled Concrete Aggregate
                 on the Properties of Recycled Aggregate Mortar 237

        10.4.4 Bond Strength Between Recycled Aggregates and
                 New Cement Mortar. 238

10.5 Carbonation and Recycling of Recycled Concrete Fine Powder
        (RCFs, <150 um). 239

        10.5.1 Phase Assemblance and Microstructure Evolutions of

Carbonated Recycled Concrete Fines...................................239
Calcium Phase Evolution in Carbonation Products..................239
Silicon Phase Evolution in Carbonated RCFs.........................240
Aluminium Phase Evolution in Carbonated RCFs...................242
Microstructure Development..243
Potential Application of Carbonated RCFs as
High-Reactive SCMs..245
10.5.2   Synthesis of Amorphous Nano-Silica from Recycled Concrete Fines...245
Design of Two-Step Carbonation Process for Synthesizing
Amorphous Nano-Silica..245
Compositions of Carbonation Products............................246
Microstructure of Carbonation Products..........................248
CO_2 Consumption and Yields of Carbonation Process.............250
10.6 Summary..251
10.7 Acknowledgements...252
10.8 References...252

Chapter 11

**Large-Scale Structural Applications of Recycled Aggregates Concrete –
Shandong Shanghe Industrial Base** **257**

11.1 Introduction...257
11.2 Design Summary...258
11.3 Application of Recycled Concrete...........................260
      11.3.1 Mix Proportion Design..............................260
      11.3.2 Construction Process...............................260
11.4 Application Monitoring of Recycled Concrete................262
      11.4.1 Dynamic Monitoring.................................262
      11.4.2 Settlement Monitoring and Creep Monitoring.........263
      11.4.3 Deflection Monitoring for Beams....................263
      11.4.4 Shrinkage Monitoring of Recycled Concrete..........264
      11.4.5 Crack and Deflection Monitoring for Slab...........265
11.5 Summary and Prospect.......................................265
11.6 Acknowledgements...266
11.7 References...266

Chapter 12

Electric Arc Furnace Slag Aggregates in Concrete **267**

12.1 Introduction...267
12.2 Steelmaking Industry and Slags Formation...................268

12.3 Electric Arc Furnace Slag Concrete...270

    12.3.1 Characteristics of Electric Arc Furnace Slag:
          Physical Properties, Chemical Composition, and Mineralogy........270

    12.3.2 Leaching Potential of Electric Arc Furnace Slag.....................273

    12.3.3 Electric Arc Furnace Slag Concrete:
          Workability, Mechanical Properties 273

    12.3.4 Electric Arc Furnace Slag Concrete: Probabilistic Models............275

    12.3.5 Electric Arc Furnace Slag Concrete: Durability Properties...........277

    12.3.6 Electric Arc Furnace Slag Concrete: Environmental Impacts.........279

    12.3.7 Electric Arc Furnace Slag Concrete: Special Applications............280

12.4 Structural Applications...282

12.5 Conclusions 283

12.6 Acknowledgements..284

12.7 References...284

Chapter 13
Alkali-Activated Concrete Performance.................................... 289

13.1 Introduction...289

13.2 What are Alkali-Activated Cements and Concretes (ACC)?.................291

    13.2.1 Definition and Brief History....................................291

    13.2.2 Precursors and Alkaline Activators..............................291

    13.2.3 Alkaline Cement and Concrete Preparation: Nature of the
          Main Activation Products 293

13.3 Fresh and Hardened Alkali-Activated Concrete (AAC) Performance 296

13.4 Alkali-Activated Concrete Durability....................................300

    13.4.1 Sulphate Resistance...301

    13.4.2 Resistance to Acid Media.......................................303

    13.4.3 Carbonation Resistance and Reinforcement Corrosion..............304

    13.4.4 Chloride Resistance and Reinforcement Corrosion 306

    13.4.5 Heat and Fire Resistance 307

13.5 Final remarks 308

13.6 References...309

Chapter 14
Fibre-Reinforced Polymer Reinforcement Towards More Durable and
Sustainable Concrete Structures 321

14.1 Introductory Remarks 321

14.2 Fibre-Reinforced Polymer Bars for Reinforced Concrete Structures 322

    14.2.1 Constituent Materials..322

14.2.2 Geometries and Manufacturing....................................322

14.2.3 Mechanical and Physical Properties..............................324

14.2.4 Bond to Concrete 325

14.2.5 Applications 325

14.3 Structural Behaviour and Design Principles of Reinforced Concrete
Members Internally Reinforced with Fibre-Reinforced Polymer Bars.........326

14.3.1 Bending..327

14.3.2 Shear...329

14.3.3 Available Design Guidelines/Codes 330

14.4 Critical Topics for FRP-RC Structures 331

14.4.1 Fire Behaviour 331

14.4.2 Seismic Behaviour..337

14.4.3 Durability 340

14.5 Sustainability of FRP Reinforcement 345

14.5.1 Overview...345

14.5.2 LCA Studies 346

14.5.3 Future Trends...349

14.6 Concluding Remarks 349

14.7 References...350

https://doi.org/10.2749/sed020.ch01

Chapter 1

Notes for a Multiscale Approach for Low Carbon Concrete Structures

Vanderley M. John, Marco Quattrone, Pedro C.R.A. Abrão, Markus S. Rebmann, Rafael G. Pilleggi

CEMtec National Institute for Advanced Eco-efficient Cementitious Technologies, Sustainable Construction Innovation Centre CICS, Polytechnic School, University of São Paulo

1.1 Introduction

The only construction with zero environmental impact is the one that is never made. Unfortunately, we do not build to prevent environmental impacts, but to serve humankind. The hypothetical zero-impact construction cannot improve the comfort, health, longevity, meaningfulness, or enjoyment of human life. While it is undeniable that certain buildings and facilities may have a marginal or even negative impact on the quality of human life, it is crucial to urgently expand the built environment in developing countries to enable impoverished citizens to lead fulfilling lives.

Therefore, the decision of whether to build or not becomes the most consequential one when it comes to the sustainability of the built environment. Having decided that a new construction (or a retrofit) is a human need, the challenge is how to minimize negative environmental impacts without compromising the ability of the built environment to serve humanity.

The increased frequency of extreme weather events is the already visible effect of global warming, with environmental stresses over the built environment frequently above the ones it was designed for. News often report that prolonged droughts make the natural environment prone to fire or made water reservoirs that supply drinking water or hydroelectricity too small to cope with the current uncertain environment. Extreme heat waves introduce thermal stresses above design. Rains heavier than expected cause urban floods and landslides. Stronger winds present an increasing challenge to the structural stability of roofs and façades. Carbon emissions are progressively becoming a cost, not only due to increasing taxation [1, 2] but also because products and activities with high CO_2 footprints are perceived as risky, driving investment elsewhere as society moves towards a low-carbon economy. The buildup of the resilience of the built environment to cope with the inevitable effects of climate change will require more construction, which at the same time must not contribute to the carbon footprint.

Mitigating construction CO_2 emissions requires deploying solutions that are affordable and technically viable with dramatically lower greenhouse gas (GHG) emissions. This cannot be achieved if we keep acting by promoting innovation at the material scale only. The challenge requires a coordinated multiscale action: material scale, construction scale and planning scale.

1.1.1 The Implication of Net Zero CO_2 Target of the Cement Industry to Concrete Producers

The Global Cement and Concrete Association (GCCA) recently disclosed that the cement industry has set a goal (Fig. 1.1) of achieving Net Zero CO_2 emissions by 2050 [3]. The average clinker content is expected to decrease to 0.52 (kg/kg) by 2050, compared to today's average of 0.63 among GCCA members, resulting in a modest 9 % reduction in emissions, which we believe is a pessimistic view, vis-à-vis the current developments of low carbon solutions. Because it implies that clinker will make 52 % of the cement composition in the best-case scenario using renewable energy sources. It would still be an average of ~500 kg CO_2/t of clinker (or 265 kg CO_2/t of cement) from limestone decomposition, which needs to be captured and used or stored to ensure zero net emissions [4]. Carbon capture and storage or utilization (CCS/CCU) is expected to provide 36 % of the total CO_2 mitigation but it will come with a steep cost between ~US$50–100/t of CO_2. Considering the current clinker cost of approximately US$30/t and carbon capture cost of US$100/t of CO_2, the taxation or carbon capture expense will add US$88/t to clinker, resulting in a significant 200 % cost increase. A similar situation can be observed for the second most important structural material: low-carbon steel will be more costly. Unfortunately, there is no other material that is low carbon, affordable, and scalable rapidly in order to replace steel and concrete completely.

These cost increases of essential materials will have social consequences for the humans living in the developing global south, which needs to build to provide a decent and safe built environment and lacks the extra money to spare. Therefore, if the mitigation action stays confined to the materials industry scale, as it has been so far, housing and infrastructure will become more expensive.

This cost inflation, however, will motivate engineers and real estate investors to look for savings. Delivering the much-needed built environment with fewer materials, dematerialization is an obvious but not easy option. Indeed, the cement industry expects that clients will act causing a 30 % reduction of cement demand in comparison with the reference scenario, incidentally, further reducing CO_2 emissions. Concrete and mortar producers facing increase of cement cost are expected to reduce cement content in their formulation, equivalent to 11 % of the total CO_2 mitigation. However, these reductions will not be enough to totally offset cement cost increase. Therefore, real estate developers and designers are expected to reduce by 21 % the concrete needed to fulfil the clients' demand for the built environment.

Thus, affordable low-carbon cement-based construction will be a result of a multiscale, multistakeholder approach. At material scale, innovation will result in cement

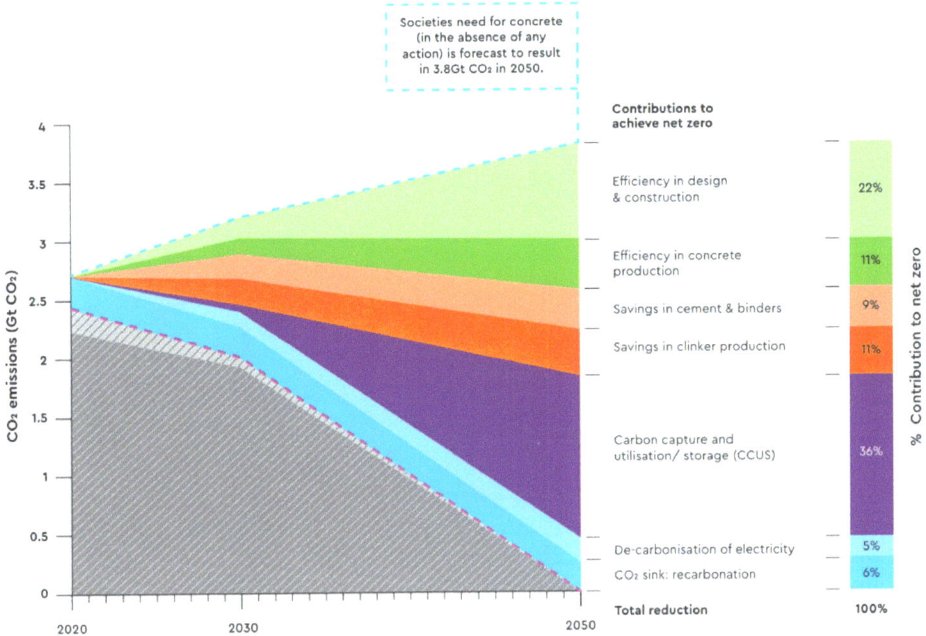

Fig. 1.1 GCCA CO_2 emission targets and the contribution of each strategy to the total mitigation [3].

with progressively less clinker. At the scale of the concrete production, concrete with less and less cement. And at design and construction scale, engineers will seek for technical solutions and designs that dematerialize construction, reducing the amount of concrete (and other cement-based materials) required and, in consequence, reducing construction costs in order to offset the cost impact of low-carbon processes.

1.1.2 Carbon Sources in Concrete and CO_2 Footprint Measuring

The GHG emissions from cement-based materials are mostly CO_2, primarily due to fossil fuel oxidation in cement kilns, which can be almost neutralized by the use of renewable fuels, and limestone decomposition in the clinker process, which is not possible to avoid. The clinker footprint varies between ~860–900 kg CO_2/t depending on the energy matrix, energy efficiency, and, to some extent, the CO_2 intensity of the raw materials. The CO_2 footprint of cement (kgCO_2/t of cement) is reduced by partially replacing clinker with supplementary cementitious materials (SCMs) that have lower CO_2 footprints (Table 1.1). Consequently, blending clinker with SCMs leads to an average direct CO_2 emission of 635 kg CO_2/t among the company's members of the GCCA – Getting the Numbers Right (GNR).

Aggregates usually possess a low carbon footprint. Depending on the energy matrix and specific details of the process, a crushed natural aggregate typically has a CO_2

footprint of 1–7 kg CO_2/t. Even when considering that the mass of aggregates is about 5–7 times higher than cement, under typical conditions, the difference in the contribution of cement and aggregates to the concrete's CO_2 footprint is around one order of magnitude.

Table 1.1 Typical emission factors for main cementitious materials (kg CO_2/t). The actual values vary widely in function of energy matrix, materials details (e.g., humidity of clay), plant detail, and operation standards. Emissions from secondary materials, such as fly ash and blast furnace slag can be higher if part of the CO_2 emissions from the process is allocated to them, which depends on rules. Data from [5]

Clinker + gypsum	Coal fly ash	Blast furnace slag	Natural pozzolans	Calcined clay	Limestone filler
~860	49	90	7	110–330	8

However, the CO_2 footprint of aggregates can increase when sophisticated processing is required. For instance, the artificial drying of aggregates can significantly impact energy demand and the CO_2 footprint, although it typically remains below 10 kg CO_2/t [6]. Marine dredged sand and gravel tend to have higher CO_2 footprints, estimated between 30–40 kg CO_2/t in the UK [7].

Transportation distance and mode also play a role. The Intergovernmental Panel on Climate Change (IPCC) estimates that the CO_2 footprint for road transportation varies between 200 g CO_2/t-km for the largest trucks to 500 g CO_2/t-km for medium trucks [8]. However, train and ship transportation result in substantially lower CO_2 footprints. The CO_2 footprint of aggregates heavily depends on transportation distances, but aggregates are widely available and affordable, leading to relatively short transportation distances. On the other hand, cement can be transported over longer distances, yet it constitutes less than 15 % of the concrete's total mass.

Regarding the processing, transportation, and placement of concrete, they require relatively little energy compared to the raw materials. However, the thermal curing of concrete requires thermal energy, and depending on the amount and heat source, it can significantly increase the CO_2 footprint of concrete. Thermal curing is often inefficient because it requires heating the aggregates, which constitute approximately 80 % of the mass. Therefore, in most of the cementitious materials, the cement used in the formulation is the main source of CO_2 emissions.

1.1.3 Objective

This chapter focuses on the facts that arise after the decision to build with concrete has been made. Its main objective is to reduce the carbon footprint of concrete by optimizing the different components (paste, mortar, and coarse aggregates) that constitute the material, with particular emphasis on the paste. For this reason, this chapter refers to the multiscale approach. Discussing how to minimize carbon emissions by improving design and construction practices are certainly essential, but outside of the scope of this chapter.

1.2 Fundamentals of Designing Low-Carbon Concretes

1.2.1 Low Carbon Concrete and Clinker Replacement

The literature on sustainable construction usually associates low carbon concrete when a large fraction of the cement (or clinker) is replaced by secondary (residues) mineral additions. It is certain that replacing clinker by low-carbon materials will reduce the CO_2 footprint of the cement ($kgCO_2$/t of cement). However, the CO_2 footprint of the concrete depends on the footprint of cement but also on the amount of cement in the mixture. The amount of cement required to achieve the required strength over time and other relevant performances of a particular concrete depends on several factors, including the fineness, shape, specific surface area and other properties of aggregates used in the mix.

SCMs have generally lower reactivity than the clinker, and in some cases their reactivity can be close to zero, especially for fillers. When blending SCMs with clinker, there is a reduction in the water chemically bound by the cement (g/g of cement) at all ages of interest. Depending on the market, sometimes a cement with little or almost no SCMs can have lower combined water. This can result from a cement particle size distribution (PSD) that includes particles coarser than 30 μm, which do not fully react at 28 days [17], eventually combined with low-quality clinker. In commercial cements, this lower reactivity leads to a reduction in the cement strength class [9], which is a proxy of the chemically combined water at 28 days when measured at a constant water/cement ratio. The lower the cement strength class, the higher the amount of cement needed to achieve a given concrete design compressive strength (Fig. 1.2). It's worth noticing that most Environmental Product Declarations (EPD) currently refer only to cement types, and do not mention the cement strength class.

The amount of cement needed also depends on the mixing water demand, which is also influenced by cement characteristics such as, specific surface area or particle size distribution. These characteristics vary significatively, particularly for blended cements [10] influencing the actual amount of cement needed to deliver a concrete with a required strength [9].

Therefore SCMs can have a significant effect on the water demand for workability and the combined water content at 28 days [11], influencing the cement consumption for a given concrete. Some SCMs may decrease the water demand: strength may be achieved with less cement. Others may increase it, and more cement will be required.
As a result, the CO_2 footprint of concrete is not solely determined by the clinker factor or the low CO_2 footprint of the cement for the amount of a given cement required to formulate a concrete is a decisive factor in the equation.

1.2.2 Defining Low Carbon Concrete

Low carbon concrete is a concrete that has lower CO_2 footprint compared to the current industrial standards in a given region at the present time. Therefore, a valid representa-

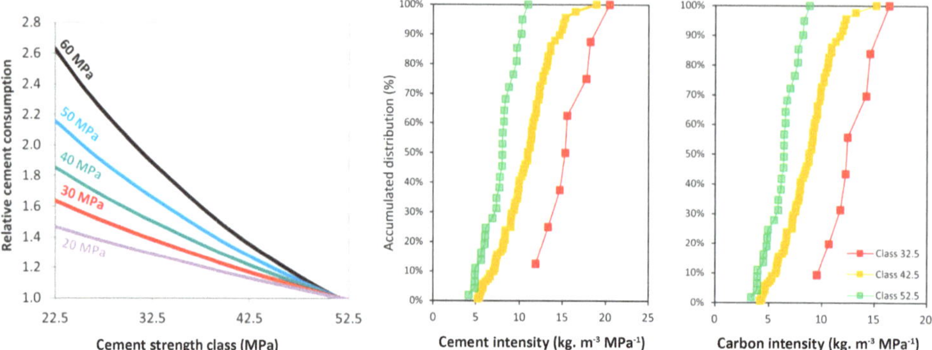

Fig. 1.2 (Left) theoretical influence on cement strength class on the relative cement consumption (kg/m³) for various design strength, all other factors remaining the same. (Right) actual frequency distribution of cement intensity and carbon intensity from literature. Figures from [9] when measured with constant *w/c* ratio, cement strength class is a proxy for combined water at 28 days.

tive baseline of the CO_2 footprint of concrete is needed for every region. As the industry progresses towards lower CO_2 emissions over time, this baseline will naturally evolve.

Usually, the functional unit for the CO_2 footprint of concrete is kg/m³. However, concrete with different design strengths will have a different performance. For that reason, we recommend including the compressive strength at 28 days in the functional unit, resulting in a unit of kg·m⁻³·MPa⁻¹ [12]. Some authors have suggested incorporating durability-related parameters (such as chloride diffusion and carbonation rate) into the functional unit. However, for environmental assessment, the crucial factor is the service life of the structure, which is highly dependent on local conditions. For example, the relevance of chloride diffusion would be negligible in chloride-free environments. Similarly, a high carbonation rate may speed up the CO_2 capture process, aiding in global warming reduction, but it might not affect the service life of reinforced concrete exposed to an indoor and dry environment or in the absence of steel reinforcement. Thus, while these parameters are important for specific applications, they might not be helpful in establishing the benchmark needed to guide the industry towards low carbon concrete production.

In the past, we produced a benchmark of binder intensity (kg·m⁻³·MPa⁻¹, at 28 days) and utilized default values for the CO_2 footprint from clinker to establish a benchmark from literature [12]. This benchmark was then used to access more recent research on low-carbon concretes [13] (Fig. 1.3). The data revealed that all concretes with SCMs exhibit a minimum CO_2 footprint of approximately 2 kg CO_2·m⁻³·MPa⁻¹ for compressive strengths above 40 MPa. As the compressive strength decreases to 20 MPa, the CO_2 footprint increases exponentially, reaching around 4 kg CO_2·m⁻³·MPa⁻¹, including when clinker is replaced by inert fillers. For ordinary Portland cement (OPC) without admixtures, the best available laboratory technology resulted in a CO_2 footprint of 4 kg CO_2·m⁻³·MPa⁻¹ for concretes with a compressive strength above 50 MPa. This value increases exponentially to 10 kg CO_2·m⁻³·MPa⁻¹ for those with a compressive strength of 20 MPa. We must keep in mind that laboratory conditions are controlled and allow for the selection of

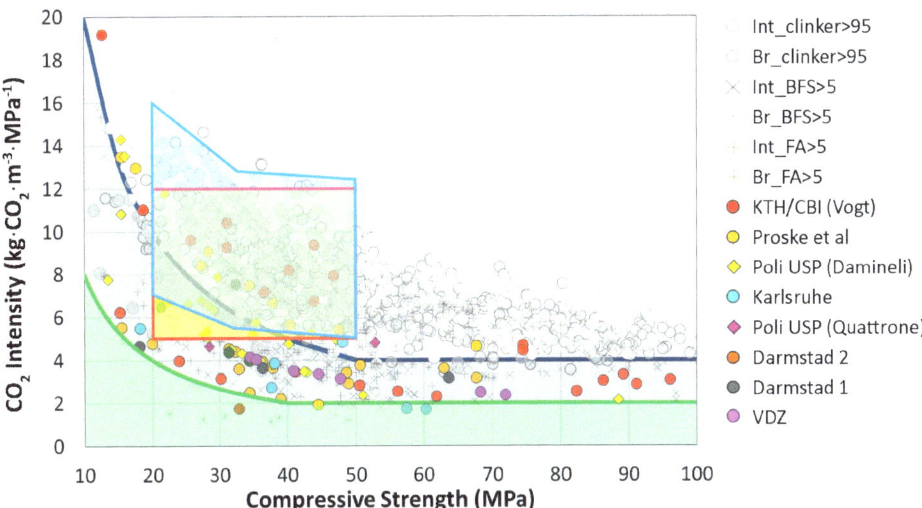

Fig. 1.3 Benchmark of CO_2 footprint from laboratory results (graph) for pure Portland cement (top white area) and blended cement (yellow), where colour dots are experimental concretes were cement is replaced by a high volume of filler (from [13]). Superimposed shapes: the blue shape is EPD derived data of Australian ready-mix concretes [15–17]; the yellow square with a red line border is from a LCA inventory of a ready-mix concrete from Brazil [18].

optimum raw materials, excellent processing machines and casting procedures, aspects that most of the time are not reproducible at industrial scale. In other words, laboratory results have a technology readiness level (TRL) up to 5 [14] showcasing the potential of a technical solution, but real-world application may require further considerations.

For a technology contributing to mitigation efforts, it must reach the market, become an innovation, and achieve production scale-up, eventually displacing high CO_2 emitting existing solutions. This process requires transforming the TRL from its initial stages of 3–5 in the laboratory to a TRL 9 market-ready solution. This transition demands considerable time and a substantially larger investment in Research, development & innovation, design and construction of industrial facilities. Additionally, it requires the development of an industrial-scale supply chain, overcoming regulation and standardization barriers, and initiating production, which will progressively scale up over time [14]. In our case, scaling up low-carbon concretes will probably require a long-time learning process.

To access what constitutes a low-carbon ready-mix concrete solution from a market perspective, it is becoming possible to produce a benchmark by combining information from EPDs from various countries, as we did with data from Australia [15–17] and from national inventories such as the one from Brazil [18]. From these limited data (Fig. 1.3) it is possible to estimate that the best available (commercial) technology of ready-mix concrete currently results in a CO_2 footprint of around 5–6 kg $CO_2 \cdot m^{-3} \cdot MPa^{-1}$. As research progresses, it is expected that new innovative low-carbon concrete solutions with even lower CO_2 footprints than the current best available technology (BAT) will enter the market. Achieving the best laboratory results involves reducing the CO_2 footprint by a factor ranging between 2 and 6. This transformation requires long-term investment, a continuous learning process, and certainly a different supply chain.

In the market, there are concretes available with lower CO_2 footprints, and some are even carbon-neutral (zero CO_2 footprint). However, this is often achieved through the vendor acquiring carbon offset credits from the market to compensate for the emissions produced during the concrete production. These credits come from officially certified projects that have reduced GHG emissions compared to the current market practice. These projects include initiatives that destroy methane from landfill, ensure native forest preservation, generation of renewable energy, and even energy efficiency projects.

1.3 Engineering the Cement Paste for a Low Carbon Concrete[1]

The cement standards are dedicated, among other things, to control the amount of chemical hydration reaction the cement will produce over time. This reaction is typically measured by the minimum cement strength at different ages or the cement strength class at 28 days, with a constant water-to-cement ratio of ~0.5. The higher the strength, the greater the fraction of cement that has reacted with the water. A 42.5 MPa cement will have a 28-days combined water around 0,22 g/g, while a 32.5 MPa cement will have 0.18 g/g. The industry manages the strength by a combination of the amount and type of SCMs and adjusting grinding fineness.

The concrete strength is a function of capillary porosity produced by the water that has not reacted with the cement. The volume of hydrated products – represented by the combined water – needed to deliver a paste with a given strength depends on the volume of mixing water, which is determined by the desired rheological behaviour. When the mixing water is reduced, the amount of hydrated products needed is also reduced.

Therefore, engineering the paste can reduce the mixing water demand and compensate for dilution caused by fillers or other SCMs with low reactivity. The use of dispersants (or superplasticizers) alone may allow to reduce 10 % to 30 % of mixing water demand [19]. This reduction in water content subsequently leads to less combined water and less cement being required, resulting in a lower CO_2 footprint for the concrete.

When the fines – cement and SCMs – are fully dispersed, it becomes possible to engineer the particle size distribution, leading to a further reduction in the mixing water demand, allowing further dilution of cement. Research data indicate that it is feasible to achieve up to 70 % reduction in mixing water by combining particle size optimization and full dispersion of fines, enabling dilution of the Portland cement by inert fillers without changing the strength, meanwhile generating a workable product [20–22].

The particle agglomeration destroys the measured particle size distribution of the fines. Therefore, particle packing optimization can only be achieved (or discussed) in formulations with fully dispersed fines. However, in the usual market, concrete formulations

---

[1]   The cement paste, or cement matrix, consists of water, admixtures, and particles below 100–150 μm, whose behaviour in the suspension depends strongly on the surface forces because their mass is very low.

rarely have fully dispersed fines in the paste, due to the cost of admixtures, concerns about segregation, the need to meet minimum cement content standards, as well as cement paste volume to ensure adequate maximum paste thickness (MPT) [20], which depends on the volume and morphology of aggregates.

Our strategy, therefore, includes: (a) select a good, reactive commercial cement; (b) select fillers or other SCMs with various particle size distributions, including ultrafine particles ($D_{90} < 5$ μm); (c) select a dispersant admixture, and for all fines – cement, fillers and SCMs – determine the optimum admixture dosage to ensure full dispersion of the fines; (d) develop a mixture between cement and SCMs to minimize the mixing water demand of the paste; (e) adjust the binder content to desired mechanical properties; (f) engineer aggregates particle size distribution to minimize concrete paste volume.

1.3.1 Procurement and Characterization of Raw Materials

Selecting a reactive (high strength class) Portland cement preferably with a low surface area is recommended. To dilute this cement at least two fillers are required: (a) an ultrafine, performance filler with $D_{90} < 5$ μm; (b) a dilution filler, which has the particle size distribution and surface area as close as possible of the original cement. Varying the proportion between the dilution filler and cement allows adjusting the chemical reactivity of the paste without significantly affecting the rheology, since both materials have similar PSD and surface area. The ultrafine improves the packing and reduces the mixing water demand. Choosing a reactive high-strength cement type CEM I is advantageous if you have access to excellent fillers.

The particle size distribution can be determined by laser diffraction, while the specific surface area should preferentially be measured by Brunauer–Emmett–Teller (BET) gas adsorption. The ratio between the BET specific surface area and the surface area estimated from the laser diffraction particle size distribution, assuming all particles are perfect spheres, represents the shape factor. A shape factor approaching 1 indicates that the particles are rounded, as observed in silica fume and some fly ashes (Table 1.2). Thermogravimetric analysis (TGA) can be used to assess the potential of cement and reactive SCMs to react with water.

1.3.2 Dispersant Selection and Dosage

Dispersants (or superplasticizers) are the simplest and more practical way to achieve up to 30 % of water reduction while maintaining similar workability [19], allowing a reduction of the combined water (or binder content) for a same given strength.

Superplasticizers interact with the chemical reactions of the cement [24], which will reduce dispersion progressively and even delay hydration affecting initial strength, as shown in Fig. 1.5. The addition of 1 wt% of a superplasticizer increased the induction period from around 2 hours (OPC-0.5-noSP) to 6 hours (OPC-0.37-1wt%SP), consequently extending the final setting time as well. Thus, achieving low-carbon cement relies on effective dispersion, requiring the use of an admixture compatible with the chosen

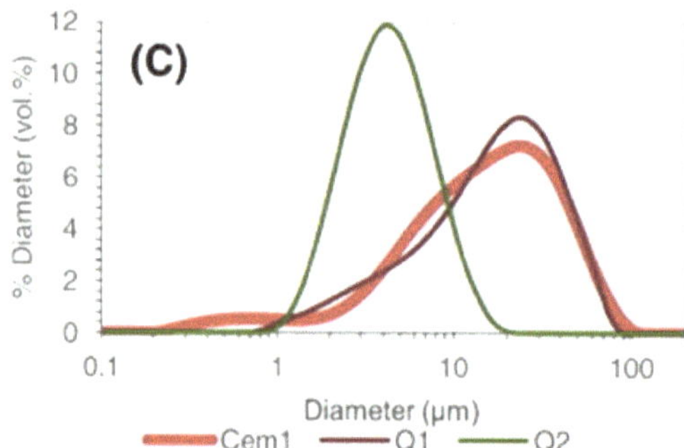

Fig. 1.4 Particle size distribution from cement and the dilution filler Q1 and the performance Q2. From [23]

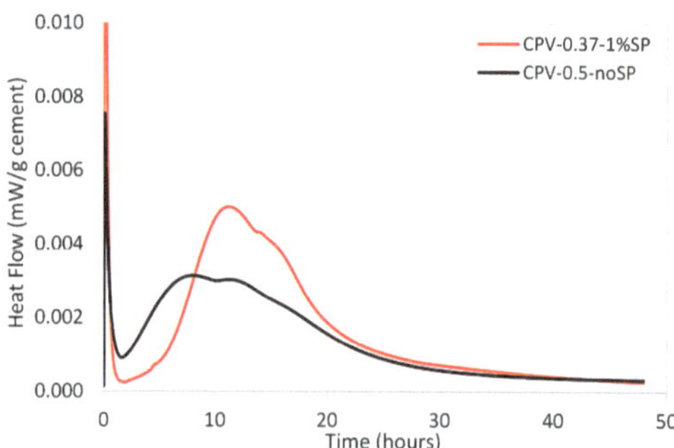

Fig. 1.5 Heat flow at 23°C of an CPV (similar to CEM I) with 0.5 w/c without superplasticizer and with 0.37 w/c and 1 wt% of a superplasticizer (polycarboxylate ether – ADVA CAST 527, GCP).

cement. Compatibility may be tested by heat of hydration and workability loss over time as shown by [25, 26].

To measure the dispersant saturation content of cement compositions, different approaches can be used, the most common ones are the mini-slump test [27] and rotational rheometry [10, 23], where the dispersant is dosed in pastes formulated with low w/c (around 0.3–0.4) to minimize segregation during the test. Fig. 1.6 provides an example of the dispersant dosage of different Brazilian commercial cements, limestone fillers, and a calcined clay [10, 28]. The pastes were subjected to rotational rheometry (Rheometer MARS 60, HAAKE) using plate-plate geometry with a diameter of 35 mm and a gap of 1 mm. The shear programme used was the stepped flow test in two cycles of acceleration and deceleration, with the shear rate varying from 0 to 50 s^{-1}. For each superplasticizer content, the yield stress was obtained at a shear rate equal to 0 s^{-1}. In this case, the saturation dosage was set as the point prior to the stabilization of the yield stress. However, as done by [23], the saturation point could be set as the point prior to the stabilization of the apparent viscosity.

From Fig. 1.6 it is observed that for the same water and dispersant content, the cement 16DE, 49DE (cement with diatomaceous earth) and the calcined clay presented higher yield stress than the others, due to the higher specific surface area of these SCMs (diatomaceous earth and calcined clay). For the analysed materials, the saturation dosage varied between 0.2 (dilution filler) and 2.55 wt% (calcined clay), demonstrating the importance of properly determining the superplasticizer dosage for each material. The amount of dispersant varies according to many factors, including the type and composition of dispersant and cement, amount of SCMs, and particle size distributions and physical characteristics of the cement [29].

Table 1.2 shows typical results from dispersant dosage estimation in the process of selecting the best raw materials. With the exception of apparent viscosity, all other properties from any combination between cement and fillers can be estimated by weighted average, including the saturation dosage of the superplasticizer (Fig. 1.7).

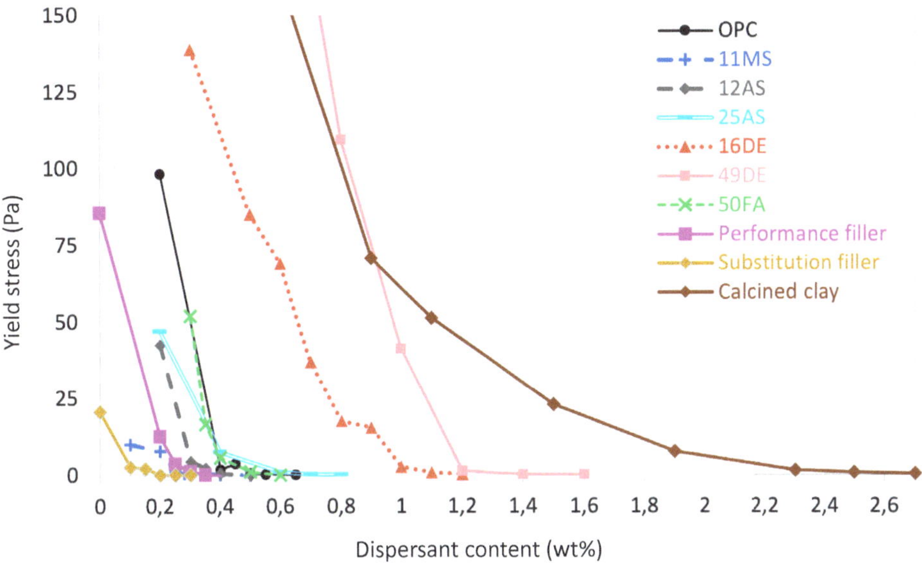

Fig. 1.6 Yield stress at 0s⁻¹ from the upwards shear rate ramp as a function of the dispersant content for different Brazilian commercial cements, two limestone fillers and a calcined clay adapted from [10, 28].

Table 1.2 Example of characterization of a commercial cement and fillers from various mineralogy.

Material	Density (g/cm³)	Shape factor	BET Surface area (m²/g)	w/s	Saturation (g/100 g)	Apparent viscosity (Nm·s)
Cement	3.2	2.6	4.6	0.275	0.375	0.49
Dolomite D1	2.82	5.91	2.26	0.4	1	0.24
Dolomite D2	2.82	4.07	3.12	0.3	0.375	0.42
Calcite C1	2.7	5.06	2.84	0.45	1	0.14
Calcite C2	2.7	2.16	1.84	0.3	0.25	0.28
Calcite C3	2.7	3.43	2.35	0.35	0.75	0.26
Quartz Q1	2.65	3.96	5.02	0.4	0.5	0.42
Quartz Q2	2.65	6.52	1.85	0.425	0.75	0.11
Granite G1	2.67	6.66	1.5	0.4	0.25	0.47
Granite G2	2.67	8.44	1.64	0.4	0.375	0.75
Granite G3	2.67	11.54	1.31	0.375	0.5	0.86
Granite G4	2.67	8.77	1.33	0.4	0.375	1.03

If the admixture consumption for full dispersion of all paste phases, whether binders or fillers, is measured, it becomes possible to estimate the admixture dosage for any combination of these phases through a simple weighted average [23].

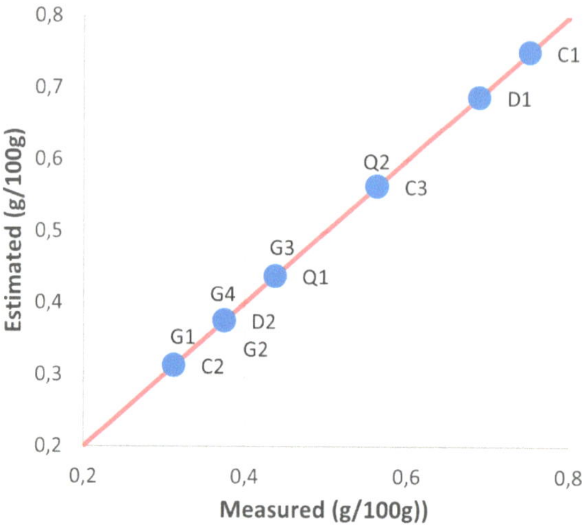

Fig. 1.7   Measured optimum dispersant content of 50-50 compositions between cement and filler versus estimated by simple weighed average of admixture demand of the cement (0.375 %) and each filler components. Labels identify the filler that have been mixed with cement. Produced from data extracted from [23]

1.4 Engineering Particle Size Distribution

1.4.1 Modelling Viscosity

It is possible to optimize a commercial cement particle packing to achieve high flowability with minimum water by combining a dispersant to ensure complete dispersion. The amount of reactive cement is an inverse function of the mixing water demand.

Complete dispersion is necessary to reproduce the particle size distribution measured in the lab (in a transparent very diluted solution) in the actual concrete, which is a much more concentrated multiphase solution including fillers such as limestone or from other minerals [30], reactive SCMs, cement, and aggregates.

However, finding the best particle size distribution for minimum water content can be laborious, especially when dealing with a high number of raw materials, which is frequently necessary to obtain better results using locally available materials. The simplest model to predict the viscosity of a suspension produced by combining various materials in different proportions is the Funk and Dinger model [20, 31], which relies on the concept of interparticle separation (IPS).

$$IPS = \frac{2}{VSA} \left[\frac{1}{V_s} - \left(\frac{1}{1-P_{of}} \right) \right] \tag{1.1}$$

Where VSA = volumetric surface area (m²/cm³), calculated by the product of specific surface area SSA (m²/g) and solid density ρ_s (g/cm³), V_s = volumetric solid fraction in suspension, and P_{of} = pore fraction at maximum packing estimated using the Westman and Hugill algorithm [32]. To improve particles mobility, the ratio between the particle size and IPS should be less than one.

The IPS concept describes the spatial distribution and distance between particles but does not directly account for the energy dissipated during liquid-particle interaction (governed by Stoke's law) or interparticle collision, which depends on the kinetic energy of each particle (a function of mass and speed). It also does not consider other aspects [20, 23]. To address these issues, the natural interference coefficient (INT_n) comes into play. For each sub-suspension, the INT_n must be estimated for various particle size classes, such as (125, 75, 45, 22.5, 11.25, 5.63, 2.37, 1.0, 0.5 and 0.25 µm). Each sub-suspension is considered to be floating in a suspension composed of particles more than 10 times smaller and water.

$$\mathrm{INT_n} = \frac{D_p^2}{IPS} \times \frac{\rho_s}{18\eta_f} \tag{1.2}$$

Where D_p is the particle diameter (µm); η_f = dynamic viscosity of the liquid (g/µm·s); ρ_s = solid density (g/cm³).

For each composition of cement and fillers, the bulk interference (INT_b) follows a linear relationship with the sum of all INT_n values in the paste. INT_b is directly correlated with the viscosity of the paste, as discussed in [20, 23].

Therefore, the mixture that flows with minimum water is the one with lowest INT_b value. The combined water and compressive strength of this formulation can be adjusted easily by varying the fraction of the dilution filler, which does not affect the particle size distribution, but requires adjusting the superplasticizer dosage.

1.5 Measuring Mixing Water Demand for Desirable Rheological Behaviour

Packing models are useful for a first estimate, but their results need to be validated by experimental measurements. These experiments can be made in paste or mortar (fixed sand and solids mix proportion), using rotational rheometry or single-point tests such as a flow table combined with careful observation of the mixing process and resulting mix behaviour. A good model, such as the one presented previously, reduces the cost and time of experimental work. However, if the number of raw materials is low, it is possible to proceed experimentally.

The water demand of cementitious material is directly associated with its processing and application route, which are not considered in the models. The study of the rheological behaviour of blended cement is complex because it depends on factors such as particle size distribution, texture, morphology, chemical and mineralogical composition, water/cement ratio, mixing conditions, presence of chemical additives, mixing and processing, temperature, measurement conditions, and others.

However, there are different methods to estimate the mixing water demand of pastes, mortars, and concretes. For pastes, De Larrard and Sedran [33] proposed a method where the water is dosed until the suspension presents an aspect of a thick paste. In our experience, it is practical to estimate the mixing water demand by the flow-table test [34–39] for mortars as it requires a smaller amount of materials than a concrete slump-test [40, 41]. In

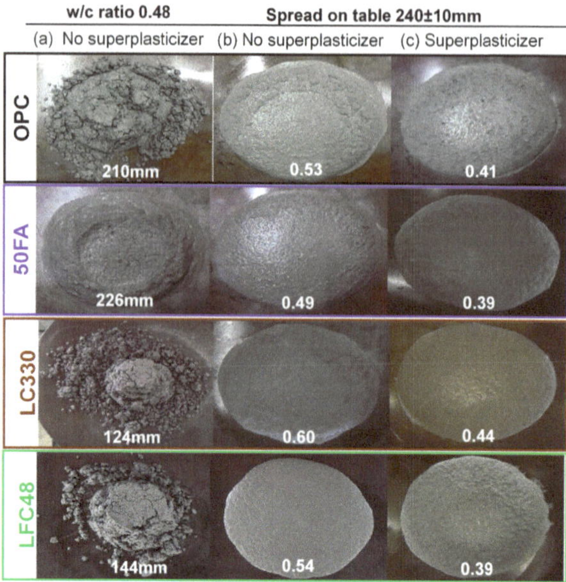

Fig. 1.8 Mortars produced with different cement (OPC, 50FA Fly ash, LC330, and LFC48 limestone filler) after the flow table test. (a) 0.48 w/c without SP; constant flow (b) without SP and (c) with SP. Images adapted from [10, 28]. [5]

both cases, the mixing water is dosed until the material achieves the requested workability, and the single-point information must be complemented by careful observation of the mix's behaviour. Fig. 1.8 shows the visual comparison between mortars (standard water/fines = 0.48) with different cement after the flow table test [10, 28]. Additionally, mortars were prepared with adjusted mixing water to achieve a spread of 240±10 mm, with or without full dispersion. During the test, observation of segregation, bleeding, and lack of cohesion can be easily made, allowing for necessary adjustments. The flow-table test can be used to dose the mixing water necessary for materials to achieve a specific rheological behaviour, confirming results from the interference model.

Rotational rheometry allows measuring the shear torque at different shear rates, leading to accurate viscosity determination [42]. As a result, it provides a direct means of validating the results obtained from the Interference model. This technique can also be employed to determine the mixing water demand by either adopting a standard torque (or shear stress) level or identifying the fluidity point in the mixing curve [43]. Moreover, the mixing water can be carefully dosed to materials achieve a desired yield stress and/or viscosity, offering greater control over the rheological behaviour of the materials.

1.6 Defining the Optimum Particle Size Distribution

From the combination of modelling and experimental validation, it becomes feasible to identify the dispersed paste particle size distribution that provides the required workability while minimizing the water demand, as demonstrated by [20, 23]

1.7 Modelling Compressive Strength of Paste Mixtures

To assess the concrete's minimum CO_2 footprint, a model to estimate its compressive strength at different ages is required. For this purpose, we use the correlation between the combined water fraction (*cwf*) and compressive strength, which we found has been considered by Powers and Brownyard as a possible alternative to gel-space ratio [44, 45]. To apply this model accurately, it is crucial to measure the combined water using a detailed methodology, such as the one presented by [45].

We defined the combined water fraction (*cwf*) as the ratio between chemically combined water (w_n) at a given time (t) and the mixing water (w_0) or water-to-cement ratio:

$$cwf(t) = \frac{w_n(t)}{w_0} \tag{1.3}$$

cwf has a linear correlation with compressive strength, for all ages (Fig. 1.9a). This approach allows adjusting the paste reactivity and its strength by varying the chemical reactivity (changing the proportion between the reactive cement and the inert dilution filler without changing the rheology) or the mixing water (and therefore changing the rheological behaviour). A given strength can be achieved by various combinations of chemical reactivity and mixing water demand.

Fig. 1.9a shows that despite the high correlation coefficient between *cwf* and compressive strength, there is significant data scattering. This dispersion is a result of different test methods used to determine the combined water, as well as variations in cement chemical composition [44]. Best results determine the correlation for each set of materials – cement, fillers and SCMs (Fig. 1.9b). For this result it is practical to produce standard 1:3 cement mortar with water/fines ratio to ensure good workability without segregation. We recommend using at least two cement-filler compositions, produced with plasticizers, and tested at two different ages, in addition to pure cement. The graph is built with all compressive strength results and combined water estimated from TGA and modelling.

1.8 Selecting a Paste with Minimum Carbon Footprint for Each Application

The best paste composition is the one that delivers the minimum mixing water for the appropriate viscosity for the application. Minimizing the mixing water demand allows minimizing the cement content of the paste formulation and, therefore, paste's carbon footprint.

The desired compressive strength and durability properties can be reached by adjusting the paste combined water at any age. If cement is the only reactive fine, this can be achieved by varying the proportion between reactive cement and the dilution filler. Because cement and dilution filler have similar PSD and surface area, this will not affect the demand of mixing water for desirable rheological behaviour [20, 30].

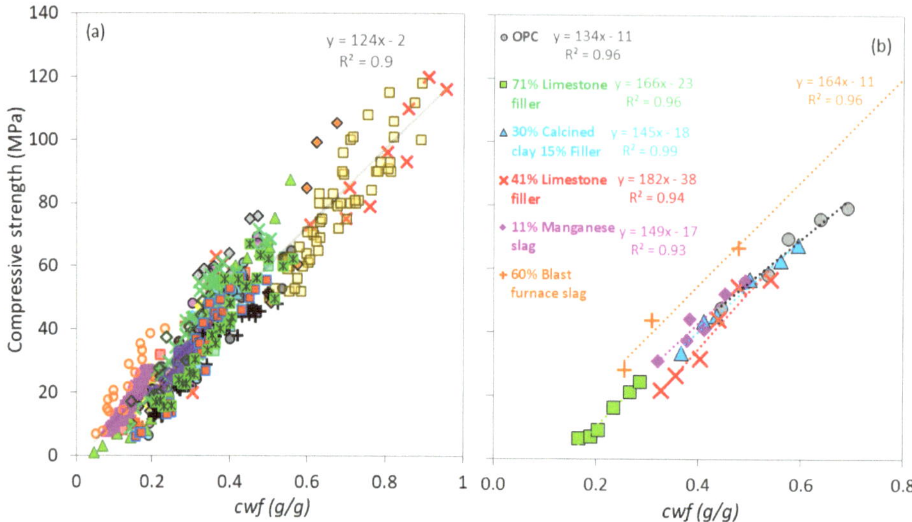

Fig. 1.9 Compressive strength vs. *cwf* of: (a) results found in 23 sources of the literature. 598 data points, including paste and mortars from 1 to 479 days, and with different binders; (b) OPC, cement diluted with limestone filler (41 and 71%), LC3 (30% calcined clay and 15% filler), cement with manganese slag and blast-furnace slag tested through a standard methodology. All data adapted from [10, 28].

1.9 Developing a Low Carbon Concrete Formulation

The carbon footprint of a concrete formulation mainly depends on the CO_2 footprint of solids in the paste and the paste volume per cubic metre. Given a fixed combination between fillers and cement, minimizing the paste volume is an effective strategy to reduce the environmental impact of concrete while maintaining the same performance of the reference concrete.

Traditional concrete mix proportioning methods, including the American Concrete Institute (ACI) and Brazilian methods, are somehow inefficient, because increasing design strength requires increasing the paste volume – more cement is added meanwhile the amount of water stays constant. The resulting cement paste has a higher volume solid concentration, higher viscosity, and higher yield stress. The only reason that slump – a practical estimation of concrete yield stress – remains constant is because the increase of paste volume increases the MPT and compensates for the changes of the paste's rheological behavior. (Fig. 1.10).

However, data shows that increasing the cement paste volume while keeping all other factors constant reduces the strength of concrete (Fig. 1.11) and also increases shrinkage and cracking [47]. There is an obvious exception to this rule, which occurs when the paste volume becomes smaller than the volume required to maintain a MPT large enough to ensure that the concrete processed with the expected casting energy will be free from large uncontrolled casting defects, that must be avoided at all costs.

Therefore, the usual proportioning models that require increasing the volume of cement paste to enhance the strength or workability are inherently inefficient. Our strategy

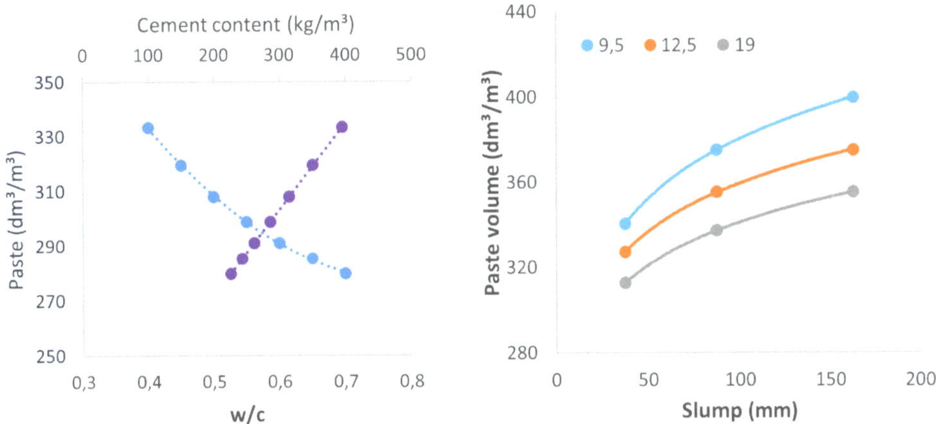

Fig. 1.10 (Left) Influence of cement content and *w/c* (proxy of strength on the volume of paste (water 178 dm³/m³ slump 7–9 cm, Dmax 9 mm). Blue markers are for w/c (primary x axis) while purple markers show the cement content (secondary x axis). (Right) Influence of slump on the paste volume for various aggregates maximum sizes (*w/c*=0.5). Data from Table 12–15 from [46] ACI concrete mix design method.

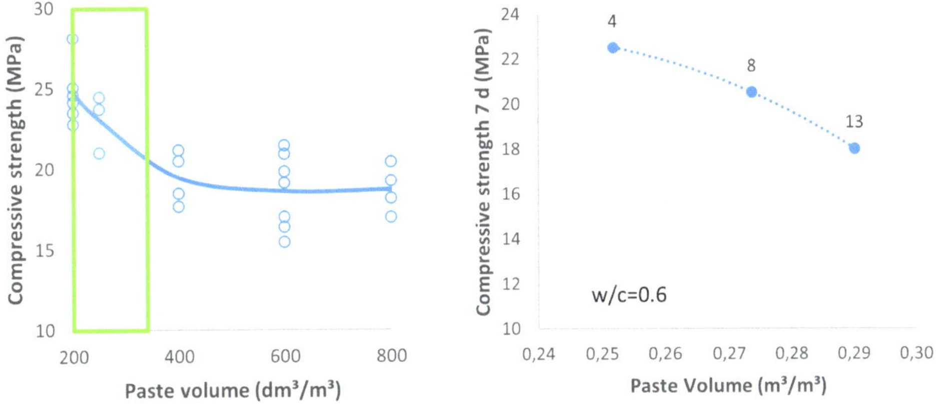

Fig. 1.11 Influence of the paste volume on the compressive strength for concretes dosed accordingly to current procedure with constant w/c ratio. (Left) A ~25% strength loss was measured by Stocks et al (1979) apud de Larrard [48]; (right) data from Metha & Monteiro [49] show a 20% strength loss when paste volume increases from 0,25 to 0,3 m³/m³ (labels are the slump values in cm).

allows adjusting the strength without significantly changing the optimum paste volume but instead by varying the proportion between dilution filler and binder (Portland cement) while maintaining the mixing water (water/solids ratio) of the paste (Table 1.3). The equation of *cwf* and compressive strength allows for a rather precise estimate of the required change. To increase strength, the binder fraction in the paste increases, and the dilution filler decreases in the same proportion, causing only marginal changes in the paste volume due to differences in density between the filler and cement, which can be easily adjusted. To vary flowability, it is necessary to adjust the water/solid ratio of the paste. Because the difference in density between powders and water is much higher, there is a more significant impact on the paste volume – perhaps up to 10 % if a substantial change in paste viscosity is needed (Table 1.3) – but this can also be easily corrected.

Table 1.3   Example of adjusting paste composition in a controlled way. Data are based on results of OPC from the paper [10]. $cwf = 163*w_n - 11{,}2$.

	Unit	Reference Mix	> Flow	< Strength
Flow table	mm	240	390	~240
Compressive strength 28d	MPa	35.0	35.0	25.0
Fillers	g	12.6	9.6	16.3
Cement	g	17.4	20.4	14
mixing water	g	12.3	14.4	12.3
Combined water	g	3.5	4.1	2.7
w/s	g/g	0.41	0.48	0.41
Combined water fraction	g/g	0.283	0.283	0.222
Paste volume	cm³	22.4	24.4	22.5
Relative paste volume		1	1.09	1.01

The minimum practical paste volumes observed with current technology and usual aggregates are of interest. The default values for mix proportioning methods, such as ACI, are understandably high: standard paste volume starts from 300 dm³/m³ for 50 mm slump concrete with maximum size aggregates (MSA) of 19 mm, reaching almost 360 dm³/m² for a 160 mm slump, and can be as high as 400 dm³/m³ for MSA of 9.5 mm (Fig. 1.10, right). However, values of 200 dm³/m³ can be found, as shown in Fig. 1.11, left.

Of course, it will depend on aggregate size, shape, resulting packing, and workability requirements, including pumpability. A report from Scandinavian countries [50] allows us to estimate the evolution of average paste volume over time, which is expected to converge in 2040 to 260 dm³/m³ (Fig. 1.12, left), around the same volume already observed in 2005 in Denmark. Our own international benchmark (Fig. 1.12, right) suggests that most of today's concretes have paste volumes above 250 dm³/m³.

All considered, we recommend reducing the paste volume below 280–300 dm³/m³ for low-carbon concretes. Increasing the maximum diameter of aggregates is an effective method to achieve this [51] and should be used whenever possible.

Proportioning low-carbon concretes should consider optimizing the different phases in concrete, i.e., the paste and the aggregate skeleton, according to application requirements (rheological behaviour, strength, and durability), thereby optimizing the binder content in the mixture. As a starting point, using models to design the paste and aggregate skeleton is recommended. Of course, the results of the model depend on the quality of the input data, thus an in-depth characterization of raw materials is needed.

Among the various models published in literature to estimate the packing density of a mixture, the Compressible Packing Model (CPM) proposed by Sedran and de Larrard [51] appears useful for an initial estimation of proportioning for both normal application concrete and low-carbon concrete mixtures. Simplifying, the model's virtual packing

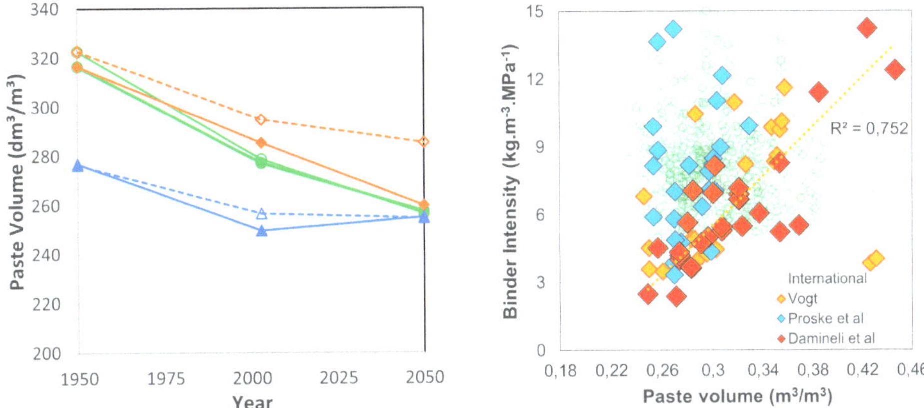

Fig. 1.12   (Left) Estimate of the evolution of paste volume from concretes in Sweden, Norway, Denmark and Iceland over time [50]; (right) Paste volume and binder intensity (amount of binder per m³ to deliver an MPa in concrete) benchmark from published papers from 29 countries plus ready mix concretes from Brazil. Large dots are from various low-carbon concrete formulations with fillers replacing binder from benchmark (see [12, 20])

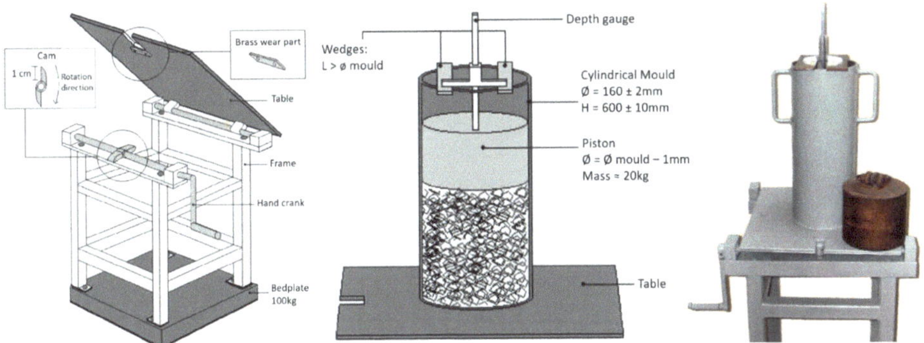

Fig. 1.13   Equipment to measure the packing density of granular fractions. Adapted from [52].

density describes the material, while the scalar compaction index K characterizes the packing process efficiency. Optimal K values lead to a packing density approaching the maximum virtual packing density. The model also accounts for *wall effect* and *loosening effect* resulting from interference between particles of different sizes and between particles and the container wall. Also, a test procedure and equipment (Fig. 1.13) to experimentally measure the packing were proposed [52].

The model results are helpful, but a few trial batches are still necessary to fine-tune the mixture to meet desired requirements. Establishing a benchmark of material properties like rheological behaviour, stability over time, strength, etc., is essential for adjusting the low-carbon concrete formulation. Because of the above-mentioned reasons, it is possible to consider that developing a low-carbon concrete formulation should pass through the general steps listed below.

1. Minimize voids in the dry mixture, i.e., maximizing solids concentration based on available raw materials, and reduce water and/or the paste volume. This is strictly related to the final application. Sometimes reaching the maximum packing density is not useful;

2. Adjust the grading by optimizing the maximum size of aggregates and coarse-to-fine grain ratio to prevent segregation of different phases in the mixture;

3. Adjust the concrete rheology using admixtures and optimize the MPT to meet application requirements, such as placeability and/or pumpability.

4. Adjust the concrete strength by adjusting binder and dilution filler content to comply with design specifications.

Since models may not fully predict important characteristics of the material at both fresh and harden state, following these steps can reduce the number of necessary trial batches during the design of new cementitious systems, particularly low-carbon ones.

1.10 Results From Practical Applications

During the past years, two real-scale applications of low-carbon concretes were successfully carried out in São Paulo, Brazil. In both cases, the above-mentioned criteria were used for proportioning and adjusting the cement pastes' rheological behaviour. Then, after optimization of the aggregate skeleton, trial batches were produced and analysed at fresh state, by rotational rheometry and standard tests, and at harden state measuring the compressive strength. Few adjustments were carried out to meet the production, transportation and placing requirements.

The first experience was related to the construction of an accessory construction (an underground rainwater reservoir) of a multistorey building. The main objectives of this case-study were a) the assessment of the low-carbon concrete manufacturing using the installed concrete production facilities, b) the understanding of the fresh concrete stability over-time along the stages of production, transportation, casting, and finishing, and c) the assessment of casting methods, i.e., placing and pumping. During this first case, the OPC replacement by engineered limestone fillers blend was about 38 %, within the limits of minimum cement content specified in the Brazilian standards. This job demonstrated that it was possible to manufacture and cast low-carbon concrete using the usual production facilities – dosing units, dry-batch concrete mixer trucks, and ordinary concrete pumps. It also showed that it is possible to produce low-carbon concrete with the same rheological behaviour and equivalent performance at hardened state of the commonly used OPC concrete for this kind of application.

The second case study was the construction of the foundation structures of the new experimental building "Sustainable Construction Innovation Centre – CICS Living lab" [53–55] at the University of São Paulo (USP). This was a challenging project due to the type of foundation – continuous flight auger (CFA) piles – and the presence of a 12-metre-long steel cage reinforcement with a geothermal pipe disposed in double-U and helicoidal closed-loop configuration (Fig. 1.14). Due to the presence of the geothermal system, the stiffness of the steel cage was higher than that normally used.

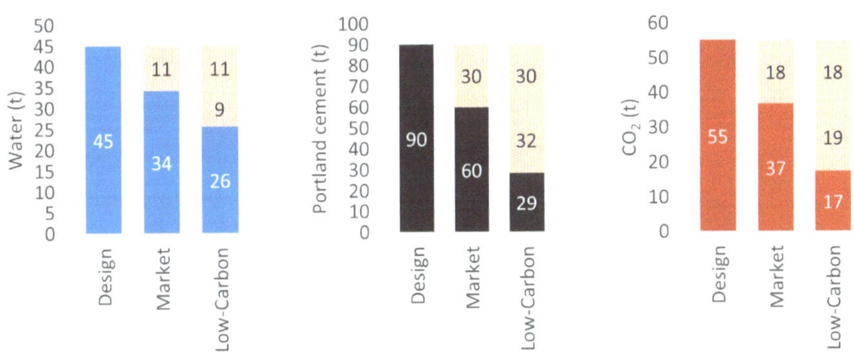

Fig. 1.14   (Left) Details of the steel cage and (right) stages of pile construction: drilling, concrete pumping, steel cage placing.

Fig. 1.15   Comparison among design specification, market practice and produced low-carbon concrete in terms of water consumption (left), Ordinary Portland Cement consumption (middle) and CO_2 emissions (right). The reduction between Market and Low-carbon formulation was 25% for water and 53% for OPC and CO_2.

For the above-mentioned reasons the low-carbon concrete had to maintain a low viscosity and avoid particles segregation after placement and during the reinforcement insertion. Also, the time window between mixing at the concrete plant and positioning the reinforcement was limited to about 3 hours in hot weather.

The results of this case study showed that is possible to reduce the OPC content by approximately 50 % while still meeting the specified rheological and mechanical requirements as per the design (Fig. 1.15). The use of limestone filler introduces a new material

Fig. 1.16   Quality control during execution by rotational planetary rheometry (Pheso from Calmetrix) was essential for adjusting the rheological behaviour.

in the process which makes the production process more complicated. However, since the limestone filler's CO_2 footprint is two orders of magnitude lower than that of OPC (Table 1.1), it contributes with less than 2 % of the total carbon footprint.

Another important result was related to the quality control at the building site. The workability of ready-mix concrete is usually controlled by the slump test. In this case study, the team of the ready-mix company followed their usual procedure, while the USP's team tested each batch by a rotational rheometer (Fig. 1.16). A slump test was performed in samples collected both at the beginning and at the end of casting.

As can be seen in Fig. 1.17, the slump test was unable to identify the low variations registered by the rotational rheometer (Fig. 1.18) and was not effective to guide adjustment of the admixture content. These adjustments, when needed, were realized following the rheometry curves, which allowed identifying limits of the rheological behaviour for an optimal execution.

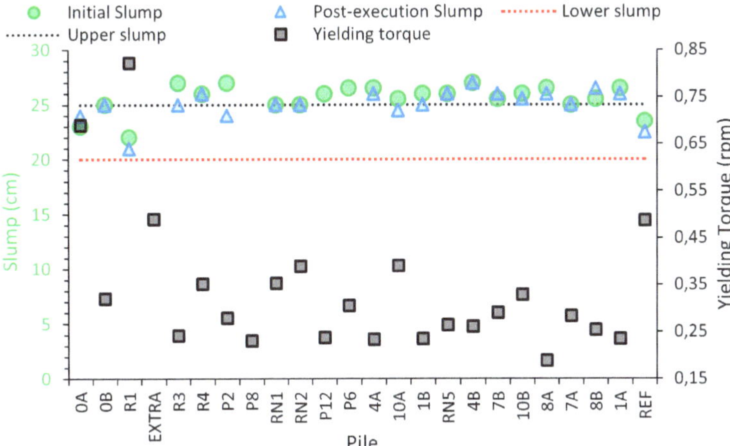

Fig. 1.17 Slump of concrete before and after the pile executions and yielding torque determined by rotational rheometry. Although the slump was always within the limit range (dashed lines) the variation of yielding torque was significant for some piles.

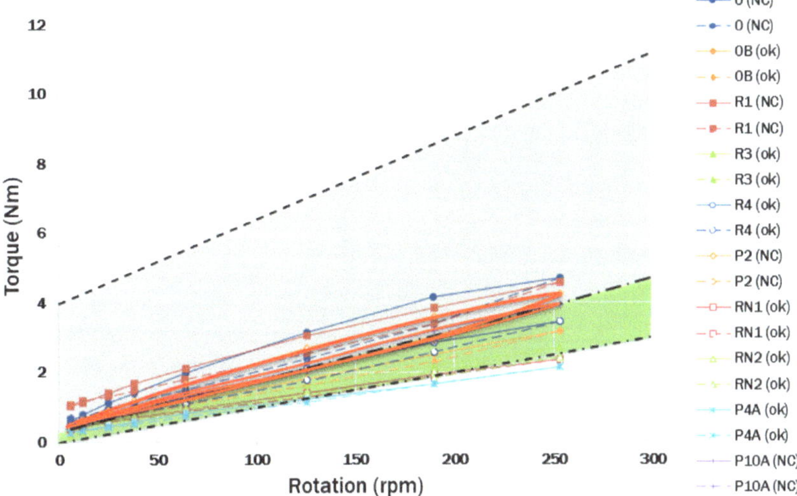

Fig. 1.18 Rotational rheometry results of CFA piles low carbon concrete. The green and grey areas represent limits of pumpable concretes, the green area represents the optimum limits of low carbon concrete for this kind of application, curves above the green area describe concretes that put up some resistance during the insertion of the steel cage.

1.11 Final Remarks

The challenge of low carbon concrete at reasonable cost cannot be delivered by the cement industry alone: it requires a multiscale effort, from materials to design and construction. We discussed in this chapter the challenge of formulating concretes with low carbon footprint.

Producing low carbon concretes requires minimizing mixing water demand at the paste scale. This requires combining particle packing optimization and dispersant admixtures ensuring full dispersion of fines. From our experience, engineering the paste is the first crucial step which is usually neglected on the mix design. There are various models and methods for it, but they are still in their infancy and much research is required.

At concrete scale it is necessary to minimize the paste volume for the given workability. A minimum volume of a paste with minimum water not only minimizes carbon footprint but also reduces shrinkage, heat of hydration and creep, potentially improving durability of the structure.

Our approach to produce low carbon concrete is mainly based on:

- Characterization and selection of raw materials with adequate characteristics in terms of reactivity, particle size and shape;
- Selection and dose of dispersants to guarantee the full dispersion of the system;
- Combining various fines to minimize mixing water for the desired rheology;
- Modelling viscosity and determining the optimum particle size distribution for a specific application;
- Modelling compressive strength by the *cwf*, a parameter linked to the chemically combined water, i.e., the reactivity of the binder, and the water demand for a set rheological behaviour;
- Optimizing aggregates packing to minimize concrete paste volume for the workability required;
- Final adjustment by some trial batches of concrete to attend production, transportation, and placing requirements.

Our understanding of the implications of low carbon concretes formulated by reducing binder content on the long-term performance in various practical situations is limited and much research is needed on this subject. However, it is known that durability of concrete is not a sole function of cement (reactive) content. Concrete porosity and the environment to which the concrete will be exposed are decisive. The actual application, particularly the need for protection of steel reinforcement, is also decisive.

We believe that in many practical situations our approach allows reducing cement content without risks. For example, changing the paste formulation influences the alkaline reserve of the concrete and its carbonation kinetics. However, carbonation is not a degradation factor for reinforced concrete that is indoor, protected from liquid water or very high relative humidities. Neither is it a problem for concrete without steel reinforcement. In those practical situations – that cover a significant fraction of the total concrete market – carbonation is desirable because it captures CO_2 from the environment and increases the concrete's mechanical properties. Also, reducing the amount of cementitious materials may influence the chloride diffusion. However, this is not relevant for applications where chloride is not expected to be present.

We hope that our simplified approach can inspire research and practitioners to engage on carbon mitigation effort, by contributing with the research and development and innovation efforts and by sharing their own practical experiences on reducing the carbon footprint.

1.12 Acknowledgements

The research was supported by CEMtec National Institute on Advanced Eco-Efficient Cement-Based Technologies (Fapesp Project nº2014/50948-3; CNPq Project nº465593/2014-3). The CICS Living lab is supported by companies, and this job was supported by InterCement, ArcelorMittal, and Tupper. P.C.R.A. Abrão's work was supported by FAPESP scholarship (19/13150-7).

Authors acknowledge the CICS and CEMtec teams, particularly Dr. Fabio A. Cardoso, Dr. Roberto Cesar O. Romano for their support on the development. We also thank the CICS living lab team, particularly Arch. Diana Csillag and Prof. Cristina Tsuha.

1.13 References

[1]   Parry I. What Is Carbon Taxation. FINANCE & DEVELOPMENT 2019;56:54–5.

[2]   IEA. South African Carbon Tax. IEA/IRENA Renewables Policies Database 2020. https://www.iea.org/policies/3041-south-african-carbon-tax.

[3]   GCCA. Concrete Future -The GCCA 2050 Cement and Concrete Industry Roadmap for Net Zero Concrete. London: Global Cement and Concrete Association; 2021.

[4]   Xi F, Davis SJ, Ciais P, Crawford-Brown D, Guan D, Pade C, et al. Substantial global carbon uptake by cement carbonation. Nature Geoscience 2016;9:880–3. https://doi.org/10.1038/ngeo2840

[5]   Miller SA, John VM, Pacca SA, Horvath A. Carbon dioxide reduction potential in the global cement industry by 2050. Cement and Concrete Research 2018;114:115–24. https://doi.org/10.1016/j.cemconres.2017.08.026

[6]   National Stone Sand & Gravel Association. THE AGGREGATES INDUSTRY GREENHOUSE GASES: LOW EMISSIONS, HIGH RESILIENCY. Alexandria: National Stone Sand & Gravel Association; 2021.

[7]   Korre A, Durucan S. Aggregates Industry Life Cycle Assessment Model: Modelling Tools and Case Studies. Banbury: WAste & Resource Action Programme; 2009.

[8]   R. Sims, Schaeffer R, F. Creutzig, X. Cruz-Núñez, M. D'Agosto, D. Dimitriu, et al. 2014: Transport. Mitigation of Climate Change. Contribution of Working Group III to the Fifth Assessment Report of the Intergovernmental Panel on Climate Change, Cambridge University Press; 2014, p. 599–670.

[9]   Reis DC, Abrão PCRA, Sui T, John VM. Influence of cement strength class on environmental impact of concrete. Resources, Conservation and Recycling 2020;163:105075. https://doi.org/10.1016/j.resconrec.2020.105075

[10] Abrão PCRA, Cardoso FA, John VM. Efficiency of Portland-pozzolana cements: Water demand, chemical reactivity and environmental impact. Construction and Building Materials 2020;247:118546. https://doi.org/10.1016/j.conbuildmat.2020.118546

[11] John VM, Quattrone M, Abrão PCRA, Cardoso FA. Rethinking cement standards: Opportunities for a better future. Cement and Concrete Research 2019;124:105832. https://doi.org/10.1016/j.cemconres.2019.105832

[12] Damineli BL, Kemeid FM, Aguiar PS, John VM. Measuring the eco-efficiency of cement use. Cement and Concrete Composites 2010;32:555–62. https://doi.org/10.1016/j.cemconcomp.2010.07.009

[13] Scrivener K, John VM, Gartner E. Eco-efficient cements: Potential, economically viable, solutions for a low-CO2, cement-based materials industry. Paris: UN Environment; 2017.

[14] Towery ND, Machek E, Thomas A. Technology Readiness Level Guidebook. Cambridge: U.S. Department of Transportation; 2017.

[15] Holcim. ECOPact LOW CARBON CONCRETE MADE EASY. (Sidney): Holcim Australia New Zealand; 2021.

[16] Boral SA Region. Pre-mix Concrete EPD. ECO Platform EPD Verified; 2021.

[17] Holcim Australia Ready-Mix Concrete. ViroDecs™ Special – Ecpáct Ramge. Australia: EPD Australasia Limited; 2022.

[18] Silva FB, Oliveira LA, Yoshida OS, John VM. Variability of environmental impact of ready-mix concrete: a case study for Brazil. IOP Conf Ser: Earth Environ Sci 2019; 323:012132. https://doi.org/10.1088/1755-1315/323/1/012132

[19] Cheung J, Roberts L, Liu J. Admixtures and sustainability. Cement and Concrete Research 2018;114:79–89. https://doi.org/10.1016/j.cemconres.2017.04.011

[20] John VM, Damineli BL, Quattrone M, Pileggi RG. Fillers in cementitious materials — Experience, recent advances and future potential. Cement and Concrete Research 2018;114:65–78. https://doi.org/10.1016/j.cemconres.2017.09.013

[21] Proske T, Hainer S, Rezvani M, Graubner C-A. Eco-friendly concretes with reduced water and cement contents — Mix design principles and laboratory tests. Cement and Concrete Research 2013;51:38–46. https://doi.org/10.1016/j.cemconres.2013.04.011

[22] Vogt C. Ultrafine particles in concrete: Influence of ultrafine particles on concrete properties and application to concrete mix design. Doctoral thesis. KTH, 2010.

[23] Damineli BL, John VM, Lagerblad B, Pileggi RG. Viscosity prediction of cement-filler suspensions using interference model: A route for binder efficiency enhancement. Cement and Concrete Research 2016;84:8–19. https://doi.org/10.1016/j.cemconres.2016.02.012

[24] Flatt RJ, Houst YF. A simplified view on chemical effects perturbing the action of superplasticizers. Cement and Concrete Research 2001;31:1169–76. https://doi.org/10.1016/S0008-8846(01)00534-8

[25] TESTING ASF, MATERIALS. ASTM C1679: Standard Practice for Measuring Hydration Kinetics of Hydraulic Cementitious Mixtures Using Isothermal Calorimetry 2014.

[26] Scrivener K, Snellings R, Lothenbach B. A practical guide to microstructural analysis of cementitious materials. Crc Press; 2016.

[27] Kantro DL. Influence of Water-Reducing Admixtures on Properties of Cement Paste – A Miniature Slump Test. CCA 1980;2:95–102. https://doi.org/10.1520/CCA10190J

[28] Abrão PCRA, Cardoso FA, Cecel RT, John VM. Evaluation of Portland cements considering chemical reactivity and mixing water demand: combined water fraction fundamentals and practical application. Cement and Concrete Composites 2022; Submitted.

[29] Ferraris CF, Obla KH, Hill R. The influence of mineral admixtures on the rheology of cement paste and concrete. Cement and Concrete Research 2001;31:245–55.

[30] Damineli BL, Pileggi RG, Lagerblad B, John VM. Effects of filler mineralogy on the compressive strength of cementitious mortars. Construction and Building Materials 2021;299:124363. https://doi.org/10.1016/j.conbuildmat.2021.124363

[31] Funk JE, Dinger D. Predictive Process Control of Crowded Particulate Suspensions: Applied to Ceramic Manufacturing. Springer Science & Business Media; 1994.

[32] Westman AER, Hugill HR. The Packing of Particles. Journal of the American Ceramic Society 1930;13:767–79. https://doi.org/10.1111/j.1151-2916.1930.tb16222.x

[33] de Larrard F, Sedran T. Optimization of ultra-high-performance concrete by the use of a packing model. Cement and Concrete Research 1994;24:997–1009. https://doi.org/10.1016/0008-8846(94)90022-1

[34] Burroughs JF, Shannon J, Rushing TS, Yi K, Gutierrez QB, Harrelson DW. Potential of finely ground limestone powder to benefit ultra-high performance concrete mixtures. Construction and Building Materials 2017;141:335–42. https://doi.org/10.1016/j.conbuildmat.2017.02.073

[35] Chindaprasirt P, Homwuttiwong S, Sirivivatnanon V. Influence of fly ash fineness on strength, drying shrinkage and sulfate resistance of blended cement mortar. Cement and Concrete Research 2004;34:1087–92. https://doi.org/10.1016/j.cemconres.2003.11.021

[36] Curcio F, DeAngelis BA, Pagliolico S. Metakaolin as a pozzolanic microfiller for high-performance mortars. Cement and Concrete Research 1998;28:803–9. https://doi.org/10.1016/S0008-8846(98)00045-3

[37] Mora EP, Paya J, Monzo J. Influence of different sized fractions of a fly ash on workability of mortars. Cement and Concrete Research 1993;23:917–24. https://doi.org/10.1016/0008-8846(93)90045-b

[38] Senff L, Labrincha JA, Ferreira VM, Hotza D, Repette WL. Effect of nano-silica on rheology and fresh properties of cement pastes and mortars. Construction and Building Materials 2009;23:2487–91. https://doi.org/10.1016/j.conbuildmat.2009.02.005

[39] Senff L, Barbetta PA, Repette WL, Hotza D, Paiva H, Ferreira VM, et al. Mortar composition defined according to rheometer and flow table tests using factorial designed experiments. Construction and Building Materials 2009;23:3107–11. https://doi.org/10.1016/j.conbuildmat.2009.06.028

[40] EN B. 12350-2, Testing fresh concrete: slump test. London: British Standards Institution 2000.

[41] American Society for Testing and Materials. ASTM-C143: Standard test method for slump of hydraulic-cement concrete 2012.

[42] Banfill PFG. Rheology of fresh cement and concrete. Rheology Reviews 2006;2006:61.

[43] Cazacliu B, Roquet N. Concrete mixing kinetics by means of power measurement. Cement and Concrete Research 2009;39:182–94. https://doi.org/10.1016/j.cemconres.2008.12.005

[44] Powers TC, Brownyard TL. Studies of the Physical Properties of Hardened Portland Cement Paste. JP 1946;43:101–32. https://doi.org/10.14359/8745

[45] John VM, Quattrone M, Abrão PCRA, Cardoso FA. Rethinking cement standards: opportunities for a better future. Cement and Concrete Research (Acepted to Publication) 2019.

[46] Kosmatka SH, Wilson ML. Design and control of concrete mixtures: the guide to applications, methods, and materials. 15th ed. Skokie, Ill: Portland Cement Association; 2011.

[47] Rozière E, Granger S, Turcry Ph, Loukili A. Influence of paste volume on shrinkage cracking and fracture properties of self-compacting concrete. Cement and Concrete Composites 2007;29:626–36. https://doi.org/10.1016/j.cemconcomp.2007.03.010

[48] de Larrard F, Sedran T. Optimization of ultra-high-performance concrete by the use of a packing model. Cement and Concrete Research 1994;24:997–1009. https://doi.org/10.1016/0008-8846(94)90022-1

[49] Mehta PK, Monteiro PJM. Concrete: microstructure, properties, and materials. 3rd ed. New York: McGraw-Hill; 2006.

[50] Jónsson G. CO2 uptake during the concrete life cycle: Information on the use of concrete in Denmark, Sweden, Norway and Iceland. Reikjavik: Icelandic Building Research Institute; 2005.

[51] de Larrard F. Concrete mixture proportioning: a scientific approach. London ; New York: E & FN Spon; 1999. https://doi.org/10.1201/9781482272055

[52] de Larrard F, Sedran T, Brochu F. Test method n°61 : Evaluation of packing density of granular fractions with a shaking table. 2004.

[53] CICS n.d. https://cics.prp.usp.br/ (accessed April 4, 2022).

[54] CICS TALKS: Concreto LEAP de baixa pegada de CO2 em fundações. 2020.

[55] Belizario-Silva F, Galimshina A, Reis DC, Quattrone M, Gomes B, Marin MC, et al. Stakeholder influence on global warming potential of reinforced concrete structure. Journal of Building Engineering 2021;44:102979. https://doi.org/10.1016/j.jobe.2021.102979

Chapter 2

Life Cycle Assessment in Support of the Sustainability of Current and Alternative Types of Structural Concrete

José D. Silvestre

CERIS, Department of Civil Engineering, Architecture and Georesources, Instituto Superior Técnico, Universidade de Lisboa, Portugal

2.1 Introduction

The rapidly growing world population has driven the highest ever urbanization rates. This has created an unusual requirement for infrastructure projects and consequently for concrete production. This unprecedented demand for concrete results in an increasing risk of global warming and high energy consumption that significantly deteriorates the environment. One way to reduce this impact is to recycle Industrial Waste Materials (IWM), such as Supplementary Cementitious Materials (SCMs) or recycled aggregates, in concrete. For that purpose, researchers have been focused on identifying the effect of integrating industrial waste as substitutes of Ordinary Portland Cement (OPC) or aggregates on the cost, environmental impact, and quality of concrete.

The use of concrete with alternative raw materials strongly relies on its potential economic and environmental advantages. As a matter of fact, this type of concrete is expected to have a similar or worse performance than the corresponding current concrete. In other words, the use of alternative raw materials in concrete production can only be justified if these potential economic and environmental advantages can be demonstrated.

This chapter is focused on the environmental impacts issue but also on the economic performance of current and alternative structural concrete mixes, based on results from many case studies. This chapter thus comprises three main parts. First, the summarized description of the environmental Life cycle assessment (LCA) methodology, and examples of its application to traditional and alternative raw materials for concrete, and to concrete use and end-of-life, are provided (Section 2.2). After that, the main results of economic and environmental LCA studies of traditional and alternative concrete mixes are described (Section 2.3). Finally, it is highlighted how LCA can support the design of concrete structures (Section 2.4).

2.2 Environmental Life Cycle Assessment of Concrete

LCA studies and the communication of their results have evolved since the 1960s towards an increasing specification and standardisation (ISO 14040 series [1, 2]), that allows more objective studies and conclusions. It allows the quantification of all relevant emissions, resources consumed, and the related potential environmental and health impacts as well as resource depletion issues that are associated with any goods or services ("products") [1]. Nowadays, LCA is a complex and data-intensive methodology that considers the potential impacts of the whole life cycle (LC) of a product, from the extraction of resources, through production, use, and recycling up to the disposal of the remaining waste. Specific standards guide LCA application to the construction sector [3].

LCA methodology comprises four analytical phases interlinked by iterative cycles [1, 2]:

1. Definition of the goal and scope of the assessment;
2. Life Cycle Inventory analysis (LCI);
3. Life Cycle Impact Assessment (LCIA);
4. Life Cycle Impact interpretation.

The definition of the goal and scope of the assessment comprises the description of the product to be assessed, the boundary of the associated system (including the definition of the cut-off criterion), and the functional unit. It also includes the selection of the environmental interventions and of the method (or methods) of environmental impact assessment, the definition of the goal of the study, its intended audience, strategy for data collection, and assumptions and limitations of the study.

Other book chapters from the author propose methods to define an equivalent functional unit for recycled aggregates concrete (RAC) [4] and for concrete with IWM [5] based on mechanical and durability properties, and that can be used in LCA.

The boundaries of the system subject to a LCA can be defined as from cradle to gate (including the extraction and processing of raw materials and the production), from cradle to grave (including also the transport, distribution and assembly, the use, maintenance and final disposal), or from cradle to cradle (also including the reuse, recovering and/or recycling potentials) (see Fig. 2.1) [6, 7]. The majority of LCA studies of concrete define a cradle to grave scope [8].

A Life Cycle Thinking approach [9] is usually considered in the construction sector by implementing an LCA. LCA is an important tool to identify, quantify, and prevent the most relevant environmental impacts over the life cycle of buildings. The application of LCA to the construction materials' industries makes manufacturers look at the processes from the production stage until the end-of-life of a product, and acknowledge and reduce their global environmental, social and/or economic impacts. Besides improving their processes, manufacturers are also challenged to choose more sustainable raw materials and energy sources, reduce transport distance, and improve their products' performance at installation, during operation, and at end-of-life.

The application of environmental LCA and economic LCA (or Life Cycle Cost – LCC) to construction are methodologies recognized and standardized at the European level [3], which are becoming more and more important, and which are being increas-

Modules	Product stage (A1-A3)	Transport to the building site (A4)	Installation in the building (A5)	Use stage		End-of-life stage - transport, processing and disposal (C2-C4)	Reuse, recovery and/or recycling potential (D)
				Maintenance, repair and replacement (B2-B4)	Energy use for heating and cooling (B6)		

Life cycle approach	Cradle to gate	Cradle to cradle	
		Gate to grave	
		Cradle to grave	

Fig. 2.1 Life cycle stages of materials, building assemblies and buildings, and boundaries of LCA studies

ingly used in this sector by experts and researchers. The LC paradigm emerges, and construction materials and buildings now represent the sum of all impacts and costs in the respective LC (no longer seen as an individual impact and cost). Standardised life cycle stages for the construction sector are presented in Fig. 2.1, along with the common boundaries of LCA studies.

The LCA of construction materials or assemblies and buildings provides quantitative data (environmental impacts) that are essential in ensuring that the designer chooses the most environmental-friendly options. However, LCA results for alternative construction products can only be compared if these products are functionally equivalent, which means that they have similar technical and functional performance. This is primordial in a design project where performance requirements are defined in advance, thus setting a comparative benchmark for all functionally equivalent alternatives evaluated for a given function. The "functional equivalent" is defined as the "quantified functional requirements and/or technical requirements for a building or an assembled system (part of works) for use as a basis for comparison" [3].

In the end, only a fair comparison between competing alternatives similar in functional terms will allow achieving, at the design project:

- The minimisation of the consumption of materials, energy and water;
- The elimination or reduction of the use of hazardous materials;
- The prevention or the minimisation of emissions and waste.

2.2.1 Life Cycle Inventory (LCI)

The LCI phase comprises the inventory, compilation, and quantification of the inputs and outputs of the system for the processes included along the life cycle of the product. This comprises the collection and treatment of the data related to these flows (of mass – quantification of materials, energy, and emissions to the air, water, or soil). The LCI of the product is presented in inventory tables that show the material and energetic balance of the system. LCI studies finish with the analysis of the inventory obtained.

The LCI of a concrete mix is divided into upstream processes, which include the energy, emissions and resource use, the latter including the extraction and transportation of all raw materials included in the mix until they reach the concrete batch plant [10]. The second group of processes is concerned with the concrete mixing process, the transport to the construction site as well as the casting, vibration, and moulding. This former group is called the core processes. The final group of processes, the downstream ones, are concerned with the maintenance, demolition, and any other impact resulting from the end-of-life phase. This means that, for the calculation of a cradle to grave LCA of a concrete mix, which is the recommended scope, there needs to be a database that includes the data concerning the resource use, energy, and emissions for all the input/output flows of the processes in the downstream, core, and upstream groups [8].

LCA Databases for Raw Materials and Construction Products

Available environmental LCA databases of raw materials and construction products at an international level include: Environmental Product Declaration (EPD) registration systems with site-specific data from each producer and generic databases developed for specific regional contexts (Fig. 2.2) [11]. It is always recommended to use specific (primary) over generic data to improve the reliability of the LCA, and also maximising the geographical representativeness of the datasets considered.

The LCA of a product system needs data related to inputs and outputs of the materials, products, or components under analysis. These data correspond to energy and mass flows within the system boundary or the life cycle. The collection of the necessary information to characterise these flows is, simultaneously, time and resources-consuming. This concern leads to the development of generic databases to model an adequate LCI, thus supporting LCA activity [12].

LCA tools are characterised by the databases included in their standard version but also by the databases that can be bought individually and additionally. The possibility of importing data included in commercial (external) databases is also an important advantage, along with the option of editing available databases and creating new ones. An LCA study can be developed using data from a single database, or from the combination of information from several databases [13]. In fact, databases are a critical issue in LCA of buildings because their availability (or absence) within the software tool allows (or avoids) the supplying of the necessary information for an adequate environmental evaluation of the building [12].

Databases for "materials and processes" include inputs and outputs of the production of materials (e.g., aluminium) and of their processing (e.g., hot lamination), as well as the corresponding data related to upstream (i.e., energy and transportation) and downstream processes (e.g., disposal or recycling), in order to allow the modelling of the system process of a given product through the association of materials and processes. The most relevant databases included in this group that can be used for the LCA of concrete are [14–15]:

- Ecoinvent version 3.0 (Switzerland) – developed by the Swiss Centre for Life Cycle Inventories, is included in the most commonly used LCA tools and comprises LCI data and extensive background reports from 4,000 industrial processes (including energy-related data – energy mix of more than 25 European countries, transportation, construction materials, metals, wood-based and chemical products, paper

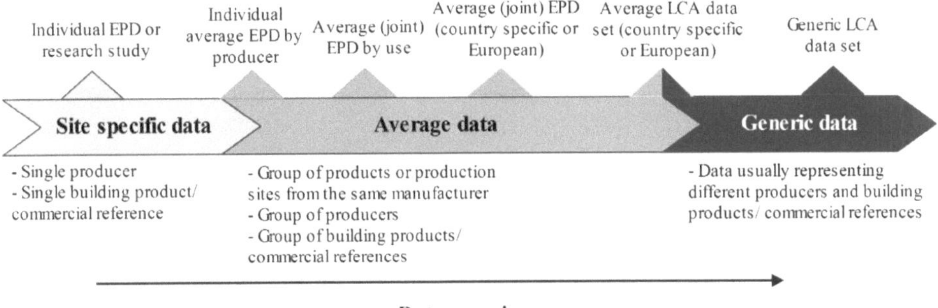

Fig. 2.2 Available environmental LCA datasets of construction products according to their genericness

and cardboard, and waste management) built with data mainly from Switzerland and Western Europe (average industry, survey or literature based, but also processes from the compilation of other databases);

– European Federation of Concrete Admixtures Associations (EFCA) – this organisation freely provides detailed eco-profiles (EPD) of six types of concrete admixtures (plasticisers, superplasticisers, retarders, accelerators, air entraining and water resisting admixtures).

2.2.2 Life Cycle Impact Assessment and Interpretation

The LCIA phase comprises the estimation and evaluation of the magnitude and significance of the potential environmental impacts of a product throughout its life cycle. This evaluation is based on the environmental interventions obtained in the LCI (e.g., Chlorofluorocarbons – CFC, CO, CO_2, Halon, and Methane – CH_4) and in the impact categories chosen (between 10 and 20) via a process of classification and characterisation. This process makes use of an Environmental Impact Assessment Method (EIAM), which converts the LCI results obtained in midpoint (e.g., potential of ozone layer depletion or greenhouse effect) or endpoint impacts (e.g., damage to human health or to the ecosystems) assuming a cause-effect relationship with different characterisation factors for each pair intervention-environmental impact. The LCIA phase can also include optional elements:

– Normalisation – calculation of the magnitude of each category indicator in relation to reference information;
– Grouping – comprising the sorting and possible ranking of the impact categories;
– Weighting – conversion, and possible aggregation across impact categories, of the results of the indicators using numerical factors based on value-choices.

The "Life Cycle Impact interpretation" phase comprises the identification, interpretation, and evaluation of the most important results obtained in LCI and LCIA phases, in accordance with the goal and scope defined for the study, to confirm whether they are

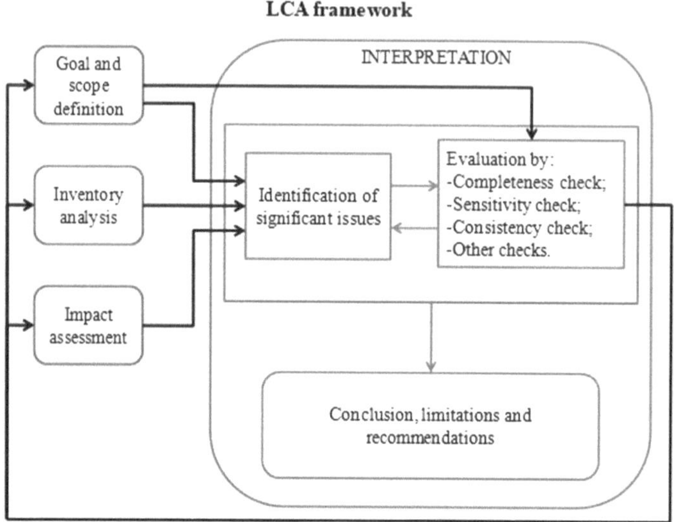

Fig. 2.3 Relationships between elements within the interpretation phase with the other phases of LCA (adapted from [2])

supported by the initial data and the model chosen (Fig. 2.3). The final report contains the main conclusions of the study and the necessary recommendations for the production and use of the product. It may also contain data on the product that are relevant for the analysis but have not been included in LCI or LCIA, such as the environmental impacts during use stage, the durability of the product or the sensitivity analysis of some of the results obtained (see Section 2.2.4).

LCA involves modelling and cannot be seen as a complete reflection of the real world. Therefore, all the decisions related to the modelling, such as data reliability and uncertainties, the effects to be included, the boundaries to be set, cut-off rules, or the interpretation, must be well documented [16]. The LCA study can also be subjected to a critical review to ensure the consistency of the study with the principles and requirements of these standards [1–2].

2.2.3 Life Cycle Assessment of Traditional and Alternative Raw Materials for Concrete

Braga *et al.* [17] compared the environmental LCA of natural aggregate concrete (NAC) and RAC mixes from cradle to gate based on previous studies. All LCI data for coarse aggregates and crushed fine aggregate, coarse aggregates recycled from concrete (CARC), and concrete production were collected from Portuguese companies. Data for river fine aggregate were collected from Marinkovic *et al.* [18]. Cement data were obtained from Blengini [19], in a study based on the production of a Portuguese company. In summary, the main conclusions of this study were (Table 2.1):

- Mixes with maximum environmental impact (EI) do not contain superplasticizer (SP) in their composition, but it is also advised to use SP to decrease cement content which is the main factor responsible for the EI;
- The water to binder ratio (w/b) reduction does not show significant improvement in EI;
- The use of CARC results in a reduction of EI. Nevertheless, transportation has a significant role in the overall impact of aggregates, depending on the location of production or processing and concrete plants.

Table 2.1 Influence of each raw material in the environmental performance of concrete

Raw materials	EI
Cement: CEM I	-
Cement: CEM II	+
River aggregate	+
Crushed aggregate (all sources)	-
Granitic coarse aggregate	+
Limestone coarse aggregate	-
CARC	++
SP +	+

Note: + represents a reduction of impact; ++ represents a significant decrease of impact; - represents an increase of impact

Kurda *et al.* [20] made a literature review on the EI of producing recycled concrete aggregates (RCA), fly ash (FA), cement, SP, and water as raw materials, and also on the effect of replacing cement and natural aggregates (NA) in the production of concrete with FA and RCA, respectively. This study mainly focuses on the cradle to gate EI of one cubic metre of concrete. The results also include the economic performance of these materials and show that the incorporation of FA in concrete significantly decreases the EI and cost of concrete. Thus, the simultaneous incorporation of FA and RCA decrease the EI, cost, use of landfill space, and natural resources extraction. Relative to FA, the incorporation of RCA does not significantly affect the EI and cost of concrete, but it significantly reduces the use of landfill space and the need of virgin materials.

Cement

The selection of the most suitable LCA databases, and corresponding elementary processes, to be used in the modelling of raw materials and related processes for concrete production, is of paramount importance and decisively conditions LCA results. The NativeLCA methodology is focused on the selection and/or adaptation of a coherent LCA dataset to produce each raw material or building product to be used as generic data for a national context [11], thus supporting the development of regional and sector-specific LCA databases.

A LCA study tested the plausibility of available national environmental datasets for the production of cement and selected reliable LCA datasets to be used as generic for a national context, using NativeLCA methodology [21]. The results include coherent LCA datasets for the cement product stage (A1–A3), based on the adaptation of available European (foreign or national) LCA datasets, that can be used as generic in LCA studies (in Portugal, in this case).

Since there are different types of cement with different compositions and strengths in the market, the aim of this study was to make available at least one LCA dataset for the production of an average cement type to be used in the modelling of the production of cement-based building products.

The selection of a coherent LCA dataset for cement starts with the description of its goal and scope. The goal was presented in the previous paragraph and the scope is defined by:

- Declared unit of the study: production of one tonne of cement;
- Characterization of the building product studied in this research: the building product corresponds to different types of Portland composite cement (CEM II), since it is the most used in Portugal, available within the European market;
- LCI flows and LCIA parameters (and corresponding EIAM): GWP (Global Warming Potential), PE-NRe (consumption of primary energy, non-renewable) and Acidification Potential (AP);
- Only the product stage (A1 to A3) is considered, without taking into account the packaging of the cement.

Table 2.2 identifies and quantifies all datasets available for the production of CEM II in the national and European context.

Table 2.2   Datasets for the production of CEM II cements identified at the European level

Product	LCA datasets						
	National (Portugal)			European			
	Site specific data from national LCA studies/individual EPD	Joint EPD	National average LCA datasets	Individual EPD	Joint EPD	Country specific or European average LCA datasets	Generic LCA dataset
Cement (CEM II)	2	0	0	7	11	0	9

Besides the quantification of the datasets available in the sample, it is also important to characterize each foreign dataset (qualitatively) by their meta-data and verify their consistency and representativeness. It was concluded that:

- All datasets were obtained using the same European LCA normative framework, although using different Product Category Rules (PCR);
- All datasets have cut-off and allocation rules defined, with the exception of generic datasets from the Ecoinvent and Ökobaudat databases, for which it was not possible to identify these rules;

- Furthermore, the EPD from EPDItaly do not clarify which are the cut-off rules and the EPD from GlobalEPD does not define the allocation rules;
- All datasets have similar system boundaries (A1-A3, excluding the packaging material);
- All EPD were subjected to external review, while generic datasets from Ecoinvent were only critically reviewed internally.

All datasets are therefore consistent in what concerns assumptions, methods, models, and data used in their calculation. The LCI flows and LCIA parameters provided in each dataset are consistent within each other but not similar between different datasets. Nevertheless, they are all in accordance with the goal and scope of this study.

A representativeness check of the selected datasets was also performed and allowed concluding that:

- All datasets are specific from a country (France, Spain, Germany, Norway, Turkey, and Italy), with the exception of the Ecoinvent datasets that have European geographical representativeness;
- Technological representativeness can be considered as "classic" in the European context for all datasets;
- All datasets include different compositions of CEM II. Therefore, the interest of separating the datasets per type of cement during the comparison of the results achieved was assessed;
- EPDItaly datasets do not identify the background LCA data used to model the electricity and transport processes. All other datasets had used European background datasets;
- Ecoinvent datasets are the oldest ones (inventory from 2005, although extrapolated to 2012) while the remaining ones resulted from studies performed in the last eight years;
- Ecoinvent datasets are the only ones that offer the possibility of modifying background data in order to provide "contextualisation".

Therefore, a "contextualisation" was made to Ecoinvent datasets (namely by changing the electricity production mix from the European to the Portuguese context) and only the "contextualised" values have been considered in the remaining steps of this study. After this step, all datasets were considered valid in terms of consistency and representativeness, even though some of them do not comply with some of the criteria defined.

18 datasets were selected for the next steps of the methodology (REVA – Reference Values quantification): seven from individual and 11 joint EPDs. Fig. 2.4 presents a general overview of the environmental impacts of these datasets (in GWP, Abiotic Depletion Potential fossil fuels – ADP (f.f.), and AP for the production of one tonne of CEM II). Taking into account the different types of cement within the sample, it was decided to calculate six different European REVA: one European REVA including all datasets of the sample (represented in Fig. 2.5, by REVA_EU) and five European REVA for the following CEM II cement types: composite, limestone, fly ash cement, slag, and pozzolana cement (represented in Fig. 2.5, respectively by REVA_CEM II (M), REVA_CEM II (L), REVA_ CEM II (V), REVA_CEM II (S), and REVA_CEM II (P)). Each REVA was calculated as

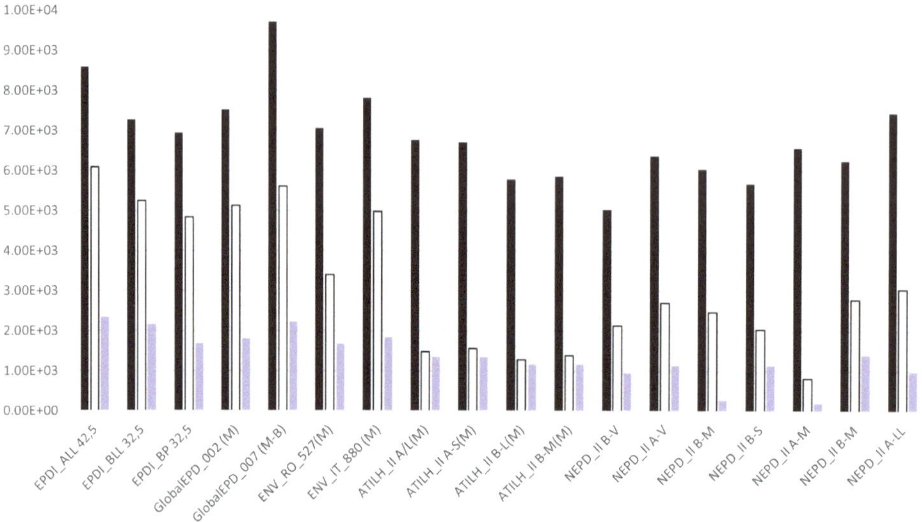

Fig. 2.4   Environmental impacts resulting from the production of one tonne of cement for all datasets available in the sample: GWP (in black, x 10^{-1} kgCO$_2$eq; ADP (f.f.) (in white, MJ); and AP (in grey, x 10^{-3} kgSO$_2$eq)

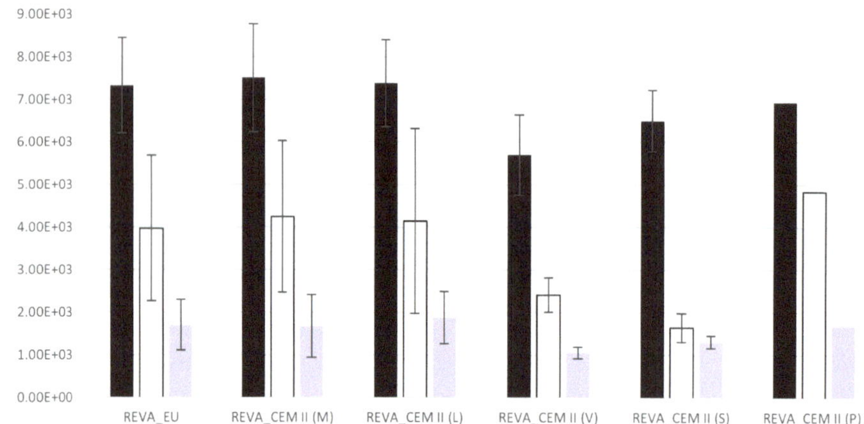

Fig. 2.5   European REVA of the environmental impacts resulting from the production of one tonne of cement (European REVA including all datasets available in the sample, CEM II (M), CEM II (L), CEM II (V), CEM II (S), and CEM II (P)): GWP (in black, x 10^{-1} kgCO$_2$eq; ADP (f.f.) (in white, MJ); and AP (in grey, x 10^{-3} kgSO$_2$eq)

a weighted arithmetic mean of all joint and individual datasets (taking into account the number of production plants included in each dataset, since not all producers declared production volumes), for each environmental indicator and for the same declared unit. Fig. 2.5 also shows the standard deviation of each REVA.

Fig. 2.5 allowed concluding that the variability of the REVA is directly dependent on the number of datasets available for each type of cement. Although REVA for pozzolan

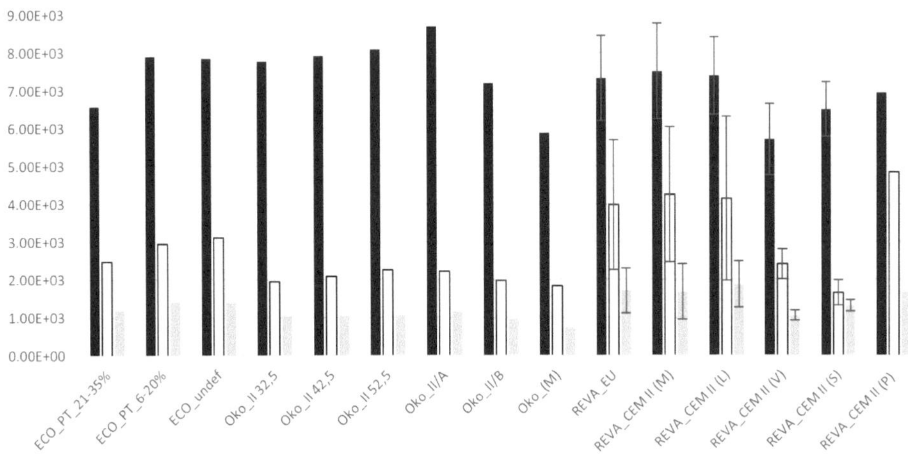

Fig. 2.6 Environmental impacts resulting from the production of one tonne of cement for European REVA, and generic datasets: GWP (in black, x 10^{-1} kgCO$_2$eq; PE-NRe (in white, MJ); and AP (in grey, x 10^{-3} kgSO$_2$eq)

cement was represented by only one dataset, the authors decided not to discard this REVA at this step of the NativeLCA methodology application, in order to guarantee the representativeness of this type of cement in the sample. Moreover, since six European REVA were calculated and there are two site-specific Portuguese datasets added to figures far from European average, the average datasets from French ATILH (Association Technique de l'Industrie des Liants Hydrauliques), Italian Global EPD (with the exception of pozzolan Portland cement), and Romanian Environdec datasets were discarded in the next steps.

Fig. 2.6 presents the comparison of REVA and generic datasets, where Ecoinvent data contextualized to Portugal are identified as ECO_PT, and the datasets from Ökobaudat database are identified by the initials Oko. It was concluded that the average generic dataset (for the Portland cement not specified) of Ökobaudat (Oko_(M)) is out of the range of the REVA_EU and REVA_CEM II (M) standard deviation in all environmental impact categories. A justification for that difference cannot be assessed due to shortage of meta-data available in generic Ökobaudat datasets. However, the proximity of figures for GWP can be related with the link between GWP and CO$_2$ release during the production process of clinker, since that relationship is similar for the whole European context for any type of cement [22, 23]. On the other hand, lower figures of PE-NRe (from Ökobaudat datasets) can be related to the increase of secondary fuel use in the production process of cement, instead of fossil fuels (in order to decrease environmental impacts from cement production in Germany). REVA values were also validated by Ecoinvent datasets, since all figures are coherent between them. Therefore, the generic dataset related to the average cement of Ökobaudat (Oko_(M)) database was discarded at this step. All other datasets validate the REVA calculated for GWP and AP, but are slightly below the REVA in ADP (f.f.).

Fig. 2.7 presents the comparison of the remaining generic datasets, European REVA, and of the Portuguese site-specific datasets obtained by [19]. From that comparison, it

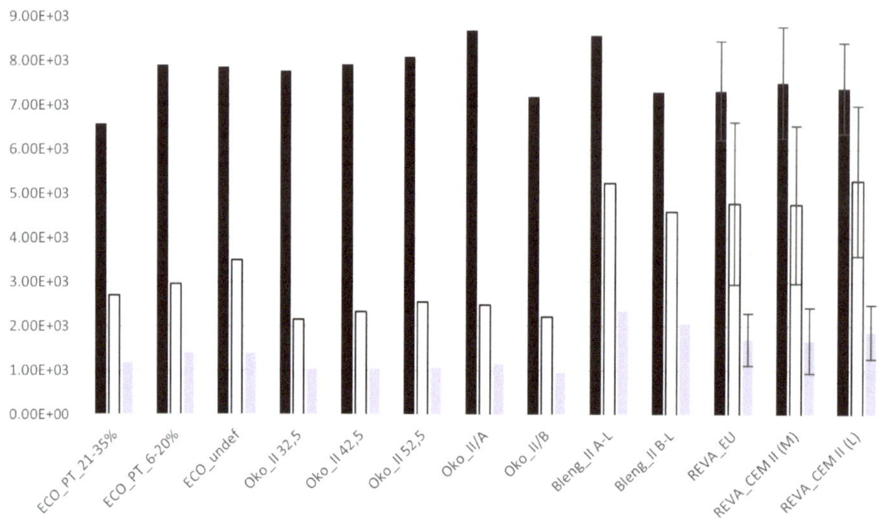

Fig. 2.7   Environmental impacts resulting from the production of one tonne of cement for European REVA, generic datasets and Portuguese site-specific datasets: GWP (in black, x 10^{-1} kgCO$_2$eq; PE-NRe (in white, MJ); and AP (in grey, x 10^{-3} kgSO$_2$eq)

was possible to validate the Portuguese values related to GWP but not to PE-NRe (environmental category chosen for the comparison since Blengini's figures did not include ADP (f.f.)) and AP, since the national figures are above those from generic datasets. These differences can be justified by the use of secondary fuel in the production processes of the Ökobaudat and Ecoinvent data, instead of non-renewable abiotic resources. Nevertheless, the comparison between site-specific data of the two Portland limestone cements (with high and low clinker content) with REVA values (REVA_EU, REVA_CEM (M), and REVA_CEM II (L)) presented in Fig. 2.8, allowed validating the Portuguese figures, since all environmental impact figures are coherent between datasets, although the national ones are higher than the mentioned REVA for the three categories assessed, but within the standard deviation of each REVA. Therefore, the Portuguese site-specific datasets were considered plausible to be used as generic data in the Portuguese context for cement production.

Considering the conclusions drawn from the analysis made, LCA datasets from Ecoinvent (contextualised for the Portuguese geography) were selected to be used as generic data within the Portuguese context for CEM II/A and CEM II/B. Its selection, instead of the Portuguese site-specific datasets obtained by Blengini, is justified by its better temporal representativeness and because the site-specific Portuguese data do not include the environmental impact category ADP (f.f.) (although its plausibility to be used had been assessed and confirmed). Furthermore, REVA_EU was selected to be used within the Portuguese context as generic data whenever the composition of the cement is not known. Table 2.3 summarises the decisions that have been made in each of the steps of the application of the NativeLCA methodology to cement production.

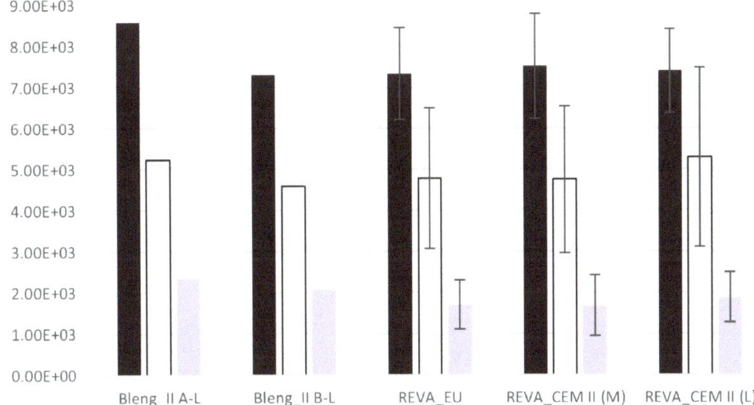

Fig. 2.8 Environmental impacts resulting from the production of one tonne of cement for REVA and Portuguese site-specific datasets: GWP (in black, x 10^{-1} kgCO$_2$eq; PE-NRe (in white, MJ); and AP (in grey, x 10^{-3} kgSO$_2$eq)

Table 2.3 Summary of the decisions made at each of the steps of the application of NativeLCA methodology to cement production

National and foreign data verification	Suitability to be used in the quantification of REVA for LCI and LCIA indicators and REVA quantification (EPD and average datasets)	Data comparison within foreign data: REVA vs generic datasets, and site-specific vs REVA vs foreign datasets	Selection of a coherent LCA dataset to be used as generic for a national context: NativeLCA
Consistency and representativeness			
None dataset discarded	Calculation of European REVA and REVA for cements CEM II type M, L, V, S and P	One generic dataset discarded because of lack of technological representativeness. Confirmed the plausibility of Portuguese site-specific datasets for Portland limestone cements	Ecoinvent generic datasets contextualized for CEM II/A and CEM II/B;
			European (REVA_EU) for Portland cements with undefined composition

Aggregates

Estanqueiro *et al.* [24] performed a LCA on RA used in the manufacture of ready-mixed concrete, and also on this structural material itself. The scenarios defined for the environmental LCA of coarse NA and RA for concrete were: 1 – the use of NA in the manufacture of concrete; 2 – the use of RA in the manufacture of concrete using a recycling fixed plant; and 3 – the use of RA in the manufacture of concrete using a recycling mobile plant. Fig. 2.9 illustrates these scenarios and the different options in the critical stages of the life cycles of NA and RA that influenced the results of this study most. These scenarios were considered in their sensitivity analysis: alternative A – use of fine RA in concrete,

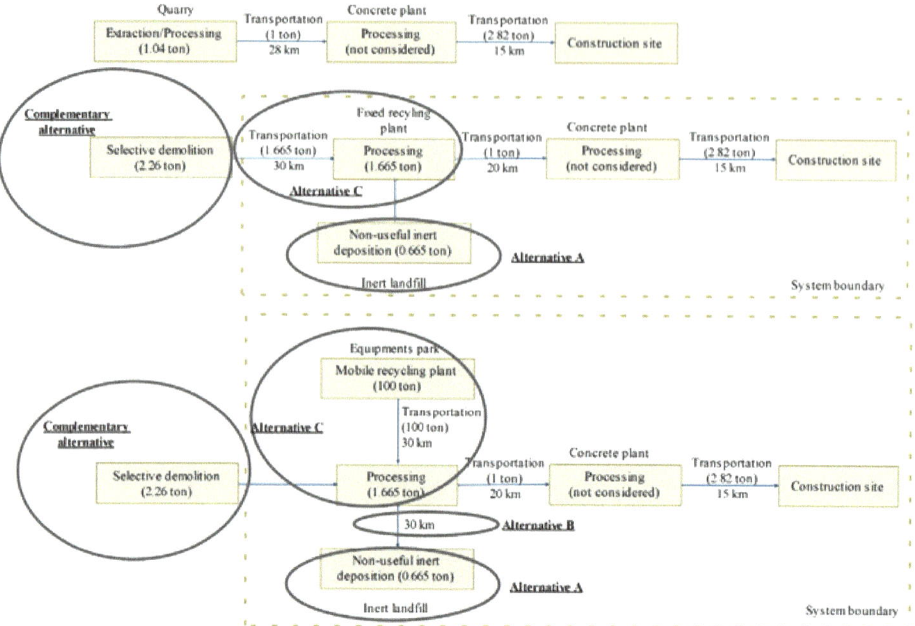

Fig. 2.9   Example of scenarios defined for the environmental LCA of coarse natural and recycled aggregates for concrete

B – landfill of no-net inert aggregates from demolition, C – recycling process of coarse aggregates not considered; D – sensitivity analysis of the transportation distances. Results showed that the assessment was very sensitive to the transportation distances.

Göswein *et al.* [25] considered a new type of non-structural bio-concrete, using invasive alien wood chips as a substitute for sand and gravel as aggregates, for future residential construction in Cape Town, comparing this new material to conventional and to earth-based materials, and benchmarking different policy scenarios. It was found that this new material, combined with a policy that promotes multi-storey buildings, offers great potential to achieve near carbon neutral cities, clearing land of invasive alien plants and thus saving annual water surface runoff.

Dias *et al.* [26] presents a review on the use of LCA methodology on NA and RCA for concrete and applies this methodology in a real context pertaining the procurement of coarse aggregates to ready-mix concrete plants in two regions in central and southern Portugal. Different scenarios for the supply of natural and recycled aggregates are studied and the scenarios for recycled aggregates procurement include different hypotheses for the installation (construction and demolition waste plant or quarry), processing the construction and demolition waste into recycled aggregates. It was found that the supply of RA produced at the construction and demolition waste plant has lower environmental impact and cost than all other scenarios, including the provision of NA, except when it is assumed that the quarry is licensed and equipped for receiving unsorted construction and demolition waste and to process it into recycled aggregates. Transport distance is a determining factor in the comparison of the impacts of the procurement of natural and recycled aggregates. Moreover, in the Portuguese context, RCA may be more costly than NA in situations in which the EI of the former are much smaller than those of the latter.

Transportation of Raw Materials

So far, environmental assessments focus mostly on the production of concrete mixes without considering the transportation-related impacts. A study by Göswein *et al.* [27] appraised the importance of transportation-related impacts of traditional and alternative raw materials for concrete production and proposes a new method combining LCA and geospatial analysis of road transportation of materials. The EI of different mixes were assessed for the specific locations of concrete plants. Traditional and alternative concrete mixes are compared to choose site-specifically the mix with lower impacts. For that purpose, two Portuguese cities were considered as case studies. This new method uncovers if and how the location of supply and demand for concrete production is important. The results show that, for traditional concrete and for mixes incorporating a low FA ratio and/or RA, transportation does not matter. However, it matters when choosing between already drastically improved mixes: then, distances from raw materials suppliers to concrete plants can tip the scales of total EI. In fact, the distances to FA and recycling plants influence the selection of the most environmentally friendly concrete mix (Fig. 2.10).

2.2.4 Life Cycle Assessment of the Use Stage

It is established that, throughout its whole service life, in the process of sequestration, concrete can absorb 13–48 % of the carbon dioxide emitted during the production phase [28]. During the "use" and "end-of-life" phases, OPC-based concrete can capture up to 47 % of its embodied carbon while SCM-based concrete can capture only up to 22 % [29]. Nevertheless, this topic deserves more attention and research effort than that currently afforded.

2.2.5 Life Cycle Assessment of the End-of-Life Stage

The construction sector was highlighted in the European Union due to the intensive use of resources and great potential for circularity. It is therefore necessary to demonstrate and disseminate the environmental and economic advantages of implementing the principles of circular economy in construction. These benefits can be attested using LCA.

A study from the author [30] proposes a mathematical and visual representation of the rules from European standards applicable to the LCA of the end-of-life stage of construction materials and buildings, increasing the LCA's contribution to closing the materials' life cycle.

Fig. 2.11 shows the assignment of the environmental impacts and benefits to the end-of-life stage (C) and to module D for two scenarios. Potential benefits from use in the next product system of energy (i.e. heat and power – Ie in common practice in Fig. 2.11)

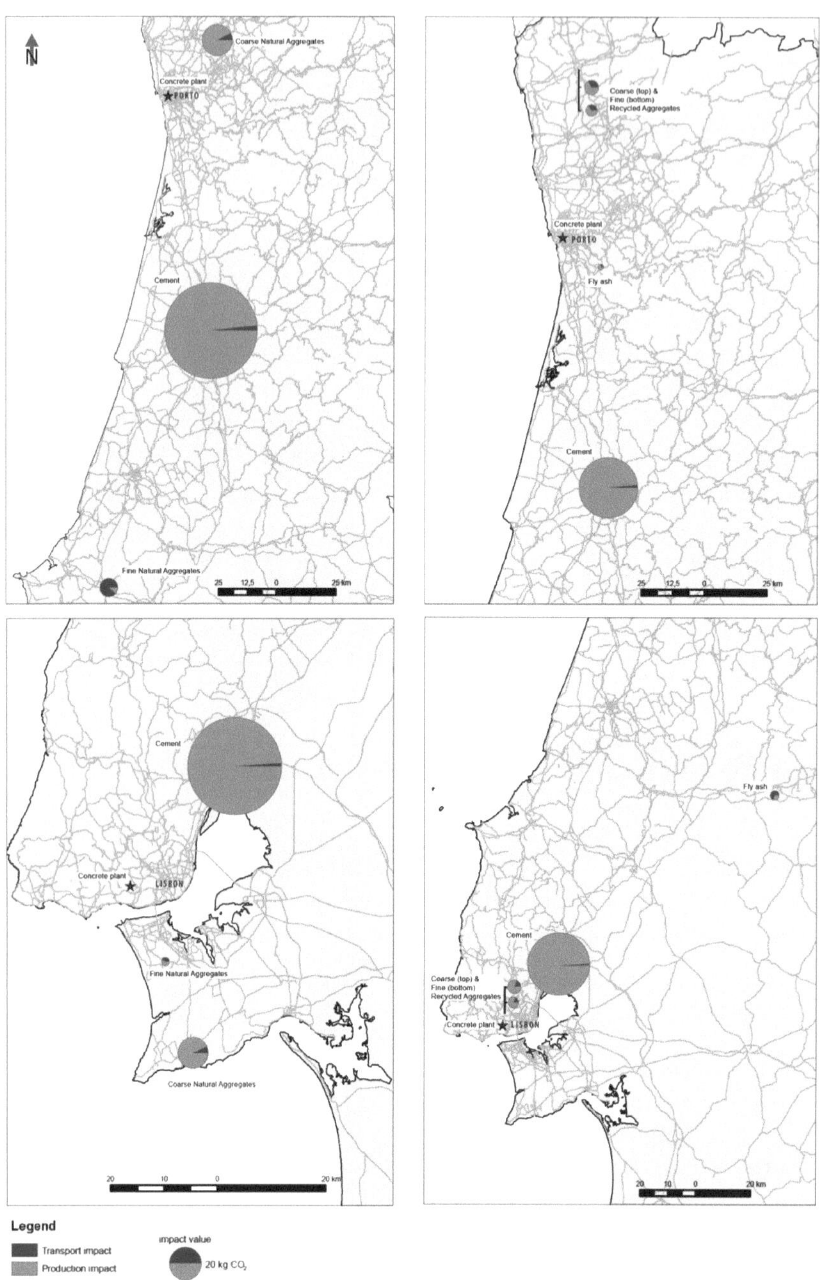

Fig. 2.10 Analysis of case studies Porto (top) and Lisbon (bottom) to compare production (A1) and transportation impacts for transportation step 1 (A2), for the raw materials necessary for 1 m³ of concrete mixes 1 (bottom and top left) and 18 (bottom and top right). The star symbol indicates the closest concrete plant to Porto and Lisbon. The size of each pie chart relates to the GWP of each raw material for A1 and A2. The centre of the pie charts are at the location of each respective plant. Note: in the images on the bottom and top right, the pie charts for coarse and fine RA are in fact at the same location but are shown at different locations due to visualization purposes

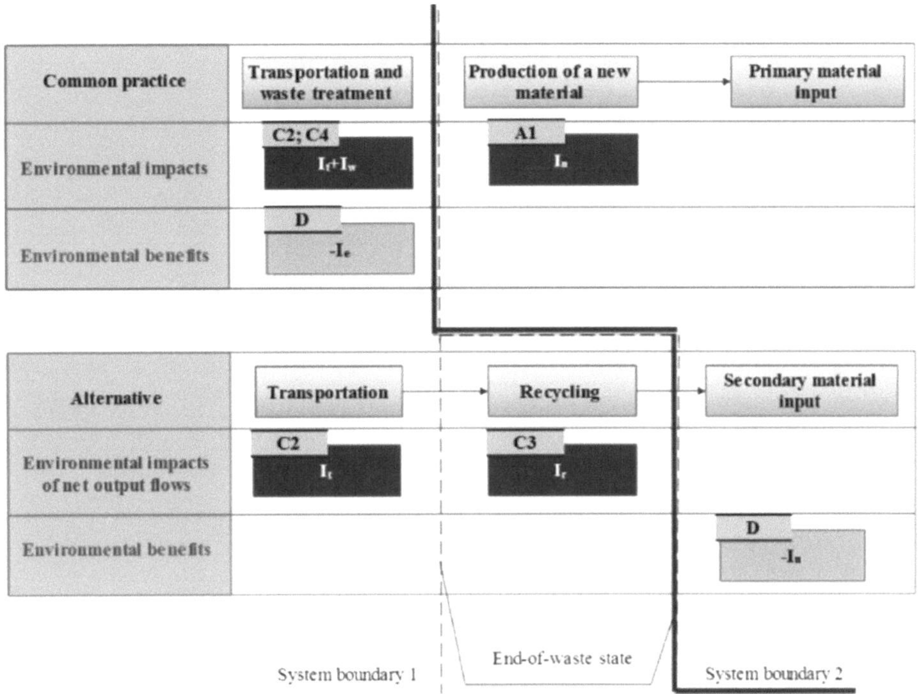

Fig. 2.11 Comparison between common practice in construction materials production (use of primary raw materials), with its alternative (use of recycled materials), and the corresponding assignment of the environmental impacts and benefits to end-of-life stage (C) and to module D

generated at sub-stage C4 (waste disposal) from waste incineration or landfill should be considered in module D, while the loads (e.g. emissions – Iw in Fig. 2.11) from waste disposal in this sub-stage are part of the product system under study, according to the 'polluter pays' principle, and should be considered at sub-stage C4 [3]. The environmental impact from waste recycling (and from the eventual transport) can be represented as: Ir+It (and assigned to the C2 and C3 sub-stages, respectively – see alternative in Fig. 2.11). This operation avoids the impacts from the production of a similar new product (potential benefit of In in module D, see alternative in Fig. 2.11) and the impacts from waste treatment (Iw – common practice in Fig. 2.11). Therefore, recycling should be promoted only if (Ir+It) < (In+Iw), because it will avoid an impact that corresponds to: (In+Iw-Ir-It) [31]. If the option is not to recycle, Iw should be included in sub-stage C4 (common practice in Fig. 2.11).

These rules should be followed on the LCA of any recycled raw materials for concrete and are exemplified for recycling IWM in concrete in another book chapter by the author [5].

It was already referred in Section 2.4 to the sequestration potential of concrete at its end-of-life stage. This capacity can be maximised by the additional specific surface produced by the crushing that accelerates carbonation of the concrete volume.

2.3 Economic and Environmental Life Cycle Assessment of Concrete

Kurda *et al.* [32] developed CONCRETop, a novel multi-criteria decision method for concrete optimization, which can be used for either conventional or non-conventional concrete mixes. This method can be used to identify the main differences between concrete mixes in three dimensions of performance: strength and durability; environmental impacts; and cost efficiency.

The characteristics of concrete fundamentally depend on its target application (e.g., the structural design project). In other words, it depends on the requested characteristics demanded by the consumer. However, this method allows optimizing concrete mixes based on the three mentioned dimensions even if the specific characteristics of concrete for the selected application are unknown. Five scenarios (business as usual, green, strength, service life, and cost) were proposed with different threshold values, in order to cover most of concrete applications – see Fig. 2.12.

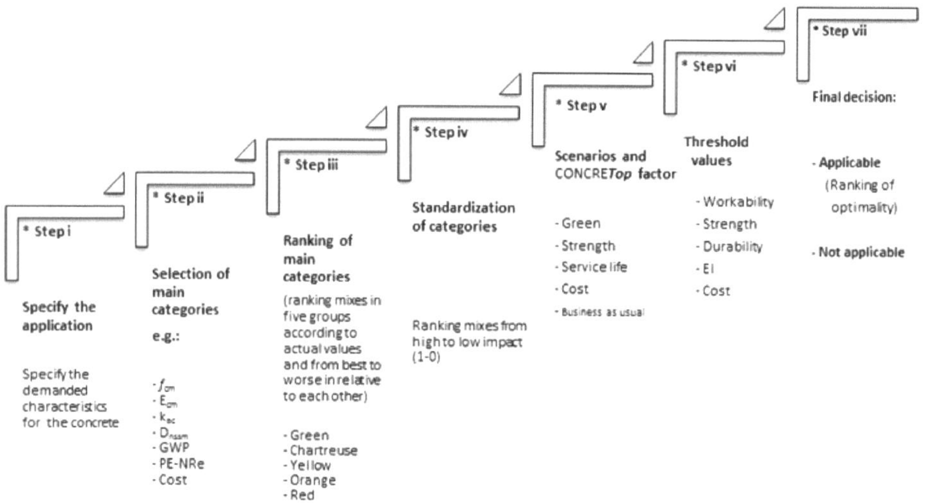

Fig. 2.12   Profile of the multi-criteria decision method for concrete optimization – CONCRETop

The major outcome of this method is to provide researchers with a validated approach to support the development of traditional and non-conventional concrete mixes with optimized technical performance and minimum cost, environmental impact, and natural resources' consumption.

Kurda *et al.* [33] applied the CONCRETop method to optimize concrete mixes containing various amounts of RCA and/or FA for different applications in the construction sector, namely high-rise building, sustainable residential housing, economical residential housing, and residential housing close to (Table 2.4) or far from the sea. The results show that, for all applications and scenarios that users may demand, the optimum concrete mixes in terms of concrete characteristics, cost, and environmental impact are the ones produced with both FA and RCA incorporation, rather than their individual incorporation.

Table 2.4 Applicable concrete mixes for residential housing close to the sea according to the "Service life" scenario

No.	CM	Fine RCA (%)	Coarse RCA (%)	FA (%)	SP (%)	CONCRETop factor	Strength			Carbonation resistance	Chloride resistance	Cost
1	M14sp	50	100	30	1	0.715	45/55	-	36	Fair	Excellent	Medium
2	M5sp	50	0	30	1	0.693	50/60	-	37	Very good	Excellent	High
3	M18sp	100	100	60	1	0.628	30/37	-	33	Fair	Excellent	Low
4	M13	0	100	30	0	0.598	35/45	-	34	Fair	Very high	Low
5	M10sp	0	100	0	1	0.590	45/55	-	36	Very good	Very high	Very low
6	M14	50	100	30	0	0.584	30/37	-	33	Fair	Very high	Low
7	M4	0	0	30	0	0.560	35/45	-	34	Fair	Very high	Medium
8	M15	100	100	30	0	0.556	30/37	-	33	Fair	Very high	Low
9	M16	0	100	60	0	0.554	25/30	-	31	Fair	Very high	Very low
10	M1sp	0	0	0	1	0.550	55/67	-	38	Very good	Very high	Very high
11	M5	50	0	30	0	0.547	30/37	-	33	Fair	Very high	Medium
12	M16sp	0	100	60	1	0.546	35/45	-	34	Fair	Excellent	Medium
13	M17	50	100	50	0	0.535	20/25	-	30	Fair	Very high	Very low
14	M9sp	100	0	60	1	0.530	30/37	-	33	Fair	Excellent	Medium
15	M18	100	100	60	0	0.528	25/30	-	31	Fair	Very high	Medium
16	M6	100	0	30	0	0.495	30/37	-	33	Fair	Very high	Low
17	M1	0	0	0	0	0.482	40/50	-	35	Good	Very high	High
18	M8	50	0	60	0	0.481	25/30	-	31	Fair	Very high	Low
19	M7	0	0	60	0	0.479	25/30	-	31	Fair	Very high	Low
20	M3sp	100	0	0	1	0.477	35/45	-	34	Fair	High	Very low
21	M9	100	0	60	0	0.466	25/30	-	31	Fair	Very high	Very low
22	M7sp	0	0	60	1	0.465	30/37	-	33	Fair	Excellent	Medium
23	M2	50	0	0	0	0.414	30/37	-	33	Good	High	High

A research work by Duarte *et al.* [34] presents a comparative assessment of the sustainability of a concrete-filled square steel tubular (CFSST) column, using rubberized concrete (RuC), thus designated 'RuCFSST'. RuCFSST columns are structural members that take advantage of the optimized structural performance of CFSST columns and of the improved ductility of RuC compared to that of NAC. Besides the two most obvious environmental advantages of using RuC instead of NAC in CFSST columns – reductions of tyre disposal into landfills and of extraction of natural aggregates – it is shown that a more complete assessment of the sustainability of RuC and RuCFSST columns is required. Hence, the embodied energy (EE), thermal performance, cost, expected durability, and recycling potential of the RuCFSST

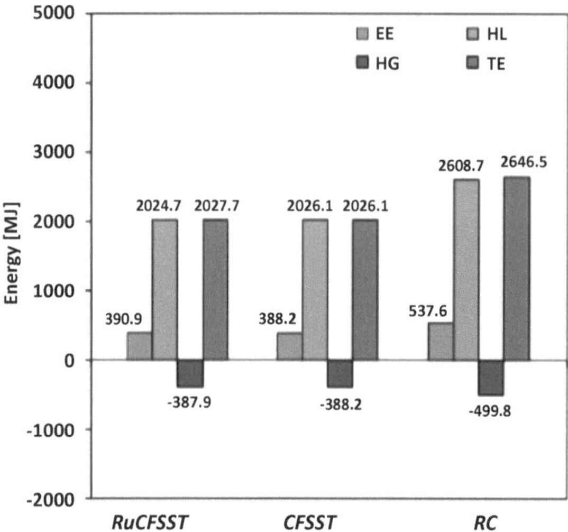

Fig. 2.13 Total energy of each column for a 50 years life span (total energy (TE): sum of (i) EE (ii), heat losses (HL) during heating seasons, and (iii) heat gains (HG) during cooling seasons)

column analysed are assessed and compared with those of a CFSST column and a reinforced concrete (RC) column, all three having a similar mechanical performance. It is shown that the analysed RuCFSST column has a marginal higher EE and cost than the CFSST column – the difference being less than 1% (Fig. 2.13). Conversely, the thermal transmittance coefficient of the former is lower than that of the latter, which improves its behaviour concerning thermal bridges. Compared to the composite solutions, the RC column possesses circa 35% more EE but is about 25% less expensive. Overall, the RuCFSST and CFSST columns incorporate a similar total energy (including the contribution of the EE and thermal performance during service life), which is 25% lower than that of the RC column.

2.4 Life Cycle Assessment-Driven Structural Design of Concrete

Braga *et al.* [17] compared the environmental and economic LCA of NAC and RAC mixes from previous studies. The functional unit of this study was 1 m³ of ready-mixed NAC and RAC per strength class defined by its 28 days' compressive strength. The cradle to gate life cycle stages considered are detailed in Fig. 2.14:

- Production/extraction of all raw materials needed for the production of each concrete (A1);
- Transportation of raw materials to the concrete plant (A2);
- Concrete production at the plant (A3).

The results were analysed by strength class and considering the influence of the cement content, SP incorporation and w/c ratio. In summary, the main conclusions of this study in terms of the relation between LCA and strength were:

- Higher concrete strengths do not necessarily lead to a higher EI;
- The use of CARC can significantly reduce the EI and costs, proving that cement is the main contributor to both impacts. The concrete mixes with best mechanical results use CARC with better characteristics (low water absorption and porosity, higher density and specific mass), usually corresponding to lower EI and costs.

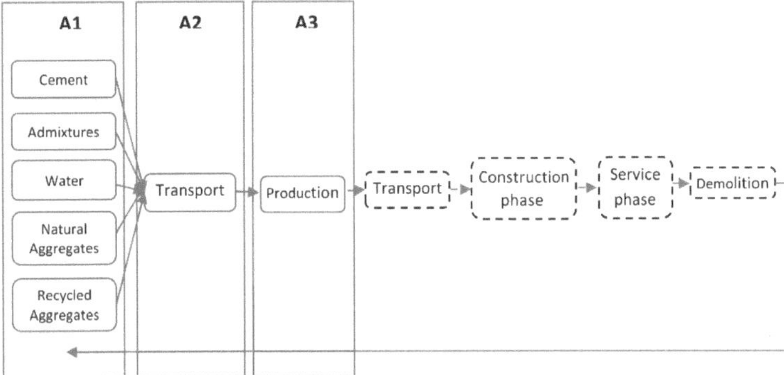

Fig. 2.14   Concrete life cycle

The results for GWP by strength class are presented in Fig. 2.15 (0 % CARC) and Fig. 2.16 (100 % CARC). It was confirmed that GWP is strongly related with the average amount of cement of each strength class – more cement corresponds to larger EI [19]. This conclusion is independent from the type of aggregates used (natural or recycled). For each strength class, and regardless of the amount of cement, the incorporation of CARC contributes to a GWP reduction [18]. The same behaviour occurs for the other categories of EI, except for ADP.

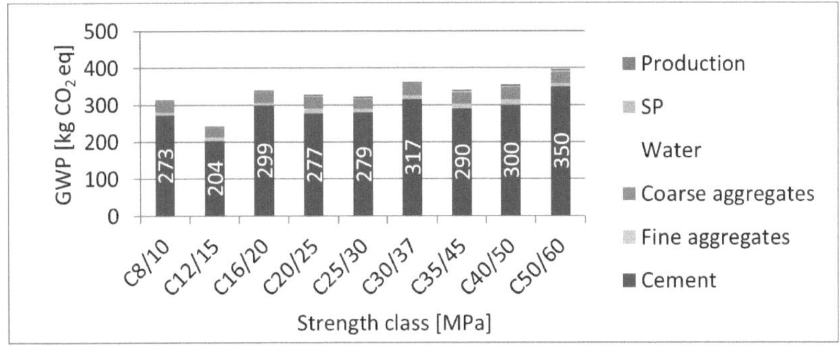

Fig. 2.15   Average GWP per m³ of mixes without CARC by strength class

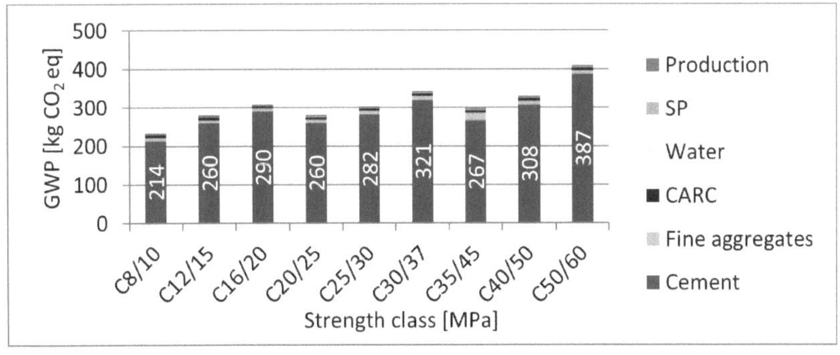

Fig. 2.16   Average GWP per m³ of mixes with 100% of CARC by strength class

The same behaviour occurs for the Pe-NRe that is significantly associated with the average amount of cement independent of strength class. Pe-NRe is lower in 100 % CARC mixes (Fig. 2.18) than in 0 % CARC mixes (Fig. 2.17), since this EI is significantly higher for natural coarse aggregates than for CARC.

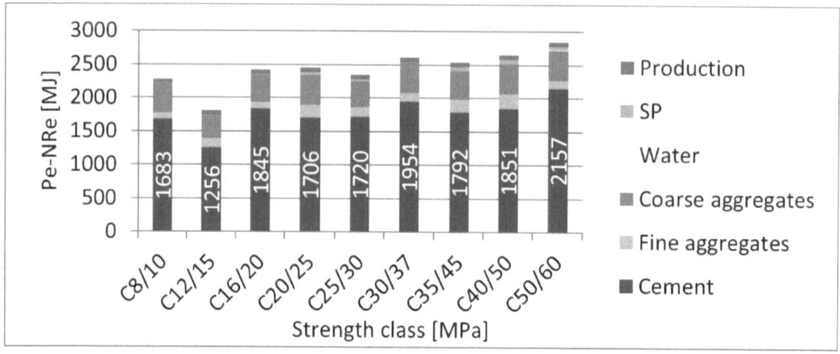

Fig. 2.17   Average Pe-NRe per m³ of mixes without CARC by strength class

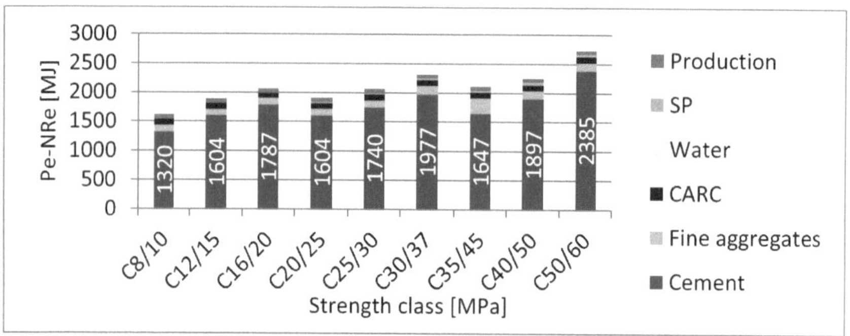

Fig. 2.18   Average Pe-NRe per m³ of mixes with 100% of CARC by strength class

In the parametric study of the w/c ratio, the selected ranges were: less than 0.55; from 0.55 (included) to 0.6; and equal to or greater than 0.6. Regarding each strength class, it is possible to conclude that an increase of the w/c ratio results in a decrease of the associated GWP. This trend is confirmed in other environmental categories.

Transport distances considered for the A2 stage were actual data from a Portuguese scenario. It was found that the impacts from this stage did not significantly influence the total impacts of the concrete mixes.

A study from cradle to gate of the technical and environmental performance of 1 m³ of concrete mixes made with high incorporation ratios of FA and RCA, individually and jointly, with and without SP ([35–38] – Fig. 2.19), found that:

1. The optimum mixes in terms of cost, strength and environmental impacts are the ones made with incorporation of both RCA and FA rather than single incorporation (Fig. 2.20);

2. The strength to GWP ratio of concrete mixes depends on the FA-RCA incorporation ratio rather than the content of the individual materials;

3. The use of SP significantly increases the cost of concrete. However, the higher cost of concrete due to the use of SP can be offset by replacing cement with FA.

4. The environmental impact of concrete significantly decreases with the use of FA, regardless of the transportation scenario. There are no clear relationships between the mechanical behaviour and GWP and PE-NRe of concrete mixtures with and without RA and FA because the incorporation of each of the non-traditional materials affects each characteristic of concrete differently.

In the same research study [39], the chloride ion penetration resistance (CIPR)-related service life and EE were determined for the same mixes. The optimal solution of mixes by means of both EE and CIPR (measured by the yearly EE of the mixes relative to the reference concrete) does not necessarily consider the one that requires less EE or that has higher CIPR. According to this parameter, the best-case scenarios are always the low w/b mixes with high volumes of RCA and FA, followed by the corresponding high w/b mixes.

As already referred, other book chapters from the author propose methods to define an equivalent functional unit for RAC [4] and for concrete with IWM [5] based on mechanical and durability properties, which can be used in LCA.

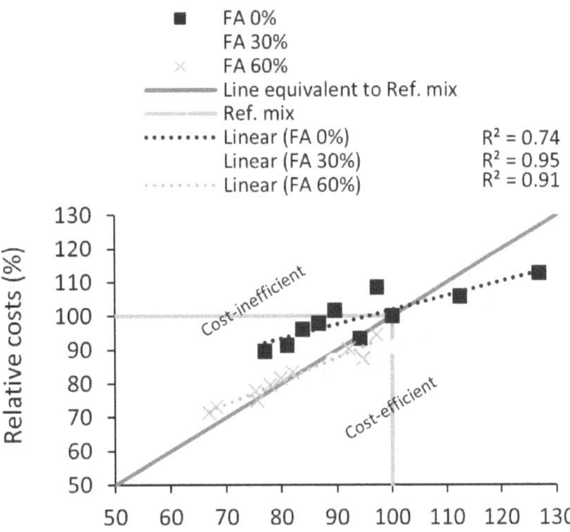

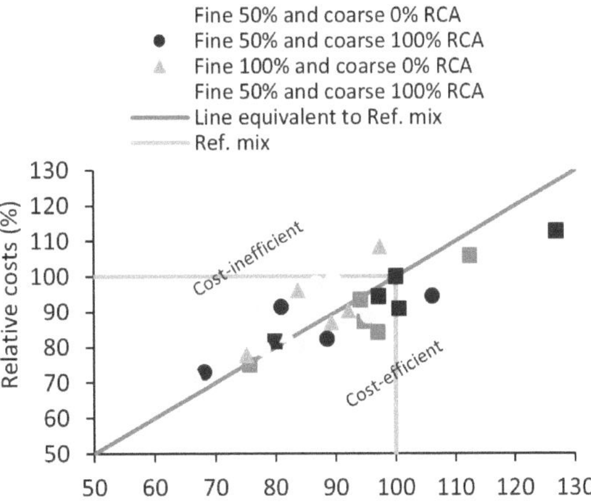

Fig. 2.19 Relationship between relative cost and mechanical performances ratio of mixes classified based on the incorporation ratio of (a) FA (regardless of the type and ratio of the aggregates) and (b) RCA (regardless of the type and ratio of the binders

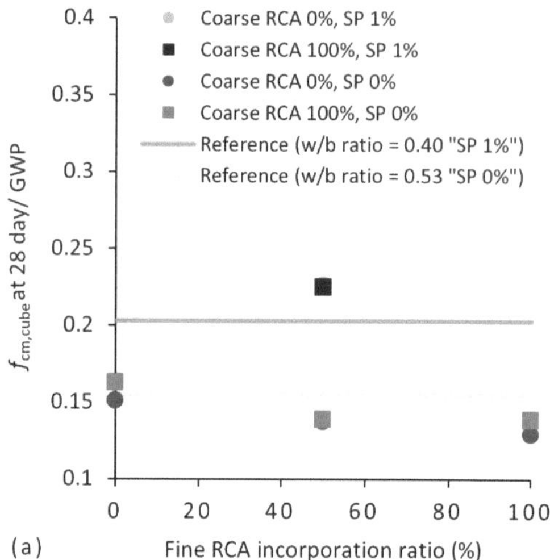

Fig. 2.20   Effect of incorporating RCA and 30% FA on the fcm/GWP ratio of concrete mixes with low ("SP 1%") and high ("SP 0%") w/b at 28 days

2.5 Conclusion

It is of outmost importance that the potential economic and environmental advantages of alternative concrete mixes are demonstrated using standardised and quantitative methodologies such as LCA and LCC, instead of being disseminated via qualitative or biased proclamations.

In this chapter, it was explained how the LCA methodology can be used in support of concrete sustainability. The research works described are focused on identifying the effect of integrating industrial waste as substitutes of OPC or aggregates on the cost, environmental impact and quality of concrete, and some of their results are summarized here. The case studies selected consider concrete production, but also upstream processes of raw materials production and transport, and downstream activities of concrete use and end-of-life. The relation between the composition, strength class and environmental and economic impacts of concrete mixes are also discussed, to motivate the inclusion of these last two dimensions of performance on the structural design of concrete on a regular basis. Lastly, it is important to highlight that an adequate equivalent functional unit based on mechanical and durability properties should always be used in LCA to guarantee a fair comparison.

2.6 References

[1]   Environmental management – Life cycle assessment – Principles and framework, ISO 14040:2006(E): International Organization for Standardization.

[2]   Environmental management – Life cycle assessment – Requirements and guidelines, ISO 14044:2006(E): International Organization for Standardization.

[3]   CEN, 2019. Sustainability of construction works. Environmental product declarations. Core rules for the product category of construction products. EN 15804:2012+A2:2019. Brussels, Belgium, Comité Européen de Normalisation.

[4]   de Brito, J.; Evangelista, L.; Silvestre, J.: "Equivalent Functional Unit in Recycled Aggregate Concrete", in Eco-Efficient Concrete: New Trends in Concrete with Recycled Aggregates, Elsevier, UK, November 2018, pp. 293–328

[5]   de Brito, J.; Hafez, H.; Kurda, R.; Silvestre, J.: 'Part Four – Life cycle assessment of concrete. Chapter 26: Calculation of the environmental impact of the integration of industrial waste in concrete using LCA', in Handbook of Sustainable Concrete and Industrial Waste Management: Recycled and Artificial Aggregate, Innovative Eco-Friendly Binders and Life Cycle Assessment, Woodhead Publishing, Elsevier, UK, December 2021, pp. 553–577

[6]   Ferrão, P. C., 2009. Industrial ecology – principles and tools (in Portuguese) (1st Ed.). Lisbon, Portugal: IST Press. 398 p.

[7]   Ortiz, O.; Castellsa, F. & Sonnemann, G., 2009. Sustainability in the construction industry: A review of recent developments based on LCA. Construction and Building Materials. 23 (1). pp. 28–39. https://doi.org/10.1016/j.conbuildmat.2007.11.012

[8]   Wu, P., Xia, B. & Zhao, X. 2014. The importance of use and end-of-life phases to the life cycle greenhouse gas (GHG) emissions of concrete – A review. Renewable and Sustainable Energy Reviews, 37, 360–369. https://doi.org/10.1016/j.rser.2014.04.070

[9]   UNEP, 2020. "Life Cycle Initiative", UN Life Cycle Initiative Retrieved March 11, 2021, from https://www.lifecycleinitiative.org/.

[10]   Del Borghi, A., 2013. LCA and communication: Environmental Product Declaration. The International Journal of Life Cycle Assessment, 18(2), p. 293. https://doi.org/10.1007/s11367-012-0513-9

[11]   Silvestre, J. D.; Lasvaux, S.; Hodkova, J.; de Brito, J.; Pinheiro, M. D., 2015a. NativeLCA – a systematic approach for the selection of environmental datasets as generic data: application to construction products in a national context. The International Journal of Life Cycle Assessment. 20 (6), pp. 731–750.

[12]   CfD, 2001. Background report LCA tools – Data and application in the building and construction industry. Australia: Centre for Design, Department of Environment and Heritage, RMIT University, 30 p.

[13]   ENSLIC. (2012). ENSLIC European project (ENergy Saving through promotion of LIfe Cycle assessment in buildings). Retrieved 2012-08-23, from http://circe.cps.unizar.es/enslic/texto/lca.htm.

[14]   EFCA, 2012. EFCA – Environmental Product Declarations (EPD). European Federation of Concrete Admixtures Associations (EFCA). Retrieved 2012-09-15, from http://www.efca.info/publications.html.

[15]   Ecoinvent, 2012. Ecoinvent LCA database. Swiss Centre for Life Cycle Inventories. Retrieved 2012-04-26, from www.ecoinvent.ch.

[16]   Blok, R.; Giarma, C. S.; Bikas, D. K.; Kontoleon, K. & Gervásio, H., 2007. Life Cycle Assessment – general methodology. First workshop COST Action C25: Sustainability of Constructions, Lisbon, Portugal. pp. 1.3–1.9.

[17] Braga, A. M.; Silvestre, J. D.; de Brito, J. (2017). Compared environmental and economic impact from cradle to gate of concrete with natural and recycled coarse aggregates. Journal of Cleaner Production. 162, pp. 529–543. https://doi.org/10.1016/j.jclepro.2017.06.057

[18] Marinkovic, S., Radonjanin, V., Malea Ev, M., Ignjatovia, I., 2010. Comparative environmental assessment of natural and recycled aggregate concrete. Waste Management. 30 (11), 2255(10). https://doi.org/10.1016/j.wasman.2010.04.012

[19] Blengini, G.A., 2006. Life Cycle Assessment Tools for Sustainable Development: Case Studies for Mining and Construction Industries in Italy and Portugal. PhD thesis in Mining Engineering. Universidade Técnica de Lisboa, Instituto Superior Técnico, Lisbon.

[20] Kurda, R.; Silvestre, J. D.; de Brito, J. (2018). Toxicity and environmental and economic performance of fly ash and recycled concrete aggregates use in concrete: A review. Heliyon. 4 (4), e00611. https://doi.org/10.1016/j.heliyon.2018.E00611

[21] Gomes, R.; Silvestre, J. D.; de Brito, J.; Lasvaux, S. (2021). Environmental datasets for cement and steel rebars to be used as generic for a national context. Journal of Cleaner Production. 316, 128003. https://doi.org/10.1016/j.jclepro.2021.128003.

[22] Josa, A., Aguado, A., Heino, A., Byars, E., Cardim, A. (2004). Comparative analysis of available life cycle inventories of cement in the EU. Cement and Concrete Research, 34 (8) 1313–1320. https://doi.org/10.1016/j.cemconres.2003.12.020

[23] Josa, A., Aguado, A., Cardim, A., Byars, E. (2007). Comparative analysis of the life cycle impact assessment of available cement inventories in the EU. Cement and Concrete Research, 37 (5) 781–788. https://doi.org/10.1016/j.cemconres.2007.02.004

[24] Estanqueiro, B.; Silvestre, J. D.; de Brito, J.; Pinheiro, M. D., 2018. Environmental life cycle assessment of coarse natural and recycled aggregates for concrete. European Journal of Environmental and Civil Engineering. 22 (4), pp. 429–449. https://doi.org/10.1080/19648189.2016.1197161

[25] Göswein, V.; Silvestre, J. D.; Gonçalves, A.; Freire, F.; Lamb, S.; Oosthuizen, D.; Lord, A.; Pittau, F.; Habert, G. (2021). Invasive alien plants as an alternative resource for concrete production – Multi-scale optimization including carbon compensation, cleared land and saved water runoff in South Africa. Resources, Conservation and Recycling. 167, 105361. https://doi.org/10.1016/j.resconrec.2020.105361

[26] Dias, A. B.; Pacheco, J. N.; Silvestre, J. D.; Martins, I. M.; de Brito, J. (2021). Environmental and economic life cycle assessment of recycled coarse aggregates: A Portuguese case study. Materials. 14 (18), 5452. https://doi.org/10.3390/ma14185452

[27] Göswein, V.; Gonçalves, A.; Silvestre, J. D.; Freire, F.; Habert, G.; Kurda, R. (2018). Transportation matters – does it? GIS-based comparative environmental assessment of concrete mixes with cement, fly ash, natural and recycled aggregates. Resources, Conservation and Recycling. 137, pp. 1–10. https://doi.org/10.1016/j.resconrec.2018.05.021

[28] Collins, F. 2010. Inclusion of carbonation during the life cycle of built and recycled concrete: influence on their carbon footprint. The International Journal of Life Cycle Assessment, 15, 549–556. https://doi.org/10.1007/s11367-010-0191-4.

[29] García-Segura, T., Yepes, V. & Alcalá, J. 2014. Life cycle greenhouse gas emissions of blended cement concrete including carbonation and durability. international journal of life cycle assessment, 19, 3–12. https://doi.org/10.1007/s11367-013-0614-0

[30] Silvestre, J., de Brito, J., Pinheiro, M., 2014. Environmental impacts and benefits of the end-of-life of building materials – calculation rules, results, and contribution to a "cradle to cradle" life cycle. Journal of Cleaner Production. 66 (1), pp. 37–45. https://doi.org/10.1016/j.jclepro.2013.10.028

[31] Peuportier, B.; Herfray, G.; Malmqvist, T.; Zabalza, I.; Staller, H.; Tritthart, W., et al. (2011). Life cycle assessment methodologies in the construction sector: the contribution of the European LORE-LCA project. SB11 Helsinki: World Sustainable Building Conference, Helsinki, Finland. pp. 110–117 – Theme four.

[32] Kurda, R.; de Brito, J.; Silvestre, J. D., 2019. CONCRETop – A multi-criteria decision method for concrete optimization. Environmental Impact Assessment Review. 74, pp. 73–85.

[33] Kurda, R.; de Brito, J.; Silvestre, J. D. (2019). CONCRETop method: Optimization of concrete with various incorporation ratios of fly ash and recycled aggregates in terms of quality performance and life-cycle cost and environmental impacts. Journal of Cleaner Production. 226, pp. 642–657. https://doi.org/10.1016/j.jclepro.2019.04.07

[34] Duarte, A.; Silvestre, N.; de Brito, J.; Júlio, E.; Silvestre, J. D. (2018). On the sustainability of rubberized concrete filled square steel tubular columns. Journal of Cleaner Production. 170, pp. 510–521. https://doi.org/10.1016/j.jclepro.2017.09.131

[35] Kurda, R.; Silvestre, J. D.; de Brito, J. (2018). Life cycle assessment of concrete made with high volume of recycled concrete aggregates and fly ash, Resources, Conservation and Recycling. 139, pp. 407–417. https://doi.org/10.1016/j.resconrec.2018.07.004

[36] Kurad, R.; Silvestre, J. D.; de Brito, J.; Ahmed, H. (2017). Effect of incorporation of high volume of recycled concrete aggregates and fly ash on the strength and global warming potential of concrete. Journal of Cleaner Production. 162, pp. 485–502. https://doi.org/10.1016/j.jclepro.2017.07.236

[37] Kurda, R., de Brito, J., Silvestre, J. D. (2020). A comparative study of the mechanical and life cycle assessment of high-content fly ash and recycled aggregates concrete. Journal of Building Engineering. 29, 101173, May. https://doi.org/10.1016/j.jobe.2020.101173

[38] Kurda, R.; de Brito, J.; Silvestre, J. D. (2018). Combined economic and mechanical performance optimization of recycled aggregate concrete with high volume of fly ash. Applied Sciences. 8 (7), p. 1189. https://doi.org/10.3390/APP8071189

[39] Kurda, R.; Silvestre, J. D.; de Brito, J.; Ahmed, H. (2018). Optimizing recycled concrete containing high volume of fly ash in terms of the embodied energy and chloride ion resistance. Journal of Cleaner Production. 194, pp. 735–750. https://doi.org/10.1016/j.jclepro.2018.05.177

Chapter 3

Optimizing the Performance of Green Concrete Using Nonconventional Recycled Materials

Rawaz Kurda[1,2,3], Jorge De Brito[3], Hawreen Ahmed[1,2,3], Hisham Hafez[4], Babar Ali[5]

1 Department of Highway Engineering, Technical, Engineering College, Erbil Polytechnic University, Erbil, Kurdistan-Region, Iraq
2 Dept. of Civil Engineering, College of Engineering, Nawroz Univ., Duhok 42001, Iraq
3 CERIS, Civil Engineering, Architecture and Georesources Department, Instituto Superior Técnico, University of Lisbon, Av. Rovisco Pais, 1049–001 Lisbon, Portugal
4 Laboratory of Construction Materials, EPFL STI IMX LMC Station 12, CH-1015 Lausanne, Switzerland.
5 Department of Civil Engineering, COMSATS University Islamabad, Sahiwal Campus, Sahiwal 57000, Pakistan.

3.1 Introduction

The negative impact of the production of cement-based materials has been mentioned in many studies. In fact, it is time to be cautious about this issue because the impact is much bigger than ever foreseen. For example, the total world production of aggregates was 48.3 billion tons [1, 2] in 2018. In volume, this amount is 32,200,000,000 m^3 (considering 1 m^3 of aggregate to be 1.5 tonnes). This volume is ≈180 times bigger than that of the volume of the Great Wall of China (180 million m^3 of raw material [3]). In other words, the volume of the aggregates mentioned above can be equal to the volume of an 80 m (height) × 10 m (width) wall around the circumference of Earth. Furthermore, the majority of the materials used in the cement-based materials production are, in *sensu stricto*, non-sustainable because their main sources are non-renewable.

Accordingly, many researchers have been focused on using construction and demolition waste instead of natural aggregates in cement-based materials, namely recycled concrete aggregate [4–7], recycled masonry aggregate [7–13], contaminated construction and demolition waste [14, 15] and mixed recycled aggregate [12, 16–22]. Nevertheless, relatively to construction and demolition waste, studies on nonconventional recycled aggregates such as agricultural wastes and aquaculture farming as aggregates, industrial wastes as aggregates, municipal wastes as aggregates and insulating aggregates are very few, especially review studies. Accordingly, this chapter will mainly focus on the effect of all nonconventional recycled aggregates mentioned above on technical performance, environmental impact, and cost of cement-based materials (mostly concrete) simultaneously. However, the focus was made only on the technical performance because studies regarding the environmental impact and cost are very limited.

3.2 Nonconventional Aggregates and Standards

The aim of this section is to show what has been identified in the worldwide standards and recommendations regarding the use of recycled aggregates as well as the gap that needs to be studied and standardized, namely for nonconventional recycled aggregates. Apart from scientific publications on recycled aggregates that have been written by researchers, there are also some recommendations or standards around the world (Fig. 3.1). For example, in

(1) Guideline for the use of recycled coarse aggregates in hydraulic binders concrete (LNEC) [23];

(2) Austria: Guideline for recycled building materials [24];

(3) Guide to use of industrial by-products and waste materials in building and civil engineering; Recycled aggregates; Concrete – complementary British Standard [25–27];

(4) Holland: Centre for Civil Engineering Research, Codes and Specifications (CUR) report [28–30];

(5) Switzerland: Instruction technique (TV) utilisation de matériaux de construction minéraux secondaires dans la construction d'abris [31];

(6) Germany: Deutsches Institut für Normung (DIN) [32];

(7) United States: American Concrete Institute (ACI) [33];

(8) Japan: Japanese Standards Association – Recycled [34–36];

(9) Denmark: Recommendations for the use of recycled aggregates for concrete in passive environmental class [37];

(10) Brazil: Aggregates from construction and demolition waste: Use on road pavements and non-structural concrete – Requirements [38];

(11) Australia: Construction and Demolition Waste [39];

(12) China: Specifications facilitating the use of recycled aggregates. Works Bureau Technical Circular (WBTC) [40];

(13) France: Guide technique pour l'utilisation des matériaux régionaux d'Ile-de-France, 2003 [41]; and globally

(14) RILEM – International Union of Testing and Research Laboratories for Materials and Structures [42].

In general, the mentioned standards and regulations are only focused on the technical characteristics of recycled aggregates such as their maximum incorporation ratio in concrete, water absorption, and density. Furthermore, most of the recommendations have been identified for recycled concrete aggregates only, especially coarse aggregates. In other words, there is not a specific standard or regulation for nonconventional recycled aggregates. Thus, similar studies to that of Silva et al. [43] need to be done in order to classify the technical properties of different aggregates to be used in standards in the future. In addition, none of the specifications have been focused on life cycle assessment (LCA). The reason is that there are very few joint studies on technical properties and LCA of recycled aggregates and recycled aggregate concrete.

3.3 Agricultural Wastes and Aquaculture Farming as Aggregates

Due to the high demand for aggregates and issues regarding the dumping process of agricultural wastes, these waste materials have been proposed to be used in concrete as aggregate. Thus, most of the studies have been made on the effect of incorporation of various agricultural and aquaculture farming wastes as partial replacement of sand (either as ash/incinerated or raw material) on the technical properties of concrete. To name a few: sawdust [44, 45], sunflower [46], seashell [47–56], plane leaf ashes [57], cork [58, 59], wild giant reed ash [60], wood ash [61], tobacco waste [62], sugarcane bagasse ash [63, 64], wheat straw [57, 65], rice husk/ash [66–68], groundnut shell [69, 70], leather [71], corn cob ash [57, 72], olive husk [73], palm tree shells [74–84] (see more details in Table 3.1). The majority of research along this line has concentrated on palm tree shells. The goal of this strategy, aside from sustainability, is to manufacture lightweight and low heat conductivity concrete [74, 85–87].

As per the summary of literature shown in Table 3.1, the most predominant feature of utilizing agricultural and aquaculture waste is the local abundance and low cost. Although the exact prices and locations were not mentioned in some papers, all wastes reviewed in the literature provide a cheaper and more environmentally friendly alternative to fine aggregates in concrete, which is the core of sustainability [88]. Moreover, apart from the rare need to incinerate the waste, which might result in toxic emissions in some cases, all wastes were recycled as fine aggregates with low or even no energy usage. As seen in Table 3.1, with the exception of bagasse sugarcane ash, reed ash, wood ash and mussel shells, as the replacement of fine aggregates with any of the agricultural wastes increase, the compressive strength of the resulting concrete mix decreases in varying degrees.

Fig. 3.1 Recommendations and standards on recycled aggregates from different countries

Table 3.1 List of the most predominant features in utilizing agricultural and aquacultural waste as aggregates in cement-based materials

Ref.	Year	Country	Type of waste	Trans-portation distance (km)	Local reserve (tonne/year)	Recycling energy (kWh/tonne)	Unit cost 7.50 %	28-day compressive strength					Water absorption		
								7.50 %	0 %	30 %	50 %	100 %	10 %	30 %	50 %
63 Sales	2010	Brazil	Sugarcane bagasse ash	150	500	No energy	-	-	-	10 % ↑	Same	-	-	-	-
64 Modani	2013	India	Sugarcane bagasse ash	locally obtained	10	No energy	-	-	-	15 % ↓	-	-	-	100 % ↑	-
69 Sada	2013	Nigeria	Ground nutshell	locally obtained	0.5 million	Low (incineration)	-	-	-	50 % ↓	70 % ↓	-	-	-	-
45 Ganiron	2014	Philip-pines	Sawdust	-	-	No energy	40 % ↓	-	-	-	-	20 % ↓	-	-	-
60 Ismail	2014	Iraq	Reed ash	-	-	No energy	-	10 % ↑	10 % ↓	-	-	-	-	-	-
61 Ottosen	2016	Denmark	Wood ash	-	-	No energy	-	Same	Same	-	-	-	-	-	-
66 Kunchari-yakun	2015	Thailand	Rice husk ash	-	5 million globally	Low (incineration)	-	-	-	-	50 % ↓	70 % ↓	-	-	-
67 Sua-iam	2013	Thailand	Rice husk ash	locally obtained	5 million globally	Low (untreated)	-	-	15 % ↓	65 % ↓	70 % ↓	95 % ↓	-	-	-
68 Chabannes	2014	France	Cork	locally obtained	20,000	Low (untreated)	-	Very low < 1 MPa so not suitable for structural purposes					-	-	-
58 Panesar	2012	Canada	Cork	locally obtained	68,000–85,000 globally	Low (oven dry at 100 °C)	-	-	-	65 % ↓	-	-	-	-	-
59 Novoa	2004	Portugal	Cork	-	-	-	-	-	-	50 % ↓	-	-	-	-	-

62	Ozturk	2005	Turkey	Tobacco waste	locally obtained	-	No energy	-	Very low < 1 MPa so not suitable for structural purposes					-	-	-
72	Memon	2019	Pakistan	Corncob ash		1 million globally	Low (incineration)	-	-	15 % ↓	-	-	-	50 % ↑	-	-
76	Shafigh	2014	Malaysia	Oil palm shell		4 million	Low (oven-dry at 100 ºC)	-	-	-	10 % ↓	20 % ↓	-	-	50 % ↑	100 % ↑
77	Muthusamy	2013					Low (crushing)	-	-	-	30 % ↓	50 % ↓	-	-	-	-
70	Gunasekaran	2008	India	Coconut shell		-		-	-	-	-	-	-	-	-	-
46	Chabannes	2015	France	Sunflower stems		-		-	Very low < 1 MPa so not suitable for structural purposes					-	-	-
47	Kuo	2013	Taiwan	Oyster shells		60,000		-	-	15 % ↓	-	-	-	50 % ↑	-	-
49	Yang	2005	South Korea			0.3 million		-	-	Same	-	-	-	-	-	-
54	Martinez-Garcia	2017	Spain	Mussel shells		25,000	Low (oven-dry at 100 ºC)	-	-	-	30 % ↓	30 % ↓	-	-	10 % ↑	20 % ↓
52	Ponnada	2016	India	Cockle shells		-	Low (crushing)	-	-	10 % ↑	Same	-	-	-	-	-
53	Varhen	2017	Peru	Scallop shells		25,000		-	-	10 % ↑	Same	-	-	-	-	-
56	Adewuyi	2008	Nigeria	Periwinkle shells		-	Low (crushing)	40 % ↓	-	-	-	35 % ↓	70 % ↓	-	-	-

Unfortunately, only compressive strength and in some cases water absorption were discussed as performance criteria for the concrete when examining the behaviour of these wastes replacing fine aggregates. Hence, more research is required to determine other fundamental characteristics such as slump, tensile strength porosity, shrinkage, abrasion, resistance to chemical attacks (e.g., chloride ingress), and carbonation.

3.4 Industrial Wastes Used as Aggregates

Industrial wastes, like agricultural and aquaculture farming wastes, are being used as a fine natural aggregate replacement in concrete. This strategy has been extensively reviewed by Rashad [86] and de Brito and Saikia [89]. The findings of these two studies demonstrate that the majority of studies has concentrated on the influence of (i) artificial pozzolan wastes [90] as sand replacement in concrete, followed by (ii) natural pozzolans [91–100], (iii) fly ash [101–109], (iv) coal bottom ash [110–117], (v) iron and steel slags [118–131], (vi) silica fume [132–134], (vii) plastic waste, (viii) rubber waste, and distantly by (ix) non-ferrous slags [130, 135–141]. As seen in Table 3.2 summarizing the findings from these studies, these types of aggregates can reduce the cost and environmental impact (EI) and enhance several properties of concrete. However, widespread reliable data are missing for the use of these aggregates in concrete.

Regarding the feasibility of coal fly ash as a replacement for fine aggregate in concrete, a study [107] shows that the compressive, flexural, shear and tensile strength were found to increase by increasing the incorporation ratio of fly ash (as fine aggregate) up to 40 % when concrete was subjected to high temperature. As shown in Table 3.2, other studies [102, 105, 108, 109] confirmed these phenomena. This is because fewer micro-cracks occur with the use of fly ash aggregates. In other words, fly ash content increases the fire resistance of concrete up to the mentioned percentage, as well as workability. In addition, this replacement also reduces the total cost of concrete.

The above trend does not occur with the use of coal bottom ash. In fact, the strength of concrete is slightly reduced or equivalent to that of conventional concrete when natural aggregates are replaced with bottom ash up to 50 % (Table 3.2). This happens because the pozzolanic reaction of bottom ash is way slower than that of the fly ash. This leads to a lower bond between the cement matrix and the aggregates [114].

Based on the data given in Table 3.2, the use of the blast-furnace slag (BFS) can be considered one of the most promising solutions to produce sustainable concrete without jeopardizing the strength and cost of concrete. A study showed that, by considering the economy index (strength/cost), the optimal replacement ratio of natural aggregates with BFS is 40–50 % [118]. Regarding electric arc furnace slag (EAFS), its use is also promising, i.e., the results show that the compressive strength of concrete increases when a low incorporation ratio of EAFS is used. For a higher incorporation ratio of EAFS (e.g., 50 %), the strength of concrete can change up to 10 % (negatively or positively), as per Table 3.2. In addition, a few studies also showed the potential of the use of copper slag and lead-zinc slag as a replacement of natural aggregates, especially in terms of compressive strength. However, they reduce the workability of concrete.

Table 3.2 List of the most predominant features in utilizing industrial waste as aggregates in cement-based materials

Ref		Year	Country	Type of waste	Cost 10%	Slump		28-day compressive strength			Water absorption
						50%	10%	30%	50%	100%	
102	Pofale	2010	India	Fly ash	20% ↓	20% ↑	-	10% ↑	20% ↑	-	-
107	Parvati	2013	India	Fly ash	-	-	-	10% ↑	10% ↑	10% ↑	-
108	Joseph	2009	India	Fly ash	-	-	-	-	20% ↑	30% ↑	-
104	Siddique	2003	USA	Fly ash	-	-	-	5% ↑	20% ↑	50% ↑	-
109	Dhir	2000	UK	Fly ash	-	20% ↓	-	10% ↑	-	50% ↑	-
110	Bai	2005	UK	Bottom ash	-	30% ↑	50% ↑	5% ↑	10% ↑	5% ↓	-
114	Aggarwal	2007	India	Bottom ash	-	10% ↓	20% ↓	-	10% ↓	20% ↓	-
116	Singh	2014	India	Bottom ash	-	20% ↓	75% ↓	Same			-
117	Singh	2016	India	Bottom ash	-	20% ↓	75% ↓	Same			-
115	Kasem-chaisiri	2008	Thailand	Bottom ash	-	20% ↓	-	5% ↓	10% ↓		-
118	Singh	2015	India	BFS	20% ↑	-	-	10% ↑	15% ↑	20% ↑	-
119	Miyanoto	2015	Japan	BFS	↓	-	-	-	-	10% ↑	-
120	Binici	2012	Turkey	BFS	-	-	-	5% ↑	10% ↑	-	-
122	Ozbakka-loglu	2016	Australia	BFS	-	-	10% ↓	-	-	10% ↓	-
125	Manso	2003	Spain	EAFS	-	-	-	-	-	10% ↓	20% ↓
126	Alizadeh	2004	Iran	EAFS	-	-	-	-	-	10% ↑	-
128	Pellegrino	2013	Italy	EAFS	-	-	-	-	-	10% ↓	-
131	Qasrawi	2009	Jordan	EAFS	-	-	-	25% ↑	15% ↑	10% ↑	-
135	Sharma	2017	India	Copper slag	-	Same	Same	10% ↑	5% ↑	Same	-
136	Rajasekar	2019	India	Copper slag	-	-	-	10% ↑	15% ↑	20% ↑	10% ↓
137	Babu	2019	India	Copper slag	-	-	-	5% ↑	10% ↑	Same	-
139	Gupta	2019	India	Copper slag	-	10% ↓	50% ↓	5% ↑	5% ↑	Same	-
140	Alwaeli	2013	Poland	Lead-zinc slag	-	-	-	Same	10% ↑	20% ↑	-

Overall, the collected studies shown above have been mainly focused on concrete strength as a common indicator to explore, and rarely discussed durability properties such as sorptivity, rapid chloride penetration testing (RCPT), carbonation, shrinkage, creep. The conclusion that could be generalized is that replacing natural fine aggregates with a percentage of most types of industrial waste in concrete seems to lower slump. However, mainly due to the pozzolanic activity of the majority of them, the addition of industrial wastes as fine aggregates increases the strength of the resulting concrete, especially after 7 days of curing. The underlying premise of the utilization of industrial waste as fine aggregates is that they would contribute to a more sustainable concrete. Therefore, it is vital to explore whether it is worthwhile to utilize these industrial wastes, specifically those with pozzolanic properties like fly ash and ground granulated blast furnace slag, as replacement of sand or even better as partial replacement of Ordinary Portland cement (OPC).

3.5 Municipal Wastes as Aggregates

Municipal wastes as a raw material and ashes have been used in concrete as a partial replacement of natural aggregates, namely municipal solid waste incinerator bottom ash (MIBA), granite waste sludge and sewage sludge ash (SSA). Most of the studies have analysed the effect of municipal waste on the compressive strength and workability of concrete.

3.5.1 Municipal Solid Waste Incinerator Bottom Ash

MIBA has been used in concrete as a replacement for fine and coarse natural aggregates. Generally, the results show that slump reduces (mostly from S2 to S1) by increasing the amount of MIBA in concrete, see Fig. 3.2 [142–146]. The results show that fine MIBA aggregates reduce the workability of concrete more than coarse MIBA aggregates (Fig. 3.2, C *versus* F). This is due to lower water absorption and surface area of the coarser fraction of MIBA [147]. Thus, coarse MIBA aggregates are more recommended to be used in concrete. Furthermore, the particle shape is the most influential parameter for manufactured aggregates in terms of their effect on workability. Thus, more researches are required to find an effective method to manufacture aggregates. In addition, there is no segregation issue with the use of MIBA as aggregates and, for a low incorporation ratio (e.g. 20%), bleeding is reduced [146]. This can be related to the associated higher water retention of the ash and the higher absorptive properties [147]. Additionally, no delays in the setting times were evident with the use of MIBA as aggregates [142].

Regarding the compressive strength, the results show that the strength of concrete significantly reduces with the rising content of fine MIBA aggregates. However, chemical treatments (1 mol/l NaOH solution) and washing can improve its performance [147]. Relative to fine MIBA aggregates, coarse aggregates have fewer negative effects on the compressive strength of concrete. In addition, the w/c ratio also affects the performance of MIBA in concrete. In general, for a high incorporation ratio of the coarse oven-dried

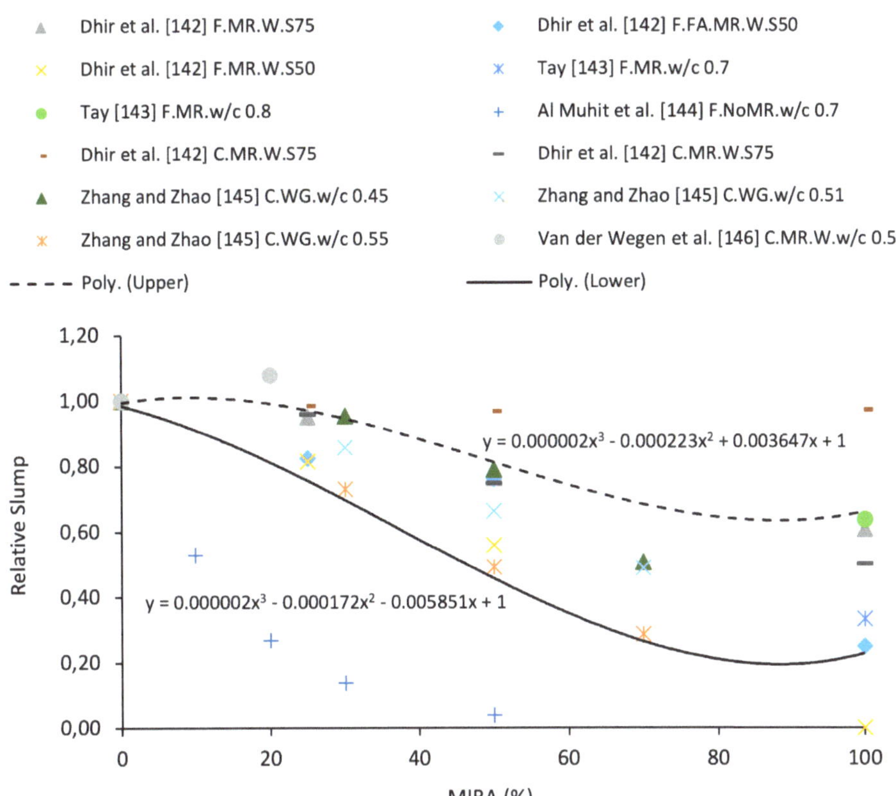

Fig. 3.2 Effect of MIBA on workability of concrete (C – coarse MIBA aggregates, F – fine MIBA aggregates, MR – metal removal, FA – fly ash, W – washed mix, no additional treatment, S – slump, W – washed mix, NoMR – no additional treatment)

MIBA, a w/c ratio (apparent) of 0.6 is optimal for concrete mixes (Fig. 3.3, 3D) [142–146, 148–150]. It is known that this value of the w/c (0.6) is a quite high value that normally brings inconvenient reductions in strength and durability. However, in this case, more water is required to get workable and applicable concrete which relates to the fact that the water absorption of the MIBA is very high. Further details regarding the strength decrements have been shown in another study [147], namely due to the high concentration of metallic aluminium in MIBA. Just like for fine MIBA aggregates, the performance of coarse MIBA aggregates in concrete can be improved with treatments [145]. Broadly speaking, MIBA replacement must be limited to low incorporation ratios in order to avoid excessive losses in strength. In general, studies regarding the effect of the MIBA on the durability of concrete are very scarce. Therefore, more studies are required to focus on this path.

Similarly to compressive strength, tensile strength [151], modulus of elasticity [142], chloride corrosion resistance [152], sulphate attack resistance [142] and carbonation resistance [146] of concrete have generally been found to reduce with increasing MIBA contents as aggregate, and the opposite can be seen for shrinkage [146], freeze thaw resistance [153] and acid attack resistance [151].

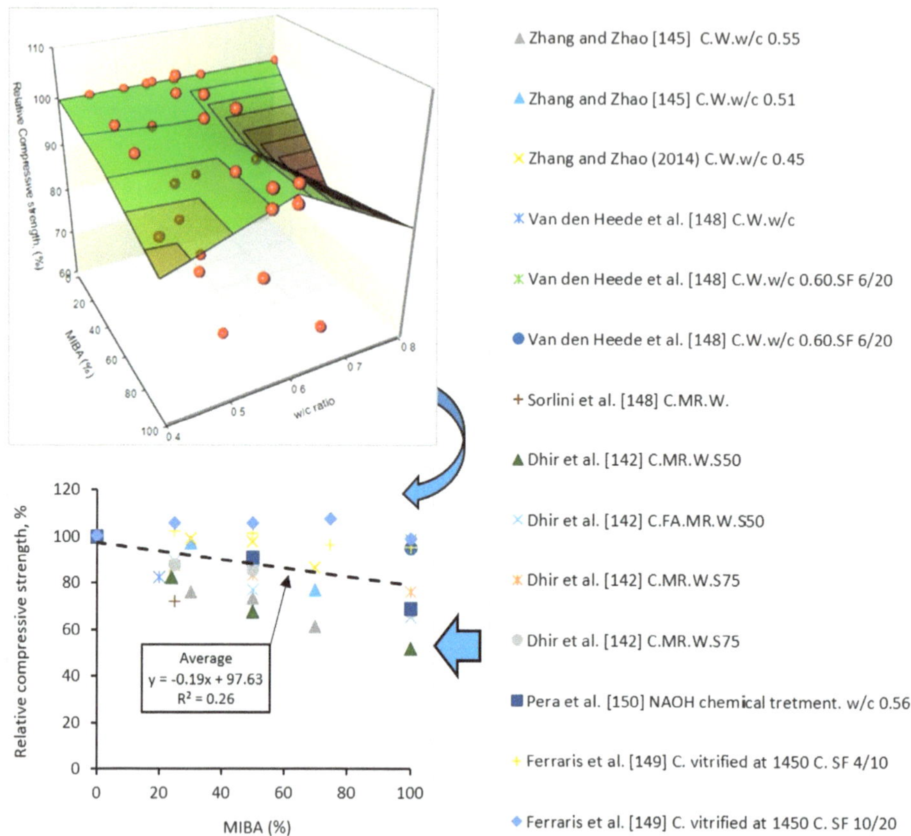

Fig. 3.3 Effect of MIBA on the compressive strength of concrete (MR – metal removal, W – washed mix, no additional treatment, S – slump, W – washed mix, NoMR – no additional treatment, SF – size fraction.)

3.5.2 Sewage Sludge Ash

SSA has been used in cement-based materials in many ways such as binder, filler [154] and sand [154–159]. Most of the studies [154–[158], [160]] have focused on the effect of low incorporation ratios of SSA on the performance of concrete and mortars, especially for slump, compressive strength, tensile strength, density, and water absorption (Fig. 3.4). In general, workability reduces with the use of SSA [156]. Fig. 3.4 reveals that there is a steep fall in the density, compressive strength, and flexural strength of concrete, and the opposite can be seen for the absorption properties. Thus, it is recommended to use only a low incorporation ratio of SSA in mortars and concrete.

Additionally, paper sludge [161] and granite waste sludge [162–164] ash can be another option to be used in concrete as aggregates. However, there are only very few such studies on this topic.

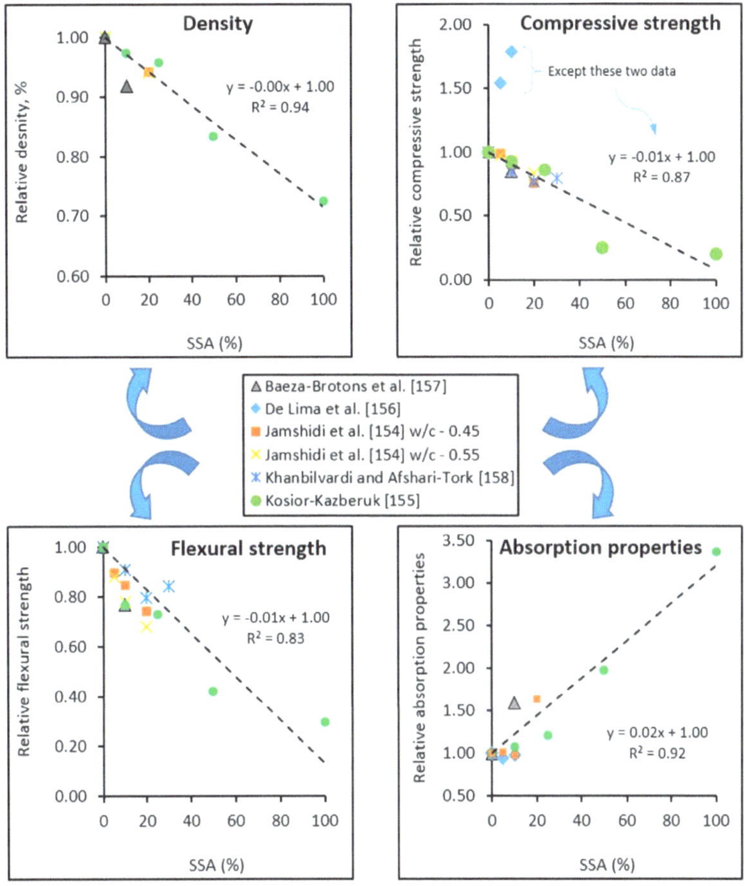

Fig. 3.4 Properties of cement based materials with sewage sludge ash as a sand replacment

3.6 Insulating Aggregates

3.6.1 Tire-Rubber Aggregates

Rubberised concrete provides a sustainable alternative that aids in reducing environmental damage associated with the disposal of waste tires. The workability of fresh concrete decreases notably with increasing replacement of natural aggregates with tire-rubber aggregates [165–167]. The workability declines as the fineness of rubber particles increases and chances of segregation are high with the utilization of rubber aggregates (Fig. 3.5). The difference in density between rubber aggregates and binder, as well as the weak bond between rubber aggregates and fresh binder, increase segregation. Compressive strength falls with the rise in tire-rubber aggregate level. This reduction can be due to the weak bond between tire-rubber aggregates and binder matrix. Due to lack of adhesion, lightweight rubber particles act as voids in the matrix of concrete [168]. The tire-rubber aggregate size also influences the degree of strength reduction.

5 mm rubber particles produce better mechanical properties than 2 mm and 10 mm particle sizes [165–166].

The effect of rubber particles is less detrimental to the tensile properties compared to their effect on compressive strength [166, 167, 169]. The incorporation of fine rubber aggregates increases the footprint of concrete as shredding the tire rubber into fine particles consumes more energy than crushing stone aggregates (Fig. 3.6). However, the use of coarse tire-rubber produces less CO_2 footprint than conventional concrete [165]. There is a deficiency in the literature on the life cycle assessment of concrete with tire waste aggregates. More focus is needed on the treatment of tire-rubber aggregates to improve their adhesion in concrete [166]. This may lead to improvement in the fresh and hardened state properties of rubberised concrete that can be used as an eco-friendly alternative to lightweight concrete. In general, studies regarding the effect of rubber particles as aggregates on the durability of concrete are very scarce. Therefore, more studies are required to focus on this path.

3.6.2 Plastic Waste Aggregate

Plastic waste aggregate can be sourced from processing of several plastic wastes, i.e., polyethylene (PE) bags waste, polyethylene terephthalate (PET) bottle waste, electronic plastic waste, or polyvinyl chloride (PVC), among others [170]. The partial replacement of fine aggregate with plastic waste aggregates does not significantly affect the workability (slump) of concrete [171]. However, chances of bleeding increased with the rising plastic aggregate content [171] and water-cement ratio [172]. The workability falls when the shape of plastic aggregates changes from spherical to flaky [173]. In general, fine and flaky plastic waste is detrimental to workability. The plasticity and consistencies are also affected negatively by partial replacement of fine aggregates with plastic waste aggregates [174]. It is established that strength declines with increasing plastic aggregates content (Fig. 3.7). The decrease in strength upon incorporation of plastic waste depends on the shape and size of aggregates and water-cement ratio.

The small particle sizes are less detrimental than large particles [175]. Round and angular particles are less harmful than flaky particles [176]. Plastic aggregates are more harmful when used in concrete with a high water-cement ratio [177, 178]. Tensile and flexural properties are affected similarly to compressive strength, owing to a weak bond between smooth plastic aggregate surfaces and binder [177, 178]. The obvious advantage of plastic aggregates concrete is in making lightweight concrete. Plastic aggregate concrete is also suitable for harsh acidic environments [179]. According to some studies, plastic aggregates decrease or insignificantly affect [171, 172, 180] the drying shrinkage of concrete. Literature still lacks information on durability and life cycle assessment of plastic aggregates concrete. The properties of concrete can be improved by increasing the surface roughness of plastic aggregates [181].

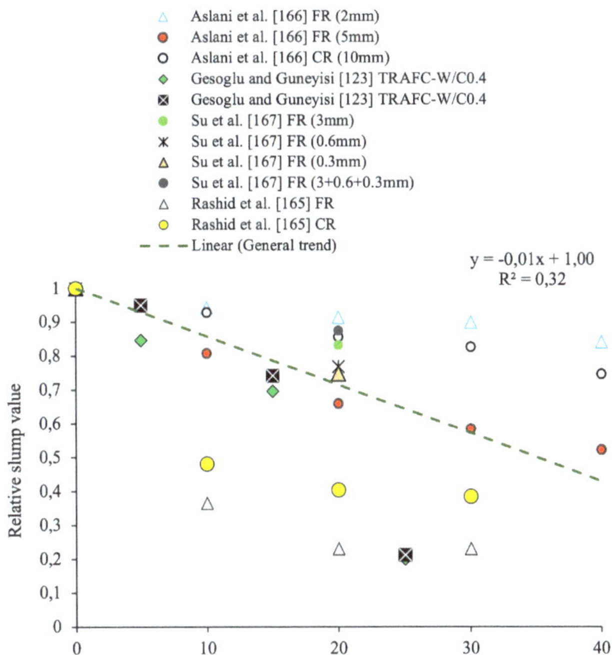

Fig. 3.5   Slump of cement-based materials with tire-rubber aggregates as an aggregate component

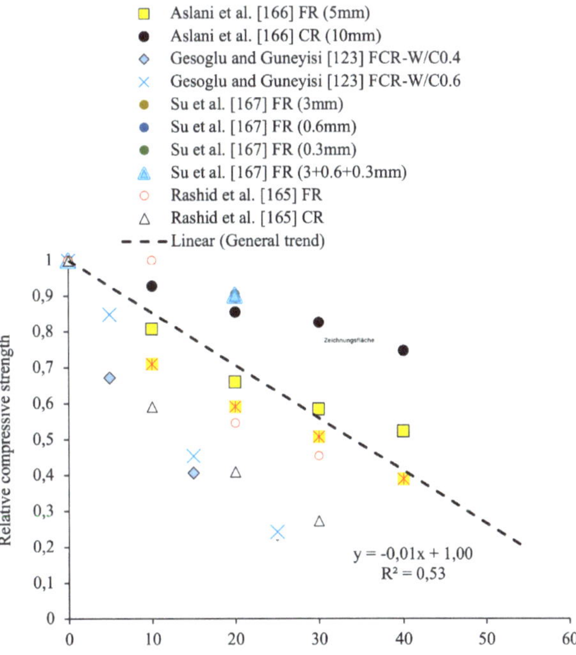

Fig. 3.6   Compressive strength of cement-based materials with tire-rubber aggregates as an aggregate component

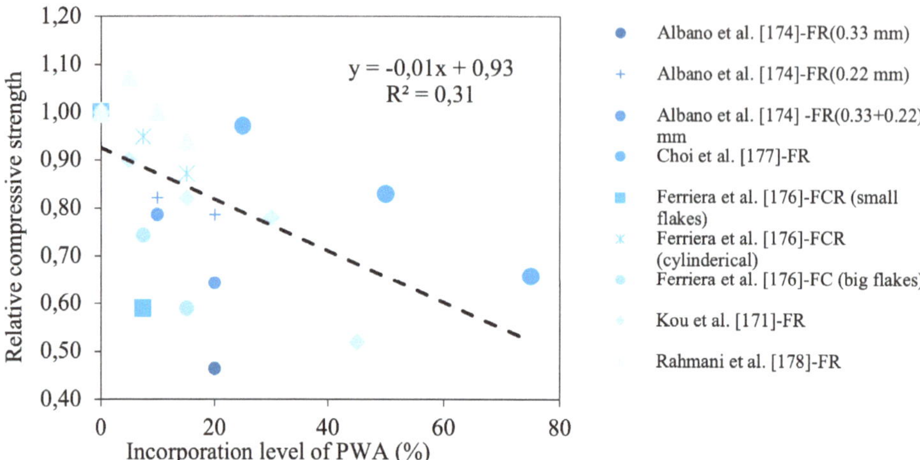

Fig. 3.7   Compressive strength of cement-based materials with plastic waste as an aggregate component

3.6.3 Glass Waste Aggregates

Glass waste aggregates can be processed from different wastes such as windows from cars and buildings, beverage bottles and glasses, medicine bottles and vials, TV screens and monitors, among others. Most related studies have investigated glass waste as a replacement of fine aggregates. Unlikesimilar reviews on plastic and rubber aggregates, high volume replacements of natural fine aggregates with glass waste aggregates have been investigated. Different to plastic aggregates, glass waste aggregates do not significantly decrease density [182], hence it can be used for normal-density concrete. Waste glass aggregates decrease the water demand of concrete owing to its low absorption capacity [183]. As shown in Figure 3.8, waste glass aggregates decrease the slump flow owing to their sharper and angular shape [182, 184, 185], while other studies reported insignificant changes [186]. For any type and size of glass waste aggregates, the mechanical properties reduce with the increasing waste glass aggregates content compared to a conventional mix (Figure 3.9). These reductions are blamed on lack of adhesion between smooth glass surface and binder. The intensity of reduction is moderate until 20-25% replacement of natural sand with glass aggregates [183, 185, 187]. With the incorporation of glass waste aggregates, the thermal conductivity of concrete reduces and improvement in residual mechanical strength is observed after exposure to elevated temperatures [188].

One major issue with the application of a high volume of glass aggregates is expansion due to an alkali-silica reaction [182, 189, 190]. Mineral admixtures like fly ash, slag and silica fume have been used to mitigate alkali-silica reactions [182, 189, 190]. A study [182] suggested that clear glass is more susceptible to alkali-silica reaction than coloured glass. Concrete with 20% screen monitor glass powder showed minimum shrinkage, while other levels of glass aggregates showed shrinkage comparable to conventional concrete [183, 186]. The literature is deficient in studies related to chloride and acid attack- and freeze-thaw-related investigations.

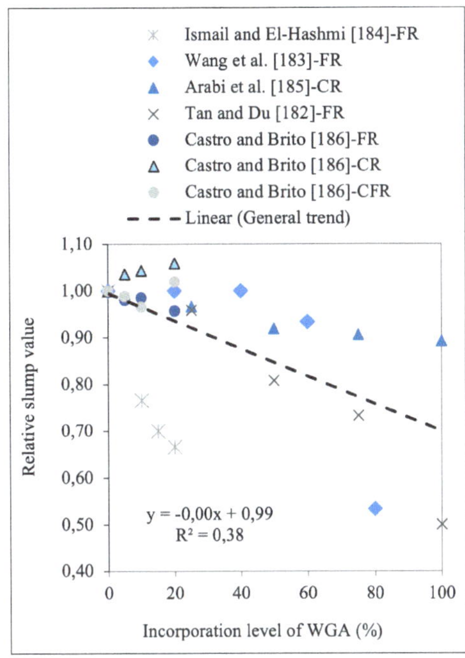
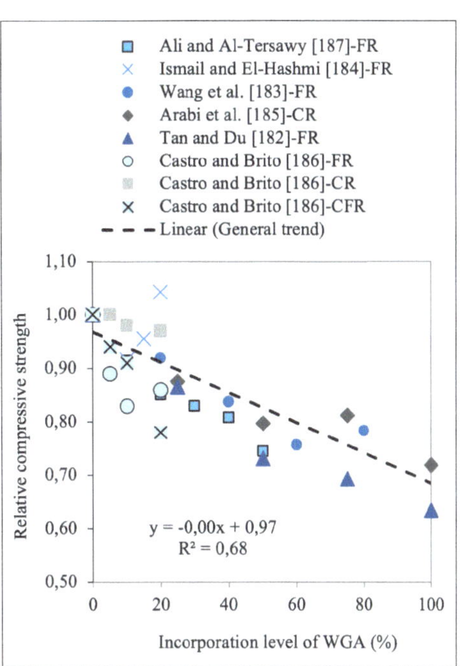

Fig. 3.8 Slump of cement-based materials made with glass waste aggregates

Fig. 3.9 Compressive strength of cement-based materials made with glass waste aggregates

3.7 Other Types of Aggregates

Apart from the nonconventional aggregates mentioned above, there are other types of aggregates that have been used in cement-based materials such as biochar aggregates [191], mine tailings [192], lead-zinc tailings [193], stone slurry [194], magnetite/hematite/ferrock [195–198], pumice stones [199, 200], ethylene-vinyl acetate [201, 202], lead-zinc tailings [193] and alkali-activated aggregates [203]. Apart from alkali-activated aggregates, studies on other types of aggregates are very scarce.

3.8 Conclusion

The main aim of this chapter was to understand the performance of nonconventional aggregates in cement-based materials in terms of technical properties, environmental impact, and cost. Thus, the authors followed the methodology illustrated in Fig. 3.10.

Based on the data collected in this study, it can be said that most categories of the technical performance of cement-based materials reduce by increasing the nonconventional aggregates' content. However, up to a content of 30%, there is a relatively small decrease in the technical performance, after which it declines increasingly. Therefore, it is recommended to use nonconventional aggregates up to 25% (replacing ¼ of conventional with nonconventional aggregates).

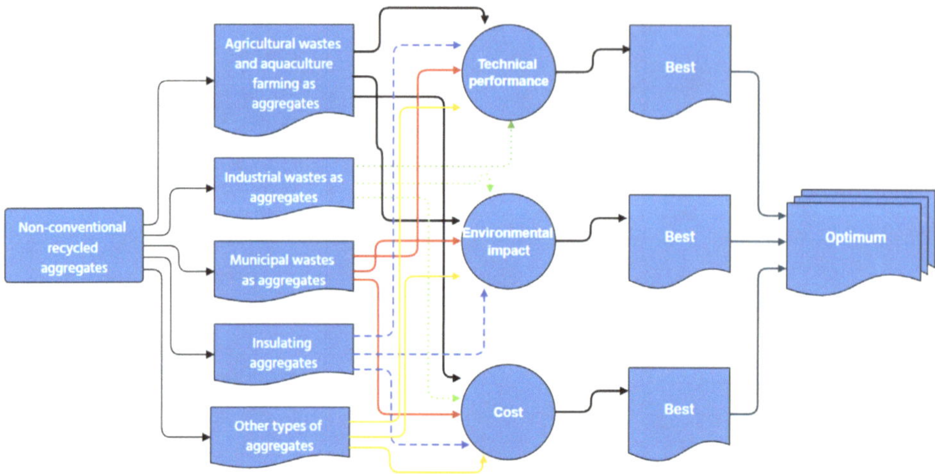

Fig. 3.10 Methodology profile to identify optimum nonconventional aggregates for concrete based on technical performance, environmental impact and cost

As future research, a meta-analysis study needs to be done according to Fig. 3.10 illustrated below (materials from each type will be compared in order to find optimum materials contents for all aspects). As future experimental work, the performance, environmental impact, and cost of each type of nonconventional recycled aggregates must be studied to decide their performance as engineering materials. In other words, the purpose of this work is to show the construction industry which types of these nonconventional materials are applicable in terms of all criteria (technical performance, environmental impact, and cost). However, a comprehensive statement can only be made in the future because, as mentioned before, studies on cost and environmental impact are very scarce. In fact, even for the technical properties, only slump, compressive strength, and water absorption have been studied. Therefore, other technical properties need to be considered in future studies.

3.9 Acknowledgements

The authors gratefully acknowledge the support of CERIS from IST, University of Lisbon, and FCT, Foundation for Science and Technology.

3.10 References

[1] IEA, *World Energy Outlook 2011*, in *International Energy Agency (IEA)*. 2011: OECD/IEA, Paris.

[2] USGS, *Mineral Commodity Summaries (Commodity statistics and information). United States Geological Survey (USGS). Mineral Yearbooks. United States*. 2018.

[3] Details, F.a. *Great wall of China*. https://factsanddetails.com/china/cat15/sub94/entry-6542.html. 2021.

[4]   Lavado, J., et al., *Fresh properties of recycled aggregate concrete.* Construction and Building Materials, 2020. 233: p. 117322. https://doi.org/10.1016/j.conbuildmat.2019.117322

[5]   Ferreira, L., J.d. Brito, and M. Barra, *Influence of the pre-saturation of recycled coarse concrete aggregates on concrete properties.* Magazine of Concrete Research, 2011. 63(8): p. 617–627.

[6]   Yang, K., H. Chung, and A. Ashour, *Influence of type and replacement level of recycled aggregates on concrete properties.* ACI Materials Journal, 2008. 105(3): p. 289–296. https://doi.org/10.14359/19826

[7]   Silva, R.V., J. de Brito, and R.K. Dhir, *The influence of the use of recycled aggregates on the compressive strength of concrete: a review.* European Journal of Environmental and Civil Engineering, 2015. 19(7): p. 825–849. https://doi.org/10.1080/19648189.2014.974831

[8]   Senthamarai, R.M. and P. Devadas Manoharan, *Concrete with ceramic waste aggregate.* Cement and Concrete Composites, 2005. 27(9): p. 910–913. https://doi.org/10.1016/j.cemconcomp.2005.04.003

[9]   Medina, C., M.I. Sánchez de Rojas, and M. Frías, *Reuse of sanitary ceramic wastes as coarse aggregate in eco-efficient concretes.* Cement and Concrete Composites, 2012. 34(1): p. 48–54. https://doi.org/10.1016/j.cemconcomp.2011.08.015

[10]   Pacheco-Torgal, F. and S. Jalali, *Compressive strength and durability properties of ceramic wastes based concrete.* Materials and Structures, 2011. 44(1): p. 155–167. https://link.springer.com/article/10.1617/s11527-021-01737-3

[11]   Bommisetty, J., et al., *Effect of waste ceramic tiles as a partial replacement of aggregates in concrete.* Materials Today: Proceedings, 2019. https://doi.org/10.1016/j.matpr.2019.08.230

[12]   Silva, R.V., J. de Brito, and R.K. Dhir, *Tensile strength behaviour of recycled aggregate concrete.* Construction and Building Materials, 2015. 83: p. 108–118. https://doi.org/10.1016/j.conbuildmat.2015.03.034

[13]   Debieb, F. and S. Kenai, *The use of coarse and fine crushed bricks as aggregate in concrete.* Construction and Building Materials, 2008. 22(5): p. 886–893. https://doi.org/10.1016/j.conbuildmat.2006.12.013

[14]   Silva, R.V., J. de Brito, and R.K. Dhir, *Use of recycled aggregates arising from construction and demolition waste in new construction applications.* Journal of Cleaner Production, 2019. 236: p. 117629. https://doi.org/10.1016/j.jclepro.2019.117629

[15]   Sormunen, P. and T. Kärki, *Recycled construction and demolition waste as a possible source of materials for composite manufacturing.* Journal of Building Engineering, 2019. 24: p. 100742. https://doi.org/10.1016/j.jobe.2019.100742

[16]   Corinaldesi, V. and G. Moriconi, *Influence of mineral additions on the performance of 100 % recycled aggregate concrete.* Construction and Building Materials, 2009. 23(8): p. 2869–2876. https://doi.org/10.1016/j.conbuildmat.2009.02.004

[17]   Gomes, M. and J. de Brito, *Structural concrete with incorporation of coarse recycled concrete and ceramic aggregates: durability performance.* Materials and Structures, 2009. 42: p. 663–675. https://doi.org/10.1617/s11527-008-9411-9

[18] Dhir, R.K. and K.A. Paine, *Performance related approach to the use of recycled aggregates*. Banbury, Oxon, UK, Waste and Resources Action Programme (WRAP) Aggregates Research Programme, 2007: p. 77.

[19] Mas, B., et al., *Influence of the amount of mixed recycled aggregates on the properties of concrete for non-structural use*. Construction and Building Materials, 2012. 27(1): p. 612–622. https://doi.org/10.1016/j.conbuildmat.2011.06.073

[20] Yang, J., Q. Du, and Y. Bao, *Concrete with recycled concrete aggregate and crushed clay bricks*. Construction and Building Materials, 2011. 25(4): p. 1935–1945. https://doi.org/10.1016/j.conbuildmat.2010.11.063

[21] Zheng, C., et al., *Mechanical properties of recycled concrete with demolished waste concrete aggregate and clay brick aggregate*. Results in Physics, 2018. 9: p. 1317–1322. https://doi.org/10.1016/j.rinp.2018.04.061

[22] Juan-Valdés, A., et al., *Paving with Precast Concrete Made with Recycled Mixed Ceramic Aggregates: A Viable Technical Option for the Valorization of Construction and Demolition Wastes (CDW)*. Materials (Basel, Switzerland), 2018. 12(1): p. 24. https://doi.org/10.3390/ma12010024

[23] LNEC E471, *Guideline for the use of recycled coarse aggregates in hydraulic binders concrete (in Portuguese)*. 2006, LNEC (National Laboratory of Civil Engineering). p. 7.

[24] BRV, *Guideline for recycled building materials*. 2007, Austrian Construction Materials Recycling Association (BRV – Österreichischer Baustoff-Recycling Verband): Austria.

[25] BS 6543, *Guide to use of industrial by-products and waste materials in building and civil engineering*. 1985, British Standards Institution (BSI). p. 40.

[26] BRE Digest 433, *Recycled aggregates*. 1998, Building research establishment (BRE): IHS BRE: Press, Walford.

[27] BS 8500-2, *Concrete – complementary British Standard to BS EN 206-1, Part 2: Specification for constituent materials and concrete*. 2002, British Standards Institution (BSI): UK.

[28] CUR 125, *Crushed concrete rubble and masonry rubble as aggregate for concrete*. 1986, Centre for Civil Engineering Research, Codes and Specifications (CUR): Holland.

[29] CUR-VB 4, *Recycled concrete aggregates for concrete use (in Dutch)*. 1984, Centre for Civil Engineering Research, Codes and Specifications (CUR): Netherlands.

[30] CUR-VB 5, *Recycled masonry aggregates as additive for concrete (in Dutch)*. 1994, Centre for Civil Engineering Research, Codes and Specifications (CUR): Netherlands.

[31] TV 70085, *Instruction technique (TV) utilisation de matériaux de construction minéraux se-condaires dans la construction d'abris*. 2006, The Federal Department of Defence, Civil Protection and Sport (DDPS): Switzerland. p. 16.

[32] DIN 4226-100, *Aggregates for concrete and mortar – Part 100: Recycled aggregates*. 2002, German Institute for Standardisation (Deutsches Institut für Normung): Germany.

[33] ACI 555R-01, *Removal and reuse of hardened concrete*. 2001, American Concrete Institute (ACI): United States. p. 26.

[34] JSA – JIS A 5021, *Recycled aggregate for concrete-class H*. 2016, Japanese Standards Association (JSA). p. 40.

[35] JSA – JIS A 5022, *Japanese language – Recycled concrete using recycled aggregate class M*. 2016, Japanese Standards Association (JSA): Japan.

[36] JSA – JIS A 5023, *Japanese language – Recycled concrete using recycled aggregate class L*. 2016, Japanese Standards Association (JSA): Japan.

[37] DCA-N.34, *Danish Recommendation for the use of recycled aggregates for concrete in passive environmental class*. 1995, Danish Concrete Association (DCA), Publication no. 34: Denmark.

[38] NBR 15.116, *Aggregates from Construction and Demolition Waste: Use on road pavements and non-structural concrete – requirements (in Portuguese)*. 2005: Brazil. p. 12.

[39] EEPL, *Construction and demolition waste guide – Recycling and re-use across the supply chain*. 2012, Edge Environment Propriety Limited (EVPL): Australia. p. 57.

[40] WBTC-N.12, *Specifications facilitating the use of recycled aggregates*. 2002, Works Bureau Technical Circular (WBTC): China.

[41] DREIF, *Guide technique pour l'utilisation des matériaux régio-naux d'Ile-de-France*. 2003, Laboratoire Régional de l'Ouest Parisien (DREIF). DRIEA / CETE Ile-de-France: France. p. 11.

[42] RILEM TC 121-DRG N. 27, *Specifications for concrete with recycled aggregates*. Materials and Structures, 1994: p. 557–559. https://doi.org/10.1007/bf02473217

[43] Silva, R.V., J. de Brito, and R.K. Dhir, *Properties and composition of recycled aggregates from construction and demolition waste suitable for concrete production*. Construction and Building Materials, 2014. 65: p. 201–217. https://doi.org/10.1016/j.conbuildmat.2014.04.117

[44] Mageswari, M. and B. Vidivelli, *The use of sawdust ash as fine aggregate replacement in concrete*. Journal of Environmental Research and Development, 2009. 3(3): p. 720–726.

[45] Ganiron, T.U., *Effect of sawdust as fine aggregate in concrete mixture for building construction*. International Journal of Advanced Science and Technology, 2014. 63: p. 73–82. https://doi.org/10.14257/ijast.2014.63.07

[46] Chabannes, M., V. Nozahic, and S. Amziane, *Design and multi-physical properties of a new insulating concrete using sunflower stem aggregates and eco-friendly binders*. Materials and Structures, 2015. 48(6): p. 1815–1829. https://doi.org/10.1617/s11527-014-0276-9

[47] Kuo, W.-T., et al., *Engineering properties of controlled low-strength materials containing waste oyster shells*. Construction and Building Materials, 2013. 46: p. 128–133. https://doi.org/10.1016/j.conbuildmat.2013.04.020

[48] Yang, E.-I., et al., *Effect of partial replacement of sand with dry oyster shell on the long-term performance of concrete*. Construction and Building Materials, 2010. 24(5): p. 758–765. https://doi.org/10.1016/j.conbuildmat.2009.10.032

[49] Yang, E.-I., S.-T. Yi, and Y.-M. Leem, *Effect of oyster shell substituted for fine aggregate on concrete characteristics: Part I. Fundamental properties*. Cement and Concrete Research, 2005. 35(11): p. 2175–2182. https://doi.org/10.1016/j.cemconres.2005.03.016

[50] Mo, K.H., et al., *Recycling of seashell waste in concrete: A review.* Construction and Building Materials, 2018. 162: p. 751–764. https://doi.org/10.1016/j.conbuildmat.2017.12.009

[51] Eo, S.-H. and S.-T. Yi, *Effect of oyster shell as an aggregate replacement on the characteristics of concrete.* Magazine of Concrete Research, 2015. 67(15): p. 833–842. https://doi.org/10.1680/macr.14.00383

[52] Ponnada, M.R., S.S. Prasad, and H. Dharmala, *Compressive strength of concrete with partial replacement of aggregates with granite powder and cockle shell.* Malaysian Journal of Civil Engineering, 2016. 28(2).

[53] Varhen, C., S. Carrillo, and G. Ruiz, *Experimental investigation of Peruvian scallop used as fine aggregate in concrete.* Construction and Building Materials, 2017. 136: p. 533–540. https://doi.org/10.1016/j.conbuildmat.2017.01.067

[54] Martínez-García, C., et al., *Performance of mussel shell as aggregate in plain concrete.* Construction and Building Materials, 2017. 139: p. 570–583. https://doi.org/10.1016/j.conbuildmat.2016.09.091

[55] Falade, F., *An investigation of periwinkle shells as coarse aggregate in concrete.* Building and Environment, 1995. 30(4): p. 573–577. https://doi.org/10.1016/0360-1323(94)00057-y

[56] Adewuyi, A. and T. Adegoke, *Exploratory study of periwinkle shells as coarse aggregates in concrete works.* ARPN Journal of Engineering and Applied Sciences, 2008. 3(6): p. 1–5.

[57] Binici, H., et al., *Effect of corncob, wheat straw, and plane leaf ashes as mineral admixtures on concrete durability.* Journal of Materials in Civil Engineering, 2008. 20(7): p. 478–483. https://doi.org/10.1061/(asce)0899-1561(2008)20:7(478)

[58] Panesar, D.K. and B. Shindman, *The mechanical, transport and thermal properties of mortar and concrete containing waste cork.* Cement and Concrete Composites, 2012. 34(9): p. 982–992. https://doi.org/10.1016/j.cemconcomp.2012.06.003

[59] Nóvoa, P.J.R.O., et al., *Mechanical characterization of lightweight polymer mortar modified with cork granulates.* Composites Science and Technology, 2004. 64(13): p. 2197–2205. https://doi.org/10.1016/s0266-3538(04)00074-0

[60] Ismail, Z.Z. and A.J. Jaeel, *A novel use of undesirable wild giant reed biomass to replace aggregate in concrete.* Construction and Building Materials, 2014. 67: p. 68–73. https://doi.org/10.1016/j.conbuildmat.2013.11.064

[61] Ottosen, L.M., et al., *Wood ash used as partly sand and/or cement replacement in mortar.* International Journal of Sustainable Development and Planning, 2016. 11(5): p. 781–791. https://doi.org/10.2495/sdp-v11-n5-781-791

[62] Ozturk, T. and M. Bayrakl, *The possibilities of using tobacco wastes in producing lightweight concrete.* International Commission of Agricultural Engineering (CIGR, Commission Internationale du Génie Rural) E-Journal, 2005. Volume 7. https://hdl.handle.net/1813/10427

[63] Sales, A. and S.A. Lima, *Use of Brazilian sugarcane bagasse ash in concrete as sand replacement.* Waste Management, 2010. 30(6): p. 1114–1122. https://doi.org/10.1016/j.wasman.2010.01.026

[64] Modani, P.O. and M.R. Vyawahare, *Utilization of Bagasse Ash as a Partial Replacement of Fine Aggregate in Concrete.* Procedia Engineering, 2013. 51: p. 25–29. https://doi.org/10.1016/j.proeng.2013.01.007

[65] Al-Akhras, N.M., K.M. Al-Akhras, and M.F. Attom, *Thermal cycling of wheat straw ash concrete.* Proceedings of the Institution of Civil Engineers – Construction Materials, 2008. 161(1): p. 9–15. https://doi.org/10.1680/coma.2008.161.1.9

[66] Kunchariyakun, K., S. Asavapisit, and K. Sombatsompop, *Properties of autoclaved aerated concrete incorporating rice husk ash as partial replacement for fine aggregate.* Cement and Concrete Composites, 2015. 55: p. 11–16. https://doi.org/10.1016/j.cemconcomp.2014.07.021

[67] Sua-Iam, G. and N. Makul, *Utilization of limestone powder to improve the properties of self-compacting concrete incorporating high volumes of untreated rice husk ash as fine aggregate.* Construction and Building Materials, 2013. 38: p. 455–464. https://doi.org/10.1016/j.conbuildmat.2012.08.016

[68] Chabannes, M., et al., *Use of raw rice husk as natural aggregate in a lightweight insulating concrete: An innovative application.* Construction and Building Materials, 2014. 70: p. 428–438. https://doi.org/10.1016/j.conbuildmat.2014.07.025

[69] Sada, B., Y. Amartey, and S. Bakoc, *An Investigation Into the Use of Groundnut as Fine Aggregate Replacement.* Nigerian Journal of Technology, 2013. 32(1): p. 54–60.

[70] Gunasekaran, K. and P. Kumar. *Lightweight concrete using coconut shell as aggregate.* in *Proceedings of the ICACC-2008. International conference on advances in concrete and construction, Hyderabad, India.* 2008.

[71] Baffa, I. and J. Akasaki. *Light-concrete with leather: Durability aspects.* in *International conference for structures. Coimbra, Portugal.* 2005.

[72] Memon, S.A., U. Javed, and R.A. Khushnood, *Eco-friendly utilization of corncob ash as partial replacement of sand in concrete.* Construction and Building Materials, 2019. 195: p. 165–177. https://doi.org/10.1016/j.conbuildmat.2018.11.063

[73] Odi, I.J.A.B., *Utilization of Olive Husk as a Replacement of Fine Aggregate in Portland Cement Concrete Mixes for Non-Structural Uses.* 2007, An-Najah National University Nablus, Palestine.

[74] Aslam, M., P. Shafigh, and M.Z. Jumaat, *Oil-palm by-products as lightweight aggregate in concrete mixture: a review.* Journal of Cleaner Production, 2016. 126: p. 56–73. https://doi.org/10.1016/j.jclepro.2016.03.100

[75] Yap, S.P., et al., *Effect of fibre aspect ratio on the torsional behaviour of steel fibre-reinforced normal weight concrete and lightweight concrete.* Engineering Structures, 2015. 101: p. 24–33. https://doi.org/10.1016/j.engstruct.2015.07.007

[76] Shafigh, P., et al., *Structural lightweight aggregate concrete using two types of waste from the palm oil industry as aggregate.* Journal of Cleaner Production, 2014. 80: p. 187–196. https://doi.org/10.1016/j.jclepro.2014.05.051

[77] Muthusamy, K., N. Zulkepli, and F.M. Yahaya, *Exploratory study on oil palm shell as partial sand replacement in concrete.* Research Journal of Applied Sciences, Engineering and Technology, 2013. 5(7): p. 2372–2375. https://doi.org/10.19026/rjaset.5.4667

[78] Okpala, D.C., *Palm kernel shell as a lightweight aggregate in concrete*. Building and Environment, 1990. 25(4): p. 291–296. https://doi.org/10.1016/0360-1323(90)90002-9

[79] Ndoke, P.N., *Performance of palm kernel shells as a partial replacement for coarse aggregate in asphalt concrete*. Leonardo Electronic Journal of Practices and Technologies, 2006. 5(9): p. 145–152.

[80] Mahmud, H., M. Jumaat, and U. Alengaram, *Influence of sand/cement ratio on mechanical properties of palm kernel shell concrete*. Journal of Applied Sciences, 2009. 9(9): p. 1764–1769. https://doi.org/10.3923/jas.2009.1764.1769

[81] Alengaram, U.J., et al., *Effect of aggregate size and proportion on strength properties of palm kernel shell concrete*. International Journal of the Physical Sciences, 2010. 5(12): p. 1848–1856.

[82] Gunasekaran, K., P.S. Kumar, and M. Lakshmipathy, *Mechanical and bond properties of coconut shell concrete*. Construction and Building Materials, 2011. 25(1): p. 92–98. https://doi.org/10.1016/j.conbuildmat.2010.06.053

[83] Kaur, M. and M. Kaur, *A review on utilization of coconut shell as coarse aggregate in mass concrete*. International Journal of Applied Engineering Research, 2012. 7(11): p. 7–9.

[84] Mannan, M.A. and C. Ganapathy, *Long-term strengths of concrete with oil palm shell as coarse aggregate*. Cement and Concrete Research, 2001. 31(9): p. 1319–1321. https://doi.org/10.1016/s0008-8846(01)00584-1

[85] Shafigh, P., et al., *Agricultural wastes as aggregate in concrete mixtures – A review*. Construction and Building Materials, 2014. 53: p. 110–117. https://doi.org/10.1016/j.conbuildmat.2013.11.074

[86] Rashad, A., *Cementitious materials and agricultural wastes as natural fine aggregate replacement in conventional mortar and concrete*. Journal of Building Engineering, 2016. 5: p. 119–141. https://doi.org/10.1016/j.jobe.2015.11.011

[87] Prusty, J.K., S.K. Patro, and S.S. Basarkar, *Concrete using agro-waste as fine aggregate for sustainable built environment – A review*. International Journal of Sustainable Built Environment, 2016. 5(2): p. 312–333. https://doi.org/10.1016/j.ijsbe.2016.06.003

[88] Hafez, H., et al., *A whole life cycle performance-based ECOnomic and ECOlogical assessment framework (ECO2) for concrete sustainability*. Journal of Cleaner Production, 2021. 292: p. 126060. https://doi.org/10.1016/j.jclepro.2021.126060

[89] De Brito, J.; Saikia, N. Recycled Aggregate in Concrete; Green Energy and Technology; Springer: London, UK, 2013; ISBN 978-1-4471-4539-4

[90] Dhir, R.K., et al., *8 – Environmental Assessment*, in *Sustainable Construction Materials*, R.K. Dhir, et al., Editors. 2018, Woodhead Publishing. p. 277–330. https://doi.org/10.1016/B978-0-08-100997-0.00008-7

[91] Maia, L. and D. Neves, *Developing a Commercial Self-Compacting Concrete Without Limestone Filler and With Volcanic Aggregate Materials*. Procedia Structural Integrity, 2017. 5: p. 147–154. https://doi.org/10.1016/j.prostr.2017.07.085

[92] Bogas, J.A. and D. Cunha, *Non-structural lightweight concrete with volcanic scoria aggregates for lightweight fill in building's floors*. Construction and Building Materials, 2017. 135: p. 151–163. https://doi.org/10.1016/j.conbuildmat.2016.12.213

[93]   Juimo Tchamdjou, W.H., et al., *Mechanical properties of lightweight aggregates concrete made with cameroonian volcanic scoria: Destructive and non-destructive characterization.* Journal of Building Engineering, 2018. 16: p. 134–145. https://doi.org/10.1016/j.jobe.2018.01.003

[94]   Marra, F., et al., *Petro-chemical features and source areas of volcanic aggregates used in ancient Roman maritime concretes.* Journal of Volcanology and Geothermal Research, 2016. 328: p. 59–69. https://doi.org/10.1016/j.jvolgeores.2016.10.005

[95]   Posi, P., et al., *Pressed lightweight concrete containing calcined diatomite aggregate.* Construction and Building Materials, 2013. 47: p. 896–901. https://doi.org/10.1016/j.conbuildmat.2013.05.094

[96]   Kotwa, A., *The Replacement of the Parts of the Aggregate in Concrete with Chalcedonite Powder.* Procedia Engineering, 2017. 195: p. 183–188. https://doi.org/10.1016/j.proeng.2017.04.542

[97]   Top, S., et al., *Properties of fly ash-based lightweight geopolymer concrete prepared using pumice and expanded perlite as aggregates.* Journal of Molecular Structure, 2019: p. 127236. https://doi.org/10.1016/j.molstruc.2019.127236

[98]   Wongsa, A., et al., *Use of crushed clay brick and pumice aggregates in lightweight geopolymer concrete.* Construction and Building Materials, 2018. 188: p. 1025–1034. https://doi.org/10.1016/j.conbuildmat.2018.08.176

[99]   Öz, H.Ö., *Properties of pervious concretes partially incorporating acidic pumice as coarse aggregate.* Construction and Building Materials, 2018. 166: p. 601–609. https://doi.org/10.1016/j.conbuildmat.2018.02.010

[100]   Sallı Bideci, Ö., *The effect of high temperature on lightweight concretes produced with colemanite coated pumice aggregates.* Construction and Building Materials, 2016. 113: p. 631–640. https://doi.org/10.1016/j.conbuildmat.2016.03.113

[101]   Seo, T., et al., *Properties of drying shrinkage cracking of concrete containing fly ash as partial replacement of fine aggregate.* Magazine of Concrete Research, 2010. 62(6): p. 427–433. https://doi.org/10.1680/macr.2010.62.6.427

[102]   Pofale, A. and S. Deo, *Comparative long term study of concrete mix design procedure for fine aggregate replacement with fly ash by minimum voids method and maximum density method.* KSCE Journal of Civil Engineering, 2010. 14(5): p. 759–764. https://doi.org/10.1007/s12205-010-0911-0

[103]   Maslehuddin, M., et al., *Effect of Sand Replacement on the Early-Age Strength Gain and Long-Term Corrosion-Resisting Characteristics of Fy Ash Concrete.* ACI Materials Journal, 1989. 86(1): p. 58–62. https://doi.org/10.14359/1856

[104]   Siddique, R., *Effect of fine aggregate replacement with Class F fly ash on the abrasion resistance of concrete.* Cement and Concrete Research, 2003. 33(11): p. 1877–1881. https://doi.org/10.1016/s0008-8846(03)00212-6

[105]   Siddique, R., *Effect of fine aggregate replacement with Class F fly ash on the mechanical properties of concrete.* Cement and Concrete Research, 2003. 33(4): p. 539–547. https://doi.org/10.1016/s0008-8846(02)01000-1

[106]   Roy, D.S., *Performance of blast furnace slag concrete with partial replacement of sand by fly ash.* International Journal of Earth Sciences and Engineering, 2011. 4.

[107] Parvati, V. and K. Prakash, *Feasibility study of fly ash as a replacement for fine aggregate in concrete and its behaviour under sustained elevated temperature.* International Journal of Science & Engineering Research, 2013. 4(5): p. 87–90.

[108] Joseph, G. and K. Ramamurthy, *Influence of fly ash on strength and sorption characteristics of cold-bonded fly ash aggregate concrete.* Construction and Building Materials, 2009. 23(5): p. 1862–1870. https://doi.org/10.1016/j.conbuildmat.2008.09.018

[109] Dhir, R., M. McCarthy, and P. Tittle, *Use of conditioned PFA as a fine aggregate component in concrete.* Materials and structures, 2000. 33(1): p. 38. https://doi.org/10.1007/bf02481694

[110] Bai, Y. and P. Basheer, *Influence of furnace bottom ash on properties of concrete.* Proceedings of the Institution of Civil Engineers – Structures and Buildings, 2003. 156(1): p. 85–92. https://doi.org/10.1680/stbu.2003.156.1.85

[111] Basheer, P.M. and Y. Bai, *Strength and durability of concrete with ash aggregate.* Proceedings of the Institution of Civil Engineers – Structures and Buildings, 2005. 158(3): p. 191–199. https://doi.org/10.1680/stbu.2005.158.3.191

[112] Bai, Y., F. Darcy, and P.A.M. Basheer, *Strength and drying shrinkage properties of concrete containing furnace bottom ash as fine aggregate.* Construction and Building Materials, 2005. 19(9): p. 691–697. https://doi.org/10.1016/j.conbuildmat.2005.02.021

[113] Yuksel, I. and A. Genç, *Properties of concrete containing nonground ash and slag as fine aggregate.* ACI Materials Journal, 2007. 104(4): p. 397. https://doi.org/10.14359/18829

[114] P. Aggarwal, Y. Aggarwal, S.M. Gupta, Effect of bottom ash as replacement of fine aggregates in concrete, Asian J. Civil Eng. (Build. Hous.) 8 (2007) 49–62. https://www.sid.ir/EN/VEWSSID/J_pdf/103820070102.pdf

[115] Kasemchaisiri, R. and S. Tangtermsirikul, *Properties of self-compacting concrete in corporating bottom ash as a partial replacement of fine aggregate.* Science Asia, 2008. 34: p. 87–95. https://doi.org/10.2306/scienceasia1513-1874.2008.34.087

[116] Singh, M. and R. Siddique, *Strength properties and micro-structural properties of concrete containing coal bottom ash as partial replacement of fine aggregate.* Construction and Building Materials, 2014. 50: p. 246–256. https://doi.org/10.1016/j.conbuildmat.2013.09.026

[117] Singh, M. and R. Siddique, *Effect of coal bottom ash as partial replacement of sand on workability and strength properties of concrete.* Journal of Cleaner Production, 2016. 112: p. 620–630. https://doi.org/10.1016/j.jclepro.2015.08.001

[118] Singh, G., et al., *Study of Granulated Blast Furnace Slag as Fine Aggregates in Concrete for Sustainable Infrastructure.* Procedia – Social and Behavioral Sciences, 2015. 195: p. 2272–2279. https://doi.org/10.1016/j.sbspro.2015.06.316

[119] Miyamoto, T., et al., *Production and use of blast furnace slag aggregate for concrete.* Nippon Steel & Sumitomo Metal Technical Report, 2015. 109: p. 102–108.

[120] Binici, H., et al., *Investigation of durability properties of concrete pipes incorporating blast furnace slag and ground basaltic pumice as fine aggregates.* Scientia Iranica, 2012. 19(3): p. 366–372. https://doi.org/10.1016/j.scient.2012.04.007

[121] Senani, M., et al., *Eco-concrete with incorporation of blast furnace slag as natural aggregates replacement.* Science and Technology of Materials, 2018. 30(3): p. 144–150. https://doi.org/10.1016/j.stmat.2017.12.001

[122] Ozbakkaloglu, T., L. Gu, and A. Fallah Pour, *Normal- and high-strength concretes incorporating air-cooled blast furnace slag coarse aggregates: Effect of slag size and content on the behavior.* Construction and Building Materials, 2016. 126: p. 138–146. https://doi.org/10.1016/j.conbuildmat.2016.09.015

[123] Gesoğlu, M., et al., *Recycling ground granulated blast furnace slag as cold bonded artificial aggregate partially used in self-compacting concrete.* Journal of Hazardous Materials, 2012. 235–236: p. 352–358. https://doi.org/10.1016/j.jhazmat.2012.08.013

[124] Wang, Q., L. Bao, and P. Yan, *Research progress on converter steel slag applied for concrete.* Concrete, 2009. 2: p. 53–56.

[125] Manso, J.M., J.J. Gonzalez, and J.A. Polanco, *Electric Arc Furnace Slag in Concrete.* Journal of Materials in Civil Engineering, 2004. 16(6): p. 639–645. https://doi.org/10.1061/(ASCE)0899-1561(2004)16:6(639)

[126] Alizadeh, R., et al. *Utilization of electric arc furnace slag as aggregates in concrete-environmental issue.* in *Proceedings of the 6th CANMET/ACI international conference on recent advances in concrete technology. Bucharest, Romania.* 2003.

[127] Maharaj, D. and A. Mwasha, *Comparative analysis of the transmission factors of lead and concrete manufactured with electric arc furnace slag aggregates.* Construction and Building Materials, 2016. 112: p. 1141–1146. https://doi.org/10.1016/j.conbuildmat.2016.02.130

[128] Pellegrino, C., et al., *Properties of concretes with Black/Oxidizing Electric Arc Furnace slag aggregate.* Cement and Concrete Composites, 2013. 37: p. 232–240. https://doi.org/10.1016/j.cemconcomp.2012.09.001

[129] González-Ortega, M.A., et al., *Radiological protection and mechanical properties of concretes with EAF steel slags.* Construction and Building Materials, 2014. 51: p. 432–438. https://doi.org/10.1016/j.conbuildmat.2013.10.067

[130] Vijayaraghavan, J., A.B. Jude, and J. Thivya, *Effect of copper slag, iron slag and recycled concrete aggregate on the mechanical properties of concrete.* Resources Policy, 2017. 53: p. 219–225. https://doi.org/10.1016/j.resourpol.2017.06.012

[131] Qasrawi, H., F. Shalabi, and I. Asi, *Use of low CaO unprocessed steel slag in concrete as fine aggregate.* Construction and Building Materials, 2009. 23(2): p. 1118–1125. https://doi.org/10.1016/j.conbuildmat.2008.06.003

[132] Ismeik, M., *Environmental enhancement through utilization of silica fume as a partial replacement of fine aggregate in concrete.* Journal of Civil Engineering Research and Practice, 2010. 7(2): p. 11–21. https://doi.org/10.4314/jcerp.v7i2.63725

[133] Ghafoori, N. and H. Diawara, *Strength and wear resistance of sand-replaced silica fume concrete.* ACI Materials Journal, 2007. 104(2): p. 206. https://doi.org/10.14359/18584

[134] Ghafoori, N. and H. Diawara, *Abrasion resistance of fine aggregate-replaced silica fume concrete.* ACI Materials Journal, 1999. 96(5): p. 559–569. https://doi.org/10.14359/658

[135] Sharma, R. and R.A. Khan, *Durability assessment of self compacting concrete incorporating copper slag as fine aggregates.* Construction and Building Materials, 2017. 155: p. 617–629. https://doi.org/10.1016/j.conbuildmat.2017.08.074

[136] Rajasekar, A., K. Arunachalam, and M. Kottaisamy, *Assessment of strength and durability characteristics of copper slag incorporated ultra high strength concrete.* Journal of Cleaner Production, 2019. 208: p. 402–414. https://doi.org/10.1016/j.jclepro.2018.10.118

[137] Mahesh Babu, K. and A. Ravitheja, *Effect of copper slag as fine aggregate replacement in high strength concrete.* Materials Today: Proceedings, 2019. https://doi.org/10.1016/j.matpr.2019.07.626

[138] Lori, A.R., A. Hassani, and R. Sedghi, *Investigating the mechanical and hydraulic characteristics of pervious concrete containing copper slag as coarse aggregate.* Construction and Building Materials, 2019. 197: p. 130–142. https://doi.org/10.1016/j.conbuildmat.2018.11.230

[139] Gupta, N. and R. Siddique, *Strength and micro-structural properties of self-compacting concrete incorporating copper slag.* Construction and Building Materials, 2019. 224: p. 894–908. https://doi.org/10.1016/j.conbuildmat.2019.07.105

[140] Alwaeli, M., *Application of granulated lead-zinc slag in concrete as an opportunity to save natural resources.* Radiation Physics and Chemistry, 2013. 83: p. 54–60. https://doi.org/10.1016/j.radphyschem.2012.09.024

[141] Alwaeli, M., *Investigation of gamma radiation shielding and compressive strength properties of concrete containing scale and granulated lead-zinc slag wastes.* Journal of Cleaner Production, 2017. 166: p. 157–162. https://doi.org/10.1016/j.jclepro.2017.07.203

[142] Dhir, R., et al., *Value added recycling of incinerator ashes.* Concrete Technology Unit, University of Dundee, 2002.

[143] Tay, J.-H., *Reclamation of wastewater and sludge for concrete making.* Resources, Conservation and Recycling, 1989. 2(3): p. 211–227.

[144] Kim, J., et al., *Effect of chemical treatment of MSWI bottom ash for its use in concrete.* Magazine of Concrete Research, 2015. 67(4): p. 179–186.

[145] Zhang, T. and Z. Zhao, *Optimal use of MSWI bottom ash in concrete.* International Journal of Concrete Structures and Materials, 2014. 8(2): p. 173–182. https://doi.org/10.1007/s40069-014-0073-4

[146] van der Wegen, G., U. Hofstra, and J. Speerstra, *Upgraded MSWI bottom ash as aggregate in concrete.* Waste and Biomass Valorization, 2013. 4(4): p. 737–743. https://doi.org/10.1007/s12649-013-9255-6

[147] Lynn, C.J., R.K.D. OBE, and G.S. Ghataora, *Municipal incinerated bottom ash characteristics and potential for use as aggregate in concrete.* Construction and Building Materials, 2016. 127: p. 504–517.

[148] Sorlini, S., A. Abbà, and C. Collivignarelli, *Recovery of MSWI and soil washing residues as concrete aggregates.* Waste Management, 2011. 31(2): p. 289–297.

[149] Ferraris, M., et al., *Use of vitrified MSWI bottom ashes for concrete production.* Waste Management, 2009. 29(3): p. 1041–1047.

[150] Pera, J., et al., *Use of incinerator bottom ash in concrete.* Cement and Concrete Research, 1997. 27(1): p. 1–5.

[151] Van den Heede, P., et al., *Sustainable high quality recycling of aggregates from waste-to-energy, treated in a wet bottom ash processing installation, for use in concrete products.* Materials, 2016. 9(1): p. 9. https://doi.org/10.3390/ma9010009

[152] Del Valle-Zermeño, R., et al., *Influence of MSWI bottom ash used as unbound granular material on the corrosion behaviour of reinforced concrete.* Journal of Material Cycles and Waste Management, 2017. 19(1): p. 124–133. https://doi.org/10.1007/s10163-015-0388-5

[153] Keppert, M., et al., *Properties of concrete with municipal solid waste incinerator bottom ash.* International Proceedings of Computer Science and Information Technology, 2012. 28: p. 127–131.

[154] Jamshidi, M., et al., *Mechanical performance and capillary water absorption of sewage sludge ash concrete (SSAC).* International Journal of Sustainable Engineering, 2012. 5(3): p. 228–234. https://doi.org/10.1080/19397038.2011.642020

[155] Kosior-Kazberuk, M., *Application of SSA as partial replacement of aggregate in concrete.* Polish Journal of Environmental Studies, 2011. 20(2): p. 365–370.

[156] de Lima, F., D. Ingunza, and M. del Pilar. *Effects of Sewage Sludge Ashes Addition in Portland Cement Concretes.* In *2nd International Conference on Civil, Materials and Environmental Sciences.* 2015. Atlantis Press. https://doi.org/10.2991/cmes-15.2015.55

[157] Baeza-Brotons, F., et al., *Portland cement systems with addition of sewage sludge ash. Application in concretes for the manufacture of blocks.* Journal of Cleaner Production, 2014. 82: p. 112–124. https://doi.org/10.1016/j.jclepro.2014.06.072

[158] Khanbilvardi, R. and S. Afshari, *Sludge ash as fine aggregate for concrete mix.* Journal of Environmental Engineering, 1995. 121(9): p. 633–638. https://doi.org/10.1061/(asce)0733-9372(1995)121:9(633)

[159] Dhir, R., G. Chataora, and C. Lynn, *Sustainable construction materials: Sewage Sludge Ash.* 2017, Cambridge, United Kingdom: Elsevier Science & Technology. Woodhead Publishing Series in Civil and Structural Engineering; 1 edition.

[160] de Lima, F., D. Ingunza, and M. del Pilar. *Effects of sewage sludge ashes addition in portland cement concretes.* In *2nd International Conference on Civil, Materials and Environmental Sciences.* 2015. Atlantis Press. https://doi.org/10.2991/cmes-15.2015.55

[161] Bui, N.K., T. Satomi, and H. Takahashi, *Influence of industrial by-products and waste paper sludge ash on properties of recycled aggregate concrete.* Journal of Cleaner Production, 2019. 214: p. 403–418. https://doi.org/10.1016/j.jclepro.2018.12.325

[162] Shermale, Y. and M. Varma, *Effective use of paper sludge (hypo sludge) in concrete.* Magnesium, 2015. 1(3.33): p. 3.33.

[163] Vashistha, P., et al., *Valorization of paper mill lime sludge via application in building construction materials: A review.* Construction and Building Materials, 2019. 211: p. 371–382. https://doi.org/10.1016/j.conbuildmat.2019.03.085

[164] Toutanji HA. The use of rubber tire particles in concrete to replace mineral aggregates. Cement and Concrete Composites. 1996;18(2):135–9. doi: https://doi.org/10.1016/0958-9465(95)00010-0.

[165] Rashid, K., A. Yazdanbakhsh, and M.U. Rehman, *Sustainable selection of the concrete incorporating recycled tire aggregate to be used as medium to low strength material.* Journal of Cleaner Production, 2019. 224: p. 396–410. https://doi.org/10.1016/j.jclepro.2019.03.197

[166] Aslani, F., et al., *Experimental investigation into rubber granules and their effects on the fresh and hardened properties of self-compacting concrete.* Journal of Cleaner Production, 2018. 172: p. 1835–1847. https://doi.org/10.1016/j.jclepro.2017.12.003

[167] Su, H., et al., *Properties of concrete prepared with waste tyre rubber particles of uniform and varying sizes.* Journal of Cleaner Production, 2015. 91: p. 288–296. https://doi.org/10.1016/j.jclepro.2014.12.022

[168] Ismail, M.K. and A.A. Hassan, *Use of metakaolin on enhancing the mechanical properties of self-consolidating concrete containing high percentages of crumb rubber.* Journal of Cleaner Production, 2016. 125: p. 282–295. https://doi.org/10.1016/j.jclepro.2016.03.044

[169] Kang, J. and Y. Jiang, *Improvement of cracking-resistance and flexural behavior of cement-based materials by addition of rubber particles.* Journal of Wuhan University of Technology- Materials Science Edition, 2008. 23(4): p. 579–583. https://doi.org/10.1007/s11595-006-4579-8

[170] Sharma, R. and P.P. Bansal, *Use of different forms of waste plastic in concrete-a review.* Journal of Cleaner Production, 2016. 112: p. 473–482. https://doi.org/10.1016/j.jclepro.2015.08.042

[171] Kou, S., et al., *Properties of lightweight aggregate concrete prepared with PVC granules derived from scraped PVC pipes.* Waste Management, 2009. 29(2): p. 621–628. https://doi.org/10.1016/j.wasman.2008.06.014

[172] Frigione, M., *Recycling of PET bottles as fine aggregate in concrete.* Waste management, 2010. 30(6): p. 1101–1106. https://doi.org/10.1016/j.wasman.2010.01.030

[173] Fonseca, N., J. De Brito, and L. Evangelista, *The influence of curing conditions on the mechanical performance of concrete made with recycled concrete waste.* Cement and Concrete Composites, 2011. 33(6): p. 637–643. https://doi.org/10.1016/j.cemconcomp.2011.04.002

[174] Albano, C., et al., *Influence of content and particle size of waste pet bottles on concrete behavior at different w/c ratios.* Waste Management, 2009. 29(10): p. 2707–2716. https://doi.org/10.1016/j.wasman.2009.05.007

[175] Ávila Córdoba, L., et al., *Effects on mechanical properties of recycled PET in cement-based composites.* International Journal of Polymer Science, 2013. https://doi.org/10.1155/2013/763276

[176] Ferreira, L., J. de Brito, and N. Saikia, *Influence of curing conditions on the mechanical performance of concrete containing recycled plastic aggregate.* Construction and Building Materials, 2012. 36: p. 196–204. https://doi.org/10.1016/j.conbuildmat.2012.02.098

[177] Choi, Y.-W., et al., *Effects of waste PET bottles aggregate on the properties of concrete.* Cement and Concrete research, 2005. 35(4): p. 776–781. https://doi.org/10.1016/j.cemconres.2004.05.014

[178] Rahmani, E., et al., *On the mechanical properties of concrete containing waste PET particles.* Construction and Building Materials, 2013. 47: p. 1302–1308. https://doi.org/10.1016/j.conbuildmat.2013.06.041

[179] Araghi, H.J., et al., *An experimental investigation on the erosion resistance of concrete containing various PET particles percentages against sulfuric acid attack.* Con-

struction and Building Materials, 2015. 77: p. 461–471. https://doi.org/10.1016/j. conbuildmat.2014.12.037

[180] Silva, R.V., J. de Brito, and N. Saikia, *Influence of curing conditions on the durability-related performance of concrete made with selected plastic waste aggregates.* Cement and Concrete Composites, 2013. 35(1): p. 23–31. https://doi.org/10.1016/j. cemconcomp.2012.08.017

[181] De la Colina Martínez, A.L., et al., *Recycled polycarbonate from electronic waste and its use in concrete: Effect of irradiation.* Construction and Building Materials, 2019. 201: p. 778–785. https://doi.org/10.1016/j.conbuildmat.2018.12.147

[182] Tan, K.H. and H. Du, *Use of waste glass as sand in mortar: Part I-Fresh, mechanical and durability properties.* Cement and Concrete Composites, 2013. 35(1): p. 109–117. https://doi.org/10.1016/j.cemconcomp.2012.08.028

[183] Wang, H.-Y., H.-h. Zeng, and J.-Y. Wu, *A study on the macro and micro properties of concrete with LCD glass.* Construction and Building Materials, 2014. 50: p. 664–670. https://doi.org/10.1016/j.conbuildmat.2013.09.015

[184] Ismail, Z.Z. and E.A. Al-Hashmi, *Recycling of waste glass as a partial replacement for fine aggregate in concrete.* Waste Management, 2009. 29(2): p. 655–659. https:// doi.org/10.1016/j.wasman.2008.08.012

[185] Arabi, N., et al., *Valorization of recycled materials in development of self-compacting concrete: Mixing recycled concrete aggregates – Windshield waste glass aggregates.* Construction and Building Materials, 2019. 209: p. 364–376. https://doi. org/10.1016/j.conbuildmat.2019.03.024

[186] de Castro, S. and J. de Brito, *Evaluation of the durability of concrete made with crushed glass aggregates.* Journal of Cleaner Production, 2013. 41: p. 7–14. https:// doi.org/10.1016/j.jclepro.2012.09.021

[187] Ali, E.E. and S.H. Al-Tersawy, *Recycled glass as a partial replacement for fine aggregate in self compacting concrete.* Construction and Building Materials, 2012. 35: p. 785–791. https://doi.org/10.1016/j.conbuildmat.2012.04.117

[188] Yang, S., et al., *Influence of particle size of glass aggregates on the high temperature properties of dry-mix concrete blocks.* Construction and Building Materials, 2019. 209: p. 522–531. https://doi.org/10.1016/j.conbuildmat.2019.03.131

[189] Lam, C.S., C.S. Poon, and D. Chan, *Enhancing the performance of pre-cast concrete blocks by incorporating waste glass-ASR consideration.* Cement and Concrete Composites, 2007. 29(8): p. 616–625. https://doi.org/10.1016/j. cemconcomp.2007.03.008

[190] Du, H. and K.H. Tan, *Use of waste glass as sand in mortar: Part II – Alkali-silica reaction and mitigation methods.* Cement and Concrete Composites, 2013. 35(1): p. 118–126. https://doi.org/10.1016/j.cemconcomp.2012.08.029

[191] Akhtar, A. and A.K. Sarmah, *Novel biochar-concrete composites: Manufacturing, characterization and evaluation of the mechanical properties.* Science of The Total Environment, 2018. 616–617: p. 408–416. https://doi.org/10.1016/j. scitotenv.2017.10.319

[192] Jensen, P.E., A.M. Simonsen, and L.M. Ottosen. *Reduction of climate impact from concrete by incorporation of mine tailings.* in *Sustain Conference 2018: Creating*

Technology for a Sustainable Society. 2018. Technical University of Denmark (DTU).

[193] Wang, X., et al., *Development of a novel cleaner construction product: Ultra-high performance concrete incorporating lead-zinc tailings*. Journal of Cleaner Production, 2018. 196: p. 172–182. https://doi.org/10.1016/j.jclepro.2018.06.058

[194] Almeida, N., et al., *High-performance concrete with recycled stone slurry*. Cement and Concrete Research, 2007. 37(2): p. 210–220. https://doi.org/10.1016/j.cemconres.2006.11.003

[195] Lanuza, A., et al., *FERROCK: A life cycle comparison to ordinary Portland cement*. 2017.

[196] Kubissa, W., M.A. Glinicki, and M. Dąbrowski, *Permeability testing of radiation shielding concrete manufactured at industrial scale*. Materials and Structures, 2018. 51(4): p. 83. https://doi.org/10.1617/s11527-018-1213-0

[197] Gencel, O., et al., *Concretes Containing Hematite for Use as Shielding Barriers*. Medžiagotyra, 2010. 16.

[198] D, R., et al., *Studies on mechanical and microstructural properties of hematite modified concrete*. International Journal of ChemTech Research, 2017. 10: p. 464–472.

[199] Wang, X., et al., *An experimental study of a freeze-thaw damage model of natural pumice concrete*. Powder Technology, 2018. 339: p. 651–658. https://doi.org/10.1016/j.powtec.2018.07.096

[200] Badogiannis, E.G., K.I. Christidis, and G.E. Tzanetatos, *Evaluation of the mechanical behavior of pumice lightweight concrete reinforced with steel and polypropylene fibers*. Construction and Building Materials, 2019. 196: p. 443–456. https://doi.org/10.1016/j.conbuildmat.2018.11.109

[201] Martins, M.L.C., J. Santos, and A. Azevedo. *Production of lightweight concrete with EVA residues as recycled aggregate*. in *International RILEM conference on the use of recycled materials in buildings and structures*. Barcelona (Spain): RILEM Publications SARL. 2004.

[202] Santiago, E., et al., *Mechanical behavior of recycled lightweight concrete using EVA waste and CDW under moderate temperature*. Revista IBRACON de Estruturas e Materiais, 2009. 2(3): p. 211–221. https://doi.org/10.1590/s1983-41952009000300001

[203] de Brito, J. and R. Kurda, *The past and future of sustainable concrete: A critical review and new strategies on cement-based materials*. Journal of Cleaner Production, 2021. 281: p. 123558. https://doi.org/10.1016/j.jclepro.2020.123558

Chapter 4

Short- and Long-Term Performance of Structural Recycled Aggregate Concrete

Belén González-Fonteboa, Sindy Seara-Paz, Fernando Martínez-Abella

Department of Civil Engineering, Campus de Elviña s/n, 15071 A Coruña, Universidade da Coruña, Spain

4.1 Introduction and Objectives

The employment of recycled aggregate from construction and demolition waste is a key factor to move the field of construction towards a sustainable development. Its use as a raw material replacing conventional aggregate would reduce the consumption of natural resources. In this regard, many researchers have worked to analyse the structural behaviour of recycled concrete, that is, a concrete where the conventional aggregate has been partially replaced by recycled aggregate [1, 2]. However, and despite the numerous work that deals with this issue, the use of this material in real civil engineering applications is still scarce. It is necessary to increase the confidence of the stakeholders to spread its use, adapting the codes, where necessary.

In this scenario, recycled concrete properties have been widely studied in the last decades [3–5], however, the amount of research that deals with its structural behaviour (shear and flexural performance of structural recycled concrete beams) is not so numerous and the question of different loading stages has hardly been analysed. A concrete structure usually withstands different loading stages during its service life and consequently, the values of permanent deformations developed during these stages must be considered in structural analysis. These deformations influence the appearance of the structure, the possible damage to other non-structural members of the building, and the loss of serviceability [6]. Therefore, the performance of recycled concrete in this regard has to be taken into account in structural design, especially for structural applications of concrete with a high content of recycled aggregates, where the recycled aggregate may have a high influence.

As recycled concrete displays a different performance in terms of long-term properties (especially creep and shrinkage), deformation, bonding, and cracking, it is necessary to analyse all of these parameters together in beam elements, in order to establish the relationship between these properties and the influence of the recycled aggregate content in terms of structural response. As aforementioned, there are few works that discuss this issue as the tests to determine long-term properties in structural members are challenging: the problem of carrying out sustained loads during a long period of time and the great number of factors that must be considered are the main problems when these tests are planned.

The aim of this chapter to summarize the main experimental findings available in literature on the structural performance of reinforced concrete beams using recycled concrete aggregates in comparison with the behaviour of beams using conventional concrete.

4.2 Short-Term Behaviour

Although there are some studies on this topic, the use of recycled aggregate for structural concrete is not fully developed due to the absence of models that involved parties can use for the design phases and which can be included in codes and standards.

In general, the results regarding the structural performance of recycled concrete beams loaded incrementally up to failure indicate that the code-based proposals used to design conventional concrete can also be applied to recycled concrete [7–9]. The authors state that recycled concrete is able to fulfil strength and serviceability requirements similar to conventional reinforced concrete. However, some differences have been detected that have to be considered to design recycled concrete members with the same reliability as conventional concrete elements. Therefore, more efforts are needed to improve the understanding of the mechanisms governing recycled aggregate members and structures.

4.2.1 Flexural Performance

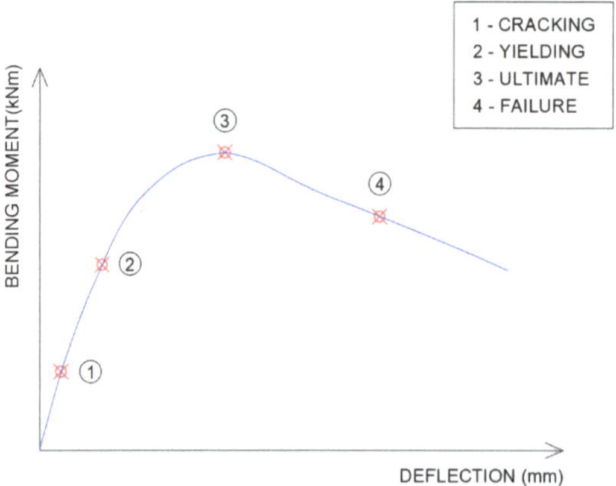

Fig. 4.1   Moment-deflection curve in the mid-span of beam

The main aspects that have to be analysed in the flexural performance of concrete beams are related to the moments and deflections developed at cracking, service, yielding and ultimate state (Fig. 4.1) [10, 11].

Seara-Paz et al. [11] developed a study with eight reinforced concrete beams using two different water to cement ratios (0.50 and 0.65) and four replacement percentages (0 %, 20 %, 50 % and 100 %). At 28 days the beams were loaded up to failure using a four-point bending test. They concluded that the cracking moment decreases as the replacement rate rises. This is due to the lower tensile splitting strength of recycled concrete that generates an earlier cracking than with conventional concrete. The same authors found drops of cracking moments of about 10 %, 30 % and 40 % when 20 %, 50 and 100 % replacement rates are used, respectively.

At serviceability, they found that bending moments and deflections are hardly influenced by the use of recycled aggregate. The bending moment for maximum deflection was similar in recycled and conventional members; only a slight decrease (under 10 %) was detected when 100 % recycled aggregate is used. In the same regard, and although it is widely known that recycled concrete shows higher deformations than conventional concrete (due to its lower modulus of elasticity), only beams with 100 % substitution rate displayed slightly higher deflections than those of conventional concrete (10 % of increment). They attributed these results to the small effect of material properties on the performance of structural members with high reinforcement ratios that lead to ductile behaviour.

In the analysis of the structural behaviour at failure, they concluded that, again, the ductile design of concrete beams (with a suitable reinforcement steel ratio, both in tension and in compression) resulted in limited concrete contribution at failure (Fig. 4.2). This led to recycled beams (even when high substitution rates are used) presenting a yielding and ultimate behaviour similar to that of conventional ones.

Fathifazl et al. [8] designed three singly reinforced beams (one beam for each longitudinal tensile reinforcement ratio) and one doubly reinforced beam (with compression steel) using two different recycled aggregate sources. The concrete mix design was developed based on the equivalent mortar volume method [12] and, therefore, the recycled concrete strength was higher than the compressive strength of the conventional concrete. They found that the performance of the recycled beams was similar to that of the conventional concrete member at both failure and serviceability states. Finally, the deflections were similar and could be predicted with the provisions suggested in the codes.

Arezoumandi et al. [7], using beams of 100 % recycled concrete, concluded that these showed a lower cracking moment, however, no important differences between conventional and recycled specimens were identified in terms of yielding and ultimate moments due to the ductile design of concrete beams.

Hamad et al. [13] studied the structural behaviour of reinforced beams using two different percentages of recycled aggregate, 40 % and 100 %. They developed a four-point bending test to analyse the flexural performance and, as a result, they found that the failure mode and crack pattern of recycled concrete beams was similar to that of the conventional beam. These authors also found that the load-deflection curves of recycled concrete elements were comparable to that of the conventional one, showing slight differences in the values of the ultimate loads. This was also seen by Pradhan et al. [4].

Evangelista and de Brito [15] analysed the behaviour of recycled beams using recycled fine aggregates. They did not find significant differences in terms of flexural behaviour, although the bearing capacity was slightly lower in the recycled concrete beams than in the conventional concrete specimens.

According to this literature review, it seems that recycled concrete beams show lower cracking moments than those of the conventional beam. Bending moments (yield, ultimate or service moments) do not show significant differences when compared to the moments obtained in conventional concrete beams. Therefore, the decrease in cracking moment and the invariability of yielding and maximum moments confirms early cracking

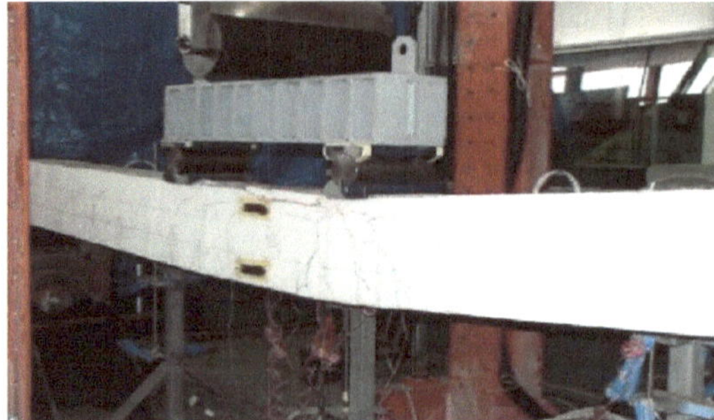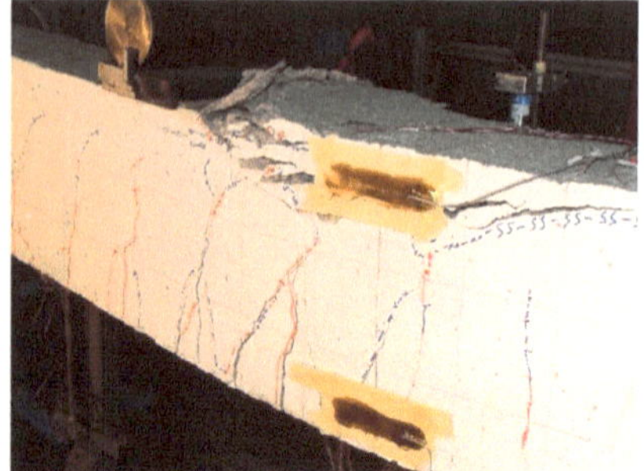

Fig. 4.2 Concrete crushing at failure [11]

development in recycled concretes. Differences in the deflections are minor, as the deflections measured in recycled concrete beams are only slightly higher than in conventional concrete beams.

The analysis of the cracking performance of recycled concrete members is also important. The higher shrinkage strain of recycled concrete (that will promote higher concrete deformation) and the lower tensile splitting strength will produce a premature cracking in recycled concrete members [16, 17]. This worse cracking behaviour [9, 18, 19] will produce a lower stiffness and therefore a lower concrete contribution after cracking.

Arezoumandi et al. [7] concluded that recycled concrete beams developed a lower cracking moment and stiffness after cracking, leading to higher ultimate deflections compared to a conventional concrete beam.

Seara-Paz et al. [11] found that recycled concrete beams after cracking developed higher strains, both in concrete and steel reinforcement, and consequently greater curvatures than those of conventional concrete. This effect is attributed to the lower concrete stiffness of the cracked cross section and its premature cracking, which is especially significant in concretes with high replacement percentages.

Regarding crack spacing and crack pattern some authors concluded that they are similar in conventional and recycled concretes [11, 20]. They state that recycled concrete beams show similar crack patterns to that of conventional ones but a slightly higher crack width. They consider that this effect is due to the lower modulus of elasticity and tensile splitting strength of recycled concrete. These two material properties influence the crack spacing and consequently the crack width. However, the lower bond strength of recycled concretes counteracts this effect and, as a result, recycled concrete showed similar crack spacing to that of conventional concrete. Consequently, this similar crack spacing and the higher strains of concrete and steel reinforcement result in greater crack width in recycled concrete compared to conventional concrete. Seara-Paz et al. [11] also detected small horizontal cracks, branched at the tensile reinforcement zone in recycled concrete beams (Fig. 4.3). These were produced by the higher strain and the lesser bond stress of recycled concrete that influences the failure mode slightly changing the crack pattern at failure.

Other authors [14] state that the overall crack pattern and morphology of conventional and recycled beams were very similar, although recycled concrete specimens showed a greater number of cracks and higher crack growth rate.

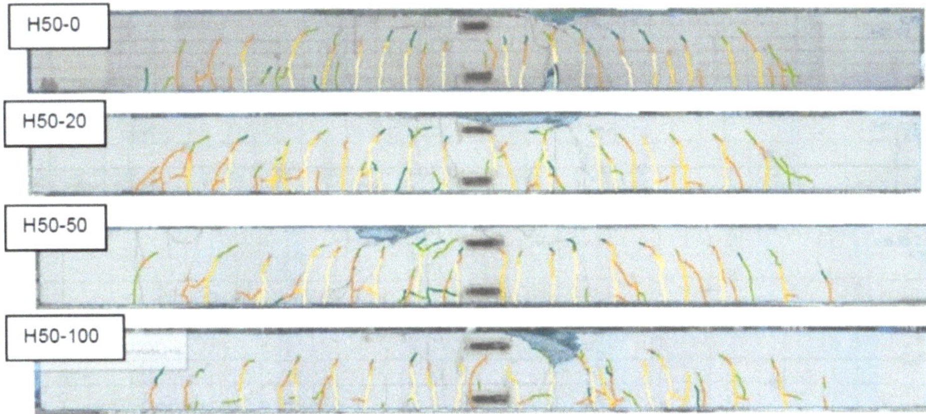

Fig. 4.3 Crack patterns at flexure failure [11]

It can be concluded, therefore, that early crack development can be expected in recycled concrete members. However, the differences in the crack pattern and crack spacing of a conventional and a recycled concrete beam seem to be small and there is not total agreement on this topic in literature.

4.2.2 Shear Performance

Several experimental studies have been developed to analyse the shear behaviour of recycled concrete beams with and without stirrups (shear reinforcement). The shear failure is brittle and therefore a suitable design has to be performed to avoid this type of failure.

Sogo et al. [18] tested ten reinforced concrete beams with recycled fine and coarse aggregates. The results indicated that shear strength of beams without stirrups decreased by 10–20 % when coarse recycled and conventional fine aggregates were used and by 10–30 % when both fractions were employed. On the other hand, the shear strength of recycled beams with stirrups is almost the same as that of conventional beams in experimental results. They also concluded that recycled concrete beams show the same cracking patterns and failure modes as those of conventional aggregate.

Arabiyat et al. [21] found that shear capacity drops with the incorporation of recycled aggregate replacing crushed conventional aggregate. However, the theoretical shear capacities calculated using the methods proposed by the concrete codes can be applicable in recycled concrete members as they are conservative. The experimental cracking patterns and the crack angles at failure were similar in recycled and conventional concrete beams.

Gonzalez-Fonteboa and Martínez-Abella [22], using a replacement ratio of 50 % of conventional crushed aggregate by recycled aggregate, found little differences in terms of deflections and ultimate load between recycled concrete beams and conventional ones. Regarding cracking behaviour, they noted that significant cracks and also premature cracking occurred in reinforced concrete beams with recycled coarse aggregates (Fig. 4.4).

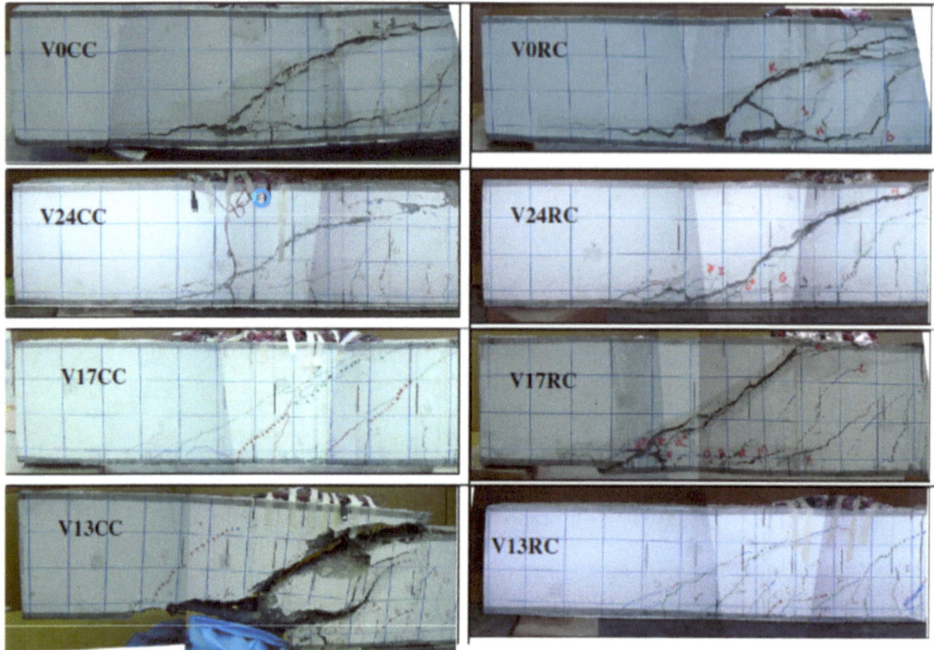

Fig. 4.4   Crack patterns at shear failure [22]

Wardeh and Ghorbel [23] analysed full-scale beams of conventional aggregates with crushed granular materials and recycled aggregates (fine and coarse) without stirrups. The authors noted that crack distribution and evolution of recycled concrete is similar to that of the conventional one. However, shear strength decreases similarly to tensile splitting strength when recycled aggregates are used.

Ignjatovic et al. [24] tested nine full-scale simply supported beams using three different replacement ratios of recycled coarse aggregate (0 %, 50 % and 100 %) and three different shear reinforcement ratios (0 %, 0.14 % and 0.19 %). They found that the shear behaviour and the shear strength of the recycled beams were very similar to that of the corresponding conventional specimens. They also concluded that the shear strength of recycled beams with and without shear reinforcement can be conservatively predicted by the codes, providing similar reliability as for the corresponding conventional beams.

Other authors [12, 25, 26] concluded that, using the equivalent mortar volume mixing method, the shear performance of recycled concrete beams can be comparable, or even superior, to that of beams made entirely with natural aggregates (rounded and crushed mixed) at both serviceability and ultimate limit states. Furthermore, the simplified methods of standards and codes were found applicable to recycled concrete members with and without stirrups.

Pradham et al. [27] tested beams with and without transverse reinforcement to examine the contribution of recycled aggregate in shear resistance mechanisms. They designed a conventional concrete (using crushed aggregates) and a 100 % recycled concrete. They also verified the applicability of the prevailing shear design provisions for recycled concrete beams. A drop of 14 % was recorded in the ultimate shear strength of recycled beams without stirrups and a theoretical relationship is proposed to predict the contribution of recycled concrete in shear resistance.

In general, it is seen that shear strength of recycled concrete beams without stirrups present slightly lower shear capacity than the shear strength of the corresponding conventional beams. In beams with shear reinforcement, no significant differences between the shear behaviour of recycled and corresponding conventional beams were reported.

The codes and standards seem to predict conservatively the shear capacity of recycled concrete beams, especially when shear reinforcement bars are used. In this case, the proposals are more conservative than in the case of shear strength without shear reinforcement.

Some differences in the cracking performance were detected in recycled concrete beams, but this has a negligible effect on the shear strength.

4.3 Long-Term Performance

Most concrete elements are subjected to different loading stages along their service life and, therefore, it is necessary to consider the impact of creep during the loading stages and also during the unloading periods. Concrete strains under loading are produced by creep; after unloading, the capacity of concrete to recover these deformations is related to creep recovery.

There are many factors affecting concrete recovery after unloading. Some authors [28] state that creep recovery does not depend on the loading age and duration while others [29, 30] consider that both these factors influence the recoverability of concrete. According to literature [31], humidity conditions or specimen size do not influence the creep recovery of concrete. In addition, it is clearly significant to consider the effect of cracking for calculations under service conditions [32]. In this situation, the structural performance of concrete structures will be influenced by the bond behaviour of the concrete and the tension stiffening in the cracked stage. The design type of concrete structures, ductile or brittle, is also a key factor to be considered in their structural analysis. Most design procedures for concrete structures include methods based on an age-adjusted stiffness or modulus, which consider the time-dependent evolution of concrete properties and different parameters related to bond performance, cracking behaviour, and stress-strain curves.

The performance of recycled concrete in terms of cracking and bond performance is worse than the behaviour of conventional concrete [11, 33, 34]. Therefore, when studying loading and unloading processes in recycled concrete members, it is necessary to consider not only the influence of recycled aggregate on mechanical strength and the deformational behaviour of concrete, but also its effect on bond behaviour and cracking performance.

4.3.1 Performance Under Sustained Load

During the last years, some researchers have studied the long-term behaviour of recycled concrete elements. Choi et al. [35] fabricated three beams, one with 100 % natural aggregate, another with 100 % recycled coarse aggregate, and the last one with 100 % conventional coarse aggregate and 50 % recycled fine aggregate. They were subjected to sustained loading at 50 % of the nominal flexural capacity for 380 days. Due to this insufficient sustained loading time, they did not get a stable state at the conclusion of the test; however, they obtained interesting results.

In terms of deflections, they indicated that the employment of recycled aggregate reduces the initial stiffness, which leads to a rise in the deflections. They also found that the long-term to instant deflection ratio of members with recycled aggregate are lower than those of conventional beams. Therefore, they presented a modified formula for assessing the long-term deflection. Due to the reduction of interfacial bonding strength between the mortar and recycled coarse aggregate, differences in the neutral axis depth of the beam have been detected with 100 % recycled aggregate. Finally, similar crack patterns were noted regardless of aggregate type, even though several cracks were seen in the specimens fabricated with recycled aggregate.

Regarding deflections, they found that an increase in the recycled aggregate content leads to an increment in the immediate and long-term deflections (Fig. 4.5). They also detected that cracking was worse in recycled concrete beams due to the lower modulus of elasticity of this material.

Seara-Paz et al. [33] analysed recycled concrete beams under sustained load for 1000 days. In this work, they noted that the use of recycled coarse aggregates highly affects long-term deformations due to the larger shrinkage strain and creep coefficient of the recycled concrete (Fig. 4.6). This higher shrinkage of recycled concrete [37–40] leads to a different flexural performance of recycled concrete structures. In addition, and contrary to the results obtained by Choi et al. [35], Seara-Paz et al. [33] detected that long-term to instant deflection ratio is higher in recycled concrete beams than in conventional ones and it increases as the recycled aggregate content rises (Fig. 4.6 and Fig. 4.7).

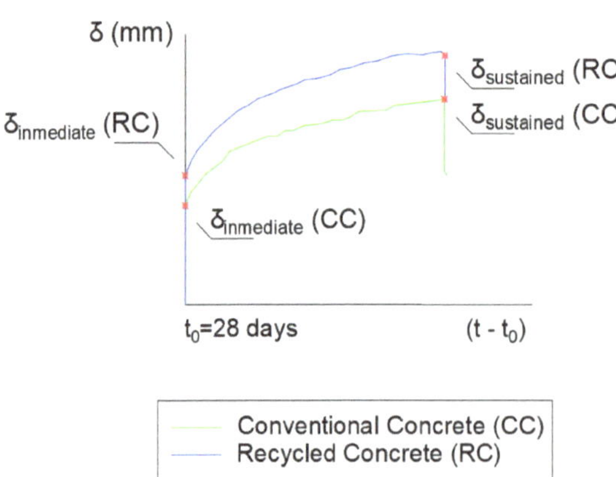

Fig. 4.5 Midspan deflections [36]

Zhu et al. [3] evaluated the long-term behaviour of recycled aggregate concrete beams for a period of 3045 days and, after the sustained load is removed, they studied their flexural performance. Three different substitution percentages (0 %, 50 % and 100 %) were employed. The results showed that the crack spacing of recycled concrete specimens is larger than that of conventional members. Compared with the baseline beam, the long-term deformations of recycled members are greater, which is attributed to the higher shrinkage and creep of recycled concrete.

Tosic et al. [6] loaded simply supported reinforced concrete beams (fabricated with conventional and 100 % recycled concrete) under sustained loads for 450 days. The cracking of the recycled concrete specimens was more noticeable compared with conventional members (crack widths were larger and crack spacing was smaller). They found that the deflections of the beams were similar, which was attributed to the high load level employed that generated significant cracking.

Most of the existing studies show that the long-term deformation of recycled concrete beams is larger than that of conventional beams due to the creep and shrinkage of recycled concrete which are more significant in recycled concrete. Regarding cracking performance, it seems that there is no agreement in literature. Some authors found differences, although they do not seem to be significant.

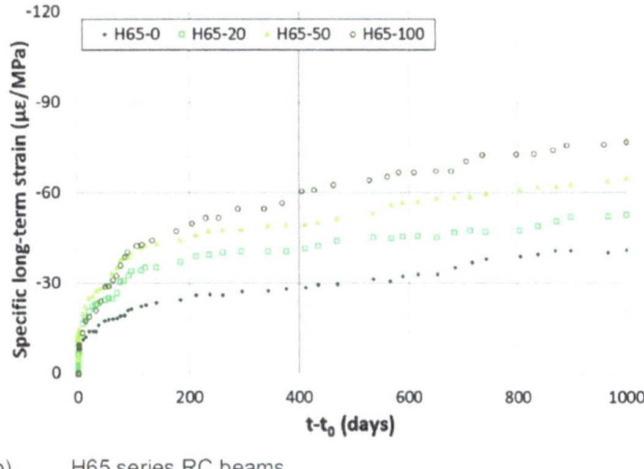

b)   H65 series RC beams

Fig. 4.6   Specific long-term strains – time [33]

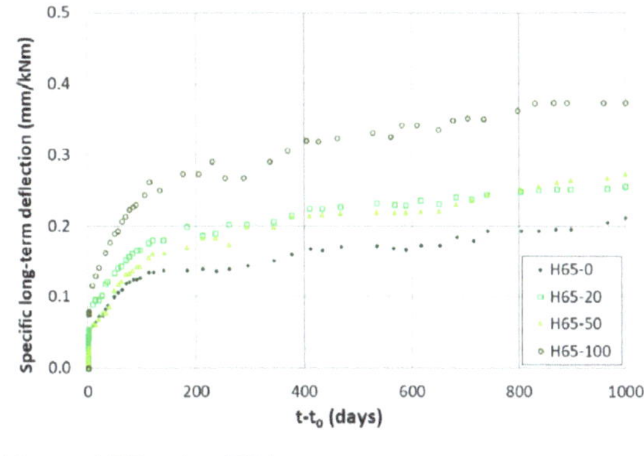

b)   H65 series RC beams

Fig. 4.7   Specific long-term deflections – time [33]

4.3.2 Performance After Sustained Loading: Recovery and Failure Behaviour

It is not easy to find studies that discuss the performance of recycled concrete beams after sustained load. The load deflection curve of a reinforced beam is represented in Fig. 4.8. The first loading process will follow 012 path, being 2 greater than the cracking load. An unloading process at this point will be developed throughout 20', producing a permanent deflection (00'). If the concrete beam is reloaded, the new curve will be the one defined by 0'2. Under sustained load the deflection will follow the path 23 and, when the sustained load is removed, the path will be 30''. The K_1, K_2 and K_3 represent the stiffness of the beam (initial, short-term reload and long-term reload, respectively). Due to the damage produced in the loading processes, these stiffnesses show the following relationship $K_1 > K_2 > K_3$.

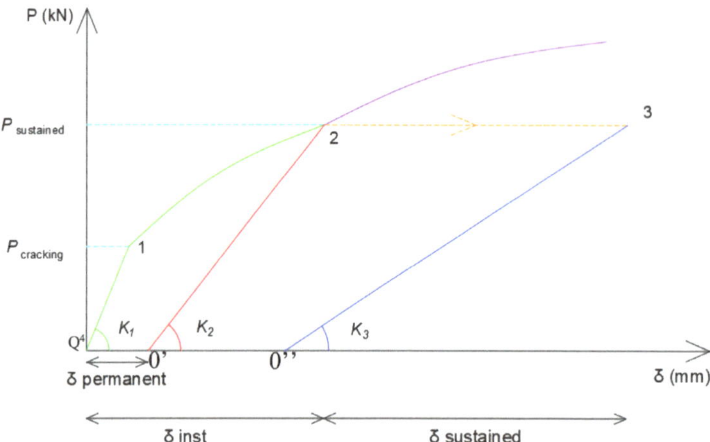

Fig. 4.8 Short and long-term response of a reinforced beam

Zhu et al. [3] investigated the long-term performance of recycled aggregate concrete beams for a period of 3045 days, and the bending behaviour of tested beams after sustained load is removed. They concluded that the recycled concrete beam shows a larger deflection recovery value after the long-term load is removed. They attributed this finding to the long-term creep effect and to the sustained load value.

These authors tested the beams after unloading. A bending test was developed, measuring the load applied and the deflection at the mid-span of the beams. In this test, they concluded that the incorporation of recycled aggregate made the crack spacing of the concrete beam larger, and the number of secondary cracks also increased. The stiffness degradation of specimens with recycled aggregate was more notable and therefore, based on their results, a formula for assessing the short-term stiffness of the recycled concrete members after being subjected to sustained loading was suggested.

Choi et al. [35] presented experimental findings on the long-term deformations of recycled aggregate concrete members for over 1 year (380 days) and flexural performance of the beams after exposure to sustained loading. They developed four-point bending tests to evaluate the flexural performance of the recycled concrete beams after sustained loading, measuring the reduction in flexural capacity. The experimental results were compared with the predictions provided by the codes and finally, a modified equation was proposed to estimate the long-term deflection in beams with recycled aggregate. These results indicated that the recovery to instant deflection ratios in recycled concrete members are higher than the ratios obtained in conventional beams.

Seara-Paz et al. [41] tested the beams under three different loading stages using four different substitution percentages, 0 %, 20 %, 50 % and 100 % (Fig. 4.9). The first stage consisted of loading and maintaining the sustained loading for over 1300 days, the second stage comprised the unloading and recovery (in this stage the deformations were recorded over a period of one year after removing the sustained load). Finally, in the last stage, the concrete beams were tested up to failure to evaluate the impact of recycled aggregates on pre-cracked concrete beams.

The strain recoverability of the concrete is affected by the concrete response after loading and by the contribution of the reinforcing steel bars. Usually, concrete structures are designed to present a ductile performance. In these cases, the influence of the concrete is limited, even after periods of loading and unloading of concrete under compression. Therefore, the experimental results indicated that all concretes show a similar strain recoverability (Fig. 4.10) regardless of the aggregate used (conventional or recycled).

When analysing the recoverability of instantaneous deflections, the difference between conventional and recycled beams is negligible. However, in terms of deflection recovery over time (considering instantaneous and long-term deformations), recycled concrete exhibits lower recoverability than conventional concrete beams (Fig. 4.11). This is due to the greater damage and cracking of recycled concrete, which reduces the capacity of recovering long-term deflections. It can be stated that recycled concrete specimens will show greater permanent deflections than conventional concrete members.

Regarding the analysis of cracked and uncracked concrete beams up to failure (Fig. 4.12), it was seen that the cracking moments decrease as the content of recycled aggregate rises. This was attributed to the reductions in tensile splitting strength of recycled concrete. In addition, once cracking occurs,

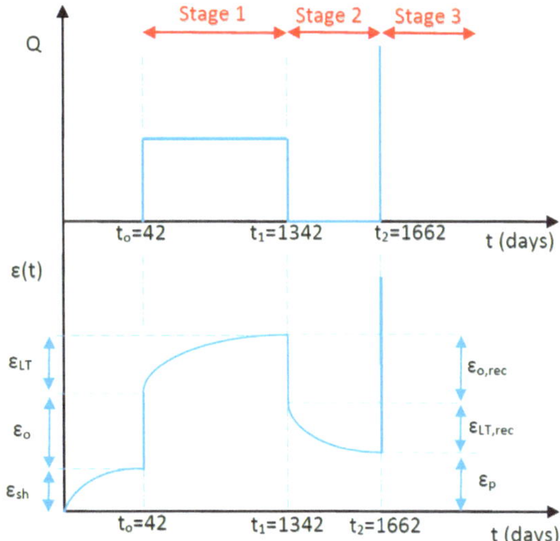

Fig. 4.9   Loading stages [41]

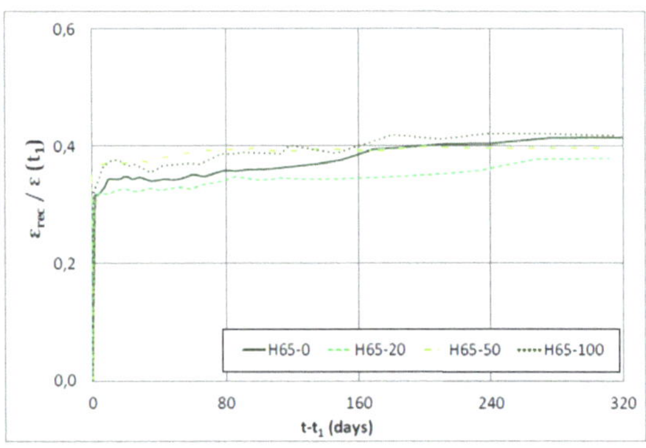

Fig. 4.10   Recoverability of strains over time [41]

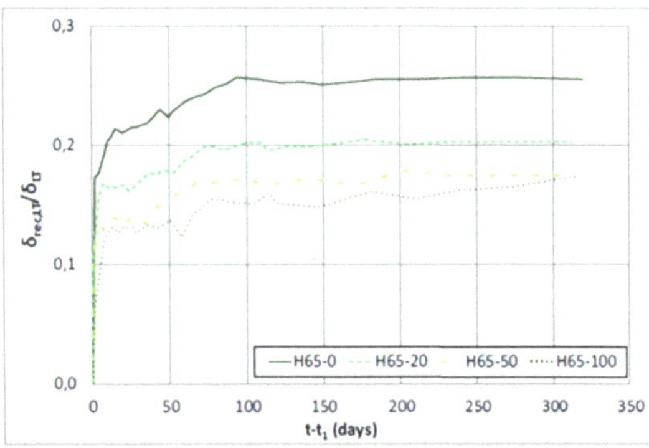

Fig. 4.11   Recoverability of deflection over time [41]

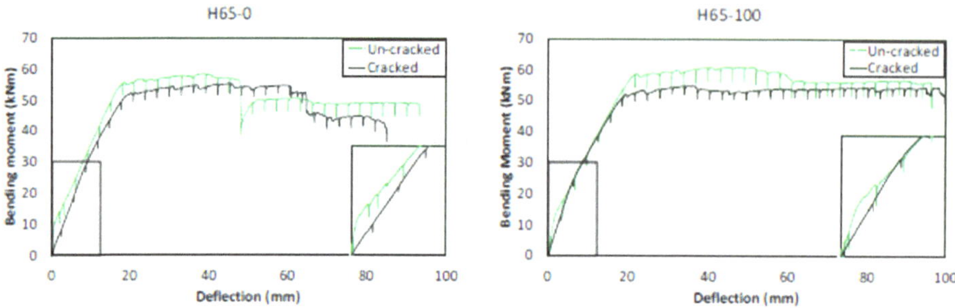

Fig. 4.12   Bending moment – deflection of uncracked and pre-cracked beams [41]

the lower bond strength of this material leads to greater crack widths and lower concrete contribution. This results in lower tension stiffening and, consequently, low effective flexural stiffness. This is especially significant when high replacement percentages are being used. Finally, at the yielding and ultimate stage, the differences between conventional and recycled concrete members are minor.

4.4 Conclusions

According to this literature review, the following conclusions can be drawn:

- The flexural performance of recycled concrete beams under short-term loads is similar to the performance of conventional concrete specimens. The bending moments (yielding, ultimate or service moments) do not show significant differences when compared to the moments measured in conventional concrete beams. In addition, the differences in the deflections are minor, being only slightly higher for the deflections measured in recycled concrete beams. Only cracking moments seem to be lower, which leads to confirm that recycled concrete beams show an early cracking development. However, the differences in the crack pattern and crack spacing of a conventional and a recycled concrete beam under flexure are negligible.
- The analysis of the shear behaviour leads to conclude that, when shear stirrups are being used, the recycled concrete beams and the conventional concrete members show a similar behaviour. Only when there is no shear reinforcement recycled concrete specimens have a slightly lower shear capacity than the shear strength of the baseline conventional beams. The cracking patterns and angles at failure are also similar in recycled and conventional concrete members.
- The long-term performance of recycled concrete members is highly related to creep performance. As the creep and shrinkage of recycled concrete are larger than that of conventional concrete, the long-term deformation of recycled concrete specimens is also larger than that of conventional members. Regarding cracking behaviour, it seems that there is no agreement in literature. Some experimental results show differences between cracking patterns of recycled and conventional concrete beams, although they do not seem to be significant.

- The performance of concrete beams after unloading has rarely been studied. According to the scarce results that can be seen in literature, it seems that recycled and conventional concrete specimens show similar strain recoverability as the strain recoverability of the concretes is highly affected by the response of the reinforcing steel bars after unloading. However, the greater damage and cracking of recycled concrete developed during loading and during long-term loading reduces the capacity of recovering long-term deflections, which leads recycled concrete specimens to show greater permanent deflections than conventional concrete members.
- Finally, the cracked performance of recycled concrete beams has shown that, once cracking occurs, the lower bond strength of recycled concrete leads to greater crack widths and lower concrete contribution, which results in lower tension stiffening. Regarding yielding and ultimate stage, no significant differences have been detected, as reinforcement contribution is highly important.

4.5 Acknowledgements

This work has been carried out within the framework of the following projects:

HACCURACEM project (BIA2017-85657-R), funded by the Ministry of Economy, Industry and Competitiveness, Spanish Programme for Research, Development and Innovation aimed at the challenges of Society, within the framework of the Spanish Plan for Scientific and Technical Research and Innovation 2013–2016, Call 2017.

Eco3DConcrete (PID2020-115433RB-I00) funded by the Ministry of Science and Innovation, Spanish Programme for Research, Development, and Innovation aimed at the challenges of Society within the framework of the Spanish Plan for Scientific and Technical Research and Innovation 2017–2020, Call 2020.

4.6 References

[1] X. Li, Recycling and reuse of waste concrete in China: Part II. Structural behaviour of recycled aggregate concrete and engineering applications, Resources, Conservation and Recycling. 53 (2009) 107–112. https://doi.org/10.1016/j.resconrec.2008.11.005

[2] B. González-Fonteboa, S. Seara-Paz, J. de Brito, I. González-Taboada, F. Martínez-Abella, R. Vasco-Silva, Recycled concrete with coarse recycled aggregate. An overview and analysis, Materiales de Construcción. 68 (2018) e151. https://doi.org/10.3989/mc.2018.13317

[3] C. Zhu, C. Liu, G. Bai, J. Fan, Study on long-term performance and flexural stiffness of recycled aggregate concrete beams, Construction and Building Materials. 262 (2020) 120503. https://doi.org/10.1016/j.conbuildmat.2020.120503

[4] K. McNeil, T.H.-K. Kang, Recycled Concrete Aggregates: A Review, International Journal of Concrete Structures and Materials. 7 (2013) 61–69. https://doi.org/10.1007/s40069-013-0032-5

[5] R. V. Silva, J. de Brito, R.K. Dhir, The influence of the use of recycled aggregates on the compressive strength of concrete: a review, European Journal of Environmental and Civil Engineering. 19 (2015) 825–849. https://doi.org/10.1080/19648189.2014.974831

[6] N. Tošić, S. Marinković, N. Pecić, I. Ignjatović, J. Dragaš, Long-term behaviour of reinforced beams made with natural or recycled aggregate concrete and high-volume fly ash concrete, Construction and Building Materials. 176 (2018) 344–358. https://doi.org/10.1016/j.conbuildmat.2018.05.002

[7] M. Arezoumandi, A. Smith, J.S. Volz, K.H. Khayat, An experimental study on flexural strength of reinforced concrete beams with 100 % recycled concrete aggregate, Engineering Structures. 88 (2015) 154–162. https://doi.org/10.1016/j.engstruct.2015.01.043

[8] A.G.R. Gholamreza Fathifazl O. Burkan Isgor Abdelgadir Abbas Benoit Fournier and Simon Foo, Flexural Performance of Steel-Reinforced Recycled Concrete Beams, ACI Structural Journal. 106 (2009). https://doi.org/10.14359/51663187

[9] Knaack Adam, Kurama Yahya, Behavior of Reinforced Concrete Beams with Recycled Concrete Coarse Aggregates, Journal of Structural Engineering. 141 (2015) B4014009. https://doi.org/10.1061/(asce)st.1943-541x.0001118

[10] S.T. Deresa, J. Xu, C. Demartino, Y. Heo, Z. Li, Y. Xiao, A review of experimental results on structural performance of reinforced recycled aggregate concrete beams and columns, Advances in Structural Engineering. 23 (2020) 3351–3369. https://doi.org/10.1177/1369433220934564

[11] S. Seara-Paz, B. González-Fonteboa, F. Martínez-Abella, J. Eiras-López, Flexural performance of reinforced concrete beams made with recycled concrete coarse aggregate, Engineering Structures. 156 (2018) 32–45. https://doi.org/10.1016/j.engstruct.2017.11.015

[12] G. Fathifazl, Structural performance of steel reinforced recycled concrete members, Library and Archives Canada = Bibliothèque et Archives Canada, 2008.

[13] B.S. Hamad, A.H. Dawi, A. Daou, G.R. Chehab, Studies of the effect of recycled aggregates on flexural, shear, and bond splitting beam structural behavior, Case Studies in Construction Materials. 9 (2018) e00186. https://doi.org/10.1016/j.cscm.2018.e00186

[14] S. Pradhan, S. Kumar, S. v Barai, Performance of reinforced recycled aggregate concrete beams in flexure: experimental and critical comparative analysis, Materials and Structures. 51 (2018) 58. https://doi.org/10.1617/s11527-018-1185-0

[15] L. Evangelista, J. de Brito, Flexural behaviour of reinforced concrete beams made with fine recycled concrete aggregates, KSCE Journal of Civil Engineering. 21 (2017) 353–363. https://doi.org/10.1007/s12205-016-0653-8

[16] A. Scanlon, P.H. Bischoff, Shrinkage Restraint and Loading History Effects on Deflections of Flexural Members, ACI Structural Journal. 105 (2008) 498–506. https://doi.org/10.14359/19864

[17] G. Kaklauskas, V. Gribniak, D. Bacinskas, P. Vainiunas, Shrinkage influence on tension stiffening in concrete members, Engineering Structures. 31 (2009) 1305–1312. https://doi.org/10.1016/j.engstruct.2008.10.007

[18] M. Sogo, T. Sogabe, I. Maruyama, R. Sato, K. Kawai, Shear behavior of reinforced recycled concrete beams, in: Proceedings of the International RILEM Conference on the Use of Recycled Materials in Buildings and Structures, Barcelona, Spain, 2004: pp. 8–11.

[19] A.B. Ajdukiewicz, A.T. Kliszczewicz, Comparative Tests of Beams and Columns Made of Recycled Aggregate Concrete and Natural Aggregate Concrete, Journal of Advanced Concrete Technology. 5 (2007) 259–273. https://doi.org/10.3151/jact.5.259

[20] I.S. Ignjatović, S.B. Marinković, Z.M. Mišković, A.R. Savić, Flexural behavior of reinforced recycled aggregate concrete beams under short-term loading, Materials and Structures. 46 (2013) 1045–1059. https://doi.org/10.1617/s11527-012-9952-9

[21] S. Arabiyat, M. Abdel Jaber, H. Katkhuda, N. Shatarat, Influence of using two types of recycled aggregates on shear behavior of concrete beams, Construction and Building Materials. 279 (2021) 122475. https://doi.org/10.1016/j.conbuildmat.2021.122475

[22] B. González-Fonteboa, F. Martínez-Abella, Shear strength of recycled concrete beams, Construction and Building Materials. 21 (2007) 887–893. https://doi.org/10.1016/j.conbuildmat.2005.12.018

[23] G. Wardeh, E. Ghorbel, Shear strength of reinforced concrete beams with recycled aggregates, Advances in Structural Engineering. 22 (2019) 1938–1951. https://doi.org/10.1177/1369433219829815

[24] I.S. Ignjatović, S.B. Marinković, N. Tošić, Shear behaviour of recycled aggregate concrete beams with and without shear reinforcement, Engineering Structures. 141 (2017) 386–401. https://doi.org/10.1016/j.engstruct.2017.03.026

[25] G. Fathifazl, A.G. Razaqpur, O.B. Isgor, A. Abbas, B. Fournier, S. Foo, Shear strength of reinforced recycled concrete beams with stirrups, Magazine of Concrete Research. 62 (2010) 685–699. https://doi.org/10.1680/macr.2010.62.10.685

[26] G. Fathifazl, A.G. Razaqpur, O.B. Isgor, A. Abbas, B. Fournier, S. Foo, Shear strength of reinforced recycled concrete beams without stirrups, Magazine of Concrete Research. 61 (2009) 477–490. https://doi.org/10.1680/macr.2008.61.7.477.

[27] S. Pradhan, S. Kumar, S. v. Barai, Shear performance of recycled aggregate concrete beams: An insight for design aspects, Construction and Building Materials. 178 (2018) 593–611. https://doi.org/10.1016/J.CONBUILDMAT.2018.05.022

[28] J.M. Illston, The components of strain in concrete under sustained compressive stress, Magazine of Concrete Research. 17 (1965) 21–28. https://doi.org/10.1680/macr.1965.17.50.21

[29] S. Qi Mei, J. Chao Zhang, Y. Feng Wang, R. Fei Zou, Creep-recovery of normal strength and high strength concrete, Construction and Building Materials. 156 (2017) 175–183. https://doi.org/10.1016/j.conbuildmat.2017.08.163

[30] H.E. Jensen, Creep and Shrinkage of High-strength Concrete, Danmarks Tekniske Højskole, 1992.

[31] A.M. Neville, Recovery of creep and observations on the mechanism of creep of concrete, Applied Scientific Research. 9 (1960) 71. https://doi.org/10.1007/bf00382191

[32] R.I. Gilbert, G. Ranzi, Time-dependent behaviour of concrete structures, CRC Press, 2011. https://doi.org/10.1201/9781482288711

[33] S. Seara-Paz, B. González-Fonteboa, F. Martínez-Abella, D. Carro-López, Long-term flexural performance of reinforced concrete beams with recycled coarse aggregates, Construction and Building Materials. 176 (2018) 593–607. https://doi.org/10.1016/j.conbuildmat.2018.05.069

[34] J. Eiras-López, S. Seara-Paz, B. González-Fonteboa, F. Martínez-Abella, Bond behavior of recycled concrete: Analysis and prediction of bond stress-slip curve, Journal of Materials in Civil Engineering. 29 (2017). https://doi.org/10.1061/(asce)mt.1943-5533.0002000

[35] W.C. Choi, H. do Yun, Long-term deflection and flexural behavior of reinforced concrete beams with recycled aggregate, Materials & Design. 51 (2013) 742–750. https://doi.org/10.1016/j.matdes.2013.04.044

[36] Adam M Knaack and Yahya C Kurama, Sustained Service Load Behavior of Concrete Beams with Recycled Concrete Aggregates, ACI Structural Journal. 112 (2015). https://doi.org/10.14359/51687799

[37] S. Seara-Paz, B. González-Fonteboa, F. Martínez-Abella, I. González-Taboada, Time-dependent behaviour of structural concrete made with recycled coarse aggregates. Creep and shrinkage, Construction and Building Materials. 122 (2016) 95–109. https://doi.org/10.1016/j.conbuildmat.2016.06.050

[38] V.W.Y. Tam, D. Kotrayothar, J. Xiao, Long-term deformation behaviour of recycled aggregate concrete, Construction and Building Materials. 100 (2015) 262–272. https://doi.org/10.1016/j.conbuildmat.2015.10.013

[39] A. Domingo, C. Lázaro, F.L. Gayarre, M.A. Serrano, C. López-Colina, Long term deformations by creep and shrinkage in recycled aggregate concrete, Materials and Structures. 43 (2010) 1147–1160. https://doi.org/10.1617/s11527-009-9573-0

[40] G. Fathifazl, A. Ghani Razaqpur, O. Burkan Isgor, A. Abbas, B. Fournier, S. Foo, Creep and drying shrinkage characteristics of concrete produced with coarse recycled concrete aggregate, Cement and Concrete Composites. 33 (2011) 1026–1037. https://doi.org/https://doi.org/10.1016/j.cemconcomp.2011.08.004

[41] S. Seara-Paz, B. González-Fonteboa, F. Martínez-Abella, J. Eiras-López, Deformation recovery of reinforced concrete beams made with recycled coarse aggregates, Engineering Structures. 251 (2022) 113482. https://doi.org/10.1016/j.engstruct.2021.113482

Chapter 5

Long-Term Deformation of Structural Recycled Aggregate Concrete

Nikola Tošić[1], Yahya Kurama[2], Jean Michel Torrenti[3]

1 Civil and Environmental Engineering Department, Universitat Politècnica de Catalunya, Jordi Girona 1–3, 08034 Barcelona, Spain
2 Department of Civil and Environmental Engineering and Earth Sciences, University of Notre Dame, Notre Dame, IN 46556, United States
2 Département Matériaux et Structures – Mast, Université Gustave Eiffel, Marne-la- Vallée, France

5.1 Introduction

Within the design of concrete structures, traditionally, more focus has been given to ultimate limit states (ULS) such as flexural, shear and punching strength. The behaviour of concrete structures in these overload situations is well described by a variety of resistance models that allow for a reliable, safe, and efficient design. At the same time, less emphasis has been given to serviceability limit states (SLS), and, in particular, long-term deformations. However, concrete structures of today are becoming increasingly governed by serviceability criteria, rather than strength requirements [1].

The deformation of concrete structures in service is a very complex issue for modelling, control, and prediction due to a large number of influencing factors and uncertainties associated with these factors [2, 3]. An important challenge is the time-dependent nature of many of the primary factors, mainly shrinkage, creep, cracking, tension stiffening, and loading history.

Nonetheless, the previous decades have seen a consistent refinement of methods for the deformation control of concrete structures, both in the tradition of the International Federation for Strucural Concrete (*fib)* and its Model Codes [4–6], as well as the American Concrete Insitute (ACI) and the Building Code 318, creep and shrinkage models of ACI 209R, and control of deflections in 435R [7–10].

All of the above-stated achievements exist for "traditional", natural aggregate concrete (NAC). However, the use of recycled aggregate concrete (RAC) as a fully accepted structural material [11, 12] requires the proper consideration of the specifics of this material from the point of view of serviceability and, in particular, long-term deformations. This is especially important because concrete properties critical to serviceability (modulus of elasticity, shrinkage, creep, deflections) are the ones in which RAC tends to differ the most from NAC.

The currently low uptake of recycled aggregate (RA) and its use for structural concrete [13, 14] is due to various barriers such as contamination during demolition, availability of competitive natural resources, and lack of integration in the value chain, but also the lack of design guidelines for RAC. Therefore, it is important to understand and be able to control and predict the time-dependent deformations of RAC as one of the critical aspects of its structural behaviour. By doing this, designers will be able to adequately consider shrinkage, creep, and tension stiffening, and their effects on cracking, deflections, and prestress losses.

For this purpose, this chapter presents a comprehensive and critical overview of the main aspects of RAC time-dependent behaviour at the material and structural levels. The sections of this chapter deal individually with the shrinkage, creep, and deflections of RAC and RAC structures. For each of these aspects, an overview of main experimental results and influencing parameters is given, as well as a summary of modelling techniques for design. Finally, conclusions are presented and recommendations for future research are given.

5.2 Shrinkage of Recycled Aggregate Concrete

When considering the long-term deformations of RAC, shrinkage should be considered first, as it is a time-dependent property that affects both structural and non-structural members, i.e., it is load-independent. Since its first observation around 135 years ago [15], knowledge on concrete shrinkage has increased immensely, to the point where there are complex physical theories for explaining and modelling it, mostly relying on capillary tension and removal of intercrystalline water from pores and diffusion [16, 17].

For all of these advances, a large number of experimental results is necessary to validate any theoretical prediction. This is helped by large databases – the most famous of which is the "NU-ITI database on concrete creep and shrinkage" assembled from 2010 to 2013 at the Northwestern University's Infrastructure Technology Institute, mainly through the support from the U.S. Department of Transportation and freely available online [18]. The NU-ITI database contains 1751 shrinkage curves (1217 'total', 417 autogenous, and 117 drying shrinkage curves).

Nonetheless, the number of influencing parameters is very large already for NAC and an accurate control of all conditions is difficult. Firstly, shrinkage is divided into autogenous and drying components (depending on whether there is water exchange between the concrete and its environment), and the experimental characterization of each component requires different approaches. Secondly, the experiments need to be under controlled environmental conditions (relative humidity and temperature) and cover varying specimen sizes (to assess the rate of diffusion). Thirdly, they need to last a sufficiently long time (years). The last point is critical – 96 % of the NU-ITI database test results do not exceed 6 years, and only two sets extend beyond 12 years. Additionally, most of the tested specimens had small thicknesses relative to common thicknesses of structural members. This all serves to illustrate the complexity of shrinkage even before RAC is considered and the additional parameters and uncertainties introduced with it.

From an engineering point of view, this has nonetheless enabled the formulation of many models for predicting shrinkage – from earlier models such as ACI 209R-92 [9] and the *fib* Model Code 1990 [5], which only considered drying shrinkage, to more advanced models such as the *fib* Model Code 2010 [6] (the basis for the new Eurocode 2-revision [11]) and the B4 model [19].

These facts highlight the challenges faced with understanding shrinkage before introducing RA and RAC into consideration. Fortunately, a lot of work has been done and is being done to illuminate this topic. In the following sub-sections, experimental results on RAC shrinkage are reviewed, and considering the main influencing parameters, modelling approaches and code proposals are given.

5.2.1 Experimental Results on the Shrinkage of Recycled Aggregate Concrete

A large number of studies report research on RAC shrinkage: for example, Lye et al. [20] report a literature search that yielded 286 publications with experimental data on RAC shrinkage. Considering such a large number of studies reporting results from a large variety of experimental setups, procedures, specimen sizes, etc., the focus of this section is on literature reviews by Lye et al. [20], Silva et al. [21], Tošić et al. [22] and Mao et al. [23], as they already summarize the literature and main findings, and facilitate discussion and identification of trends and influencing parameters.

As expected, the governing parameter specific to RAC shrinkage is the adhered mortar in RA particles [23]. Through differences in porosity, density, and water absorption, RA offers less restraint to shrinkage than natural aggregates (NA) in NAC, while the adhered mortar can exhibit shrinkage as well, depending on its age and origin [24]. The cited literature reviews provide important insights into all aspects of RAC shrinkage.

Mao et al. [23] summarize the research topics related to RAC shrinkage, Fig. 5.1. This paper is the only published study so far to separately treat autogenous and drying shrinkage of RAC. In this regard, particularly important is their observation that autogenous shrinkage of RAC is smaller than that of NAC (up to 66–83 % when 100 % of RA is used); this is attributed to the internal curing effect of RA (i.e. the gradual release of water absorbed by RA during pre-soaking or mixing with additional water) [23, 25]. Results in line with this observation were reported by Adessina [26], where RAC with 100 % of coarse RA after 90 days had one third of the autogenous shrinkage of the reference NAC. Nonetheless, autogenous shrinkage remains important only in concretes with very low water-cement ratios (i.e., high-strength concretes), and as such, is not of critical importance for typical applications of RAC. Furthermore, the number of available results that separate RAC shrinkage into autogenous and drying is very small – Mao et al. [23] report three studies [27–29], whereas Tošić et al. [22] report only one [30]. Therefore, more research is needed in this direction.

Drying shrinkage is the principal component of RAC shrinkage, constituting approximately 90 % of total RAC shrinkage [31], and much more information on results is available on it. Lye et al. [20] initially collected 286 publications with experimental results on

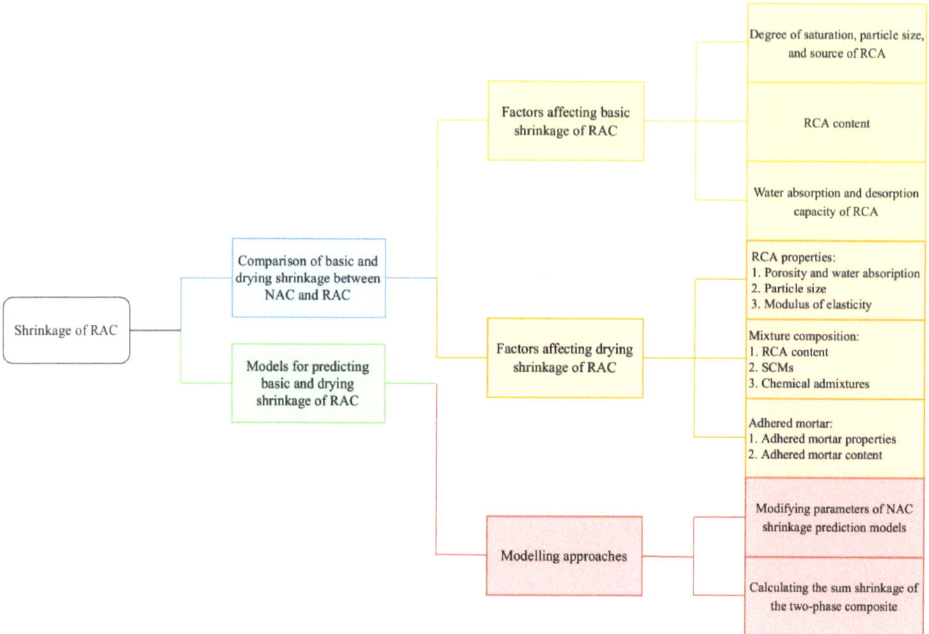

Fig. 5.1   Diagram of research topics related to RAC shrinkage [23]

RAC shrinkage. The authors considered different RA types; principally, recycled concrete aggregates (RCA), recycled masonry aggregates (RMA), and mixed recycled aggregates (MRA), which included studies with only coarse, only fine or both coarse and fine RA. The majority of literature (92 %) was on RAC produced with RCA and 75 % was on RAC produced with coarse RCA. Finally, the authors found that only 118 of the 286 publications had accessible, original experimental data with sufficient information for analysis.

Table 5.1 presents the range of the most important parameters in the studies collected by Lye et al. [20]. As can be seen from the table, the majority of the studies do not report the moisture state of the RA saturated-surface-dry (SSD, oven- or air-dried), which is a problem for drawing general conclusions. Additionally, the duration of the majority of the experiments was under 180 days.

The large dispersion in the results due to a wide range of parameter values is illustrated in Fig. 5.2, where the shrinkage of RAC with coarse RCA is given as a relative increase with regard to its "reference NAC" (prepared with the same effective w/c ratio) [20]. Nonetheless, the results (filtered for outliers) show an increase of shrinkage with increasing coarse RCA content, such that at 100 % coarse RCA, shrinkage increases by, on average, 33 % relative to NAC. This is in line with results of other reviews [22, 23], which generally report a range of 10–60 % increase.

The second most important parameter identified by the authors is the compressive strength of RAC: for RAC with 100 % of coarse RCA and f_{cm} = 30 MPa, the increase in shrinkage relative to NAC is close to 50 %, whereas for RAC with f_{cm} of 60 MPa, this increase is around 30 % [20]. Finally, lower relative humidity leads to larger differences

Table 5.1 Range of parameter values in studies collected by Lye et al. [20]

Parameter	Value	No. of studies in Lye et al. [20]
RCA specific gravity	<2.30	0/1[a]
	2.30–2.49	0/40
	2.50–2.69	39/20
	Not reported	119/133
Water absorption	<3 %	98/12
	3–5.9 %	2/102
	6–10 %	0/40
	Not reported	58/40
Moisture state of RA when mixing	Saturated surface dry	43
	Air dry	5
	Oven dry	3
	Not reported but water added	20
	Not reported	84
Relative humidity during testing	21–40 %	0
	41–60 %	88
	61–80 %	25
	81–95 %	9
	96–100 %	6
	Not reported	33
Duration (days)	<30	11
	30–100	47
	101–180	45
	>181	36
	Not reported	24

[a] Number of studies for NA/Number of studies for RCA

between RAC and NAC shrinkage: increases in shrinkage for RAC with 100 % of coarse RCA relative to NAC are around 45 % and 35 % for relative humidity values of 50 % and 70 %, respectively [20]. It is important to note that for lower amounts of coarse RCA, e.g., up to 30 %, the average increase in shrinkage remains in the range of ~15 %. This is important considering that up to such lower replacement ratios, many codes and recommendations suggest that no special consideration of RCA effect on shrinkage needs to be taken into account; therefore, such a proposal seems justified by the results.

In terms of other properties, Silva et al. [21] noted that the use of saturated surface dry RA (as opposed to oven dry RA + additional water for absorption compensation) leads to higher shrinkage. Additionally, multiple crushing during recycling, which tends to remove more adhered mortar, leads to lower shrinkage of RAC. In this regard, RAC with fine RA exhibits significantly higher shrinkage than RAC with only coarse RA [23]. The use of RMA or MRA also additionally increases RAC shrinkage [20]. Finally, in terms of supplementary cementitious materials, the results are inconclusive and more research is needed: when adding fly ash or ground granulated blast furnace slag to RAC, both

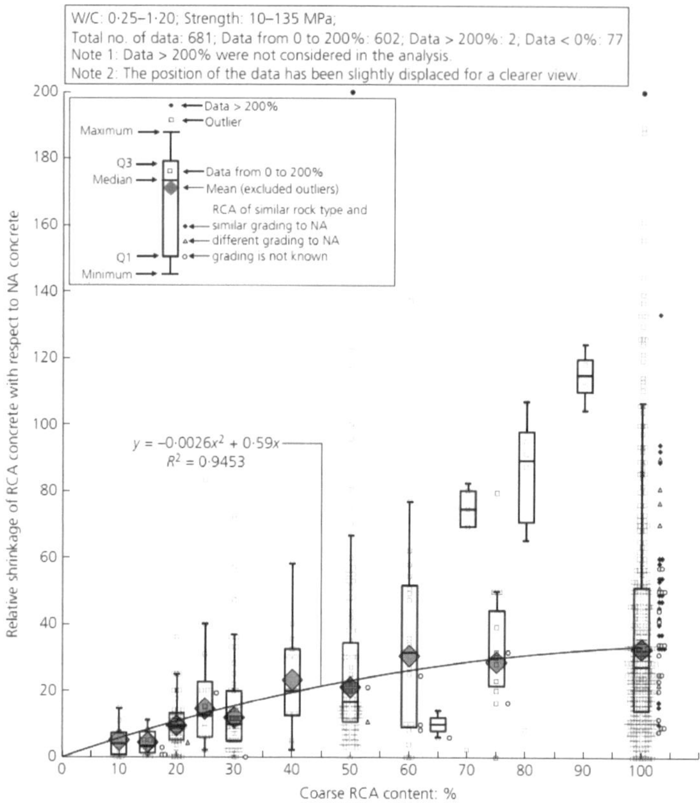

$$y = -0.0026x^2 + 0.59x$$
$$R^2 = 0.9453$$

Fig. 5.2   Shrinkage of RAC with coarse RCA relative to shrinkage of reference NAC [20]

increases and decreases in shrinkage are reported [21, 23]; in the case of silica fume and metakaolin additions, increases in RAC shrinkage have been reported (due to a higher content of C-S-H gel in the cement paste produced quickly by the pozzolanic reaction).

An important, and often overlooked, factor when comparing RAC and NAC shrinkage is the time evolution of shrinkage. Namely, the previously cited reviews [20, 21, 23] analysed the collected literature in terms of differences in RAC and NAC shrinkage for "final" reported values, at whatever time they occurred. As seen from Table 5.1, this time was often less than 180 days. However, the time evolution of RAC shrinkage might differ from that of NAC, and this needs to be analysed based on shrinkage-time curves. Such an analysis was performed by Tošić et al. [22] for 39 NAC and 86 RAC time curves. By analysing the ratio of RAC to NAC shrinkage strain, $\varepsilon_{cs,RAC}/\varepsilon_{cs,NAC}$ within individual time curves, the authors tried to identify whether there was any horizontal scaling of shrinkage, or only vertical, (the concept of vertical and horizontal scaling being illustrated in Fig. 5.3). By finding that the $\varepsilon_{cs,RAC}/\varepsilon_{cs,NAC}$ ratios were stable over time within RAC-NAC time curve pairs, the authors concluded that the time evolution of RAC and NAC shrinkage is not significantly different [22], i.e., it is sufficient to compare final values of RAC and NAC shrinkage. However, it should be noted that Zhang et al. [32] reported a slower evolution of RAC shrinkage relative to NAC. It is therefore likely that further research is needed in this direction.

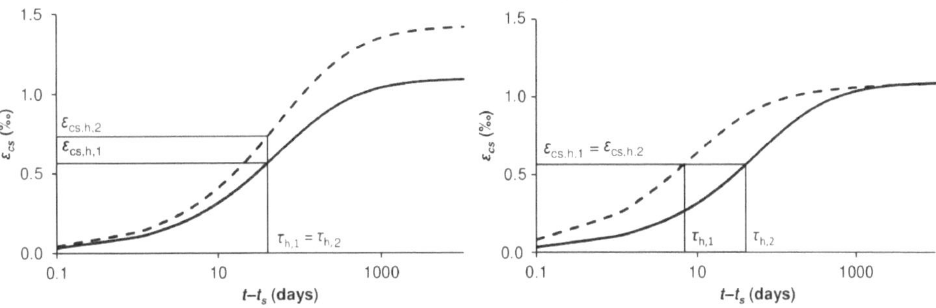

Fig. 5.3 Differences between vertical (left) and horizontal (right) scaling of shrinkage curves [22]

5.2.2 Modelling of Recycled Aggregate Concrete Shrinkage

The accurate prediction of shrinkage is a complex task due to the large number of uncertainties and the variability of input parameters. When analysing predictions of experimental results on databases, even the most advanced models can easily exhibit coefficients of variation close to 50 % [33]. Considering the added uncertainties introduced by RA, it is a legitimate question whether complex and computationally time-consuming physical models carry a real benefit over simple engineering empirical approaches.

In 2011, Fathifazl et al. [24] proposed a correction coefficient for the ACI 209R-92 shrinkage model, S_{RM}, which is meant to be multiplied by the shrinkage calculated for an NAC of equivalent characteristics:

$$S_{RM} = \left(\frac{1 - (1 - RMC) \cdot V_{RCA}}{1 - V_{RCA}} \right)^{1.45} \tag{5.1}$$

where RMC is the residual mortar content in the RA and V_{RCA} is the volume of RA in RAC. As can be seen, the method requires knowing both the mix design of RAC (V_{RCA}) and the adhered mortar content (RMC) of the RA, both of which are difficult to know in the design stage and when sourcing RA from a recycling plant.

Other researchers followed, proposing similar models that relied on inputs unknown in the design stage: density, absorption and volume of RA [29, 32 , 34–39]. These models could be considered as too complex for routine design and useful in specific cases where a detailed shrinkage analysis is necessary and the composition of the concrete mix fully characterised.

For general structural design, there are several simple code modifications to shrinkage models proposed by researchers [22, 40]. Tošić et al. [22] compiled a database of 424 data points in 125 shrinkage time curves from 19 studies and tested the applicability of the *fib* Model Code 2010 [6] shrinkage model. The authors first confirmed the suitability of the model's mathematical formulation for describing the time evolution of shrinkage, Fig. 5.4. Subsequently, the calculated versus experimental result comparison was performed for RAC and reference NAC shrinkage, Fig. 5.5. An overestimation of shrinkage by the *fib* Model Code 2010 model was found both for NAC and RAC, however, a larger overesti-

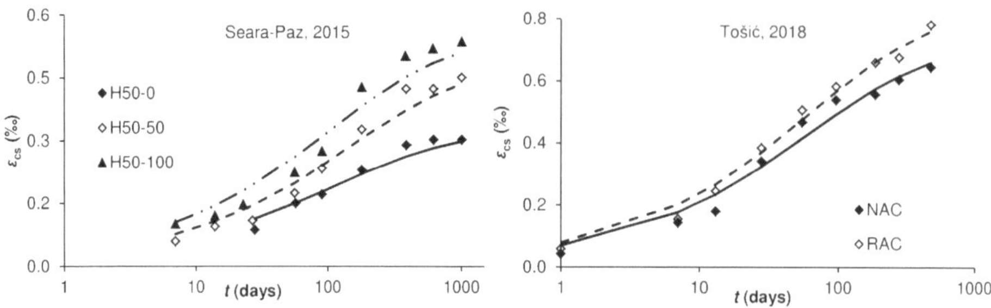

Fig. 5.4 Calibration of individual shrinkage curves from [40, 43, 22]

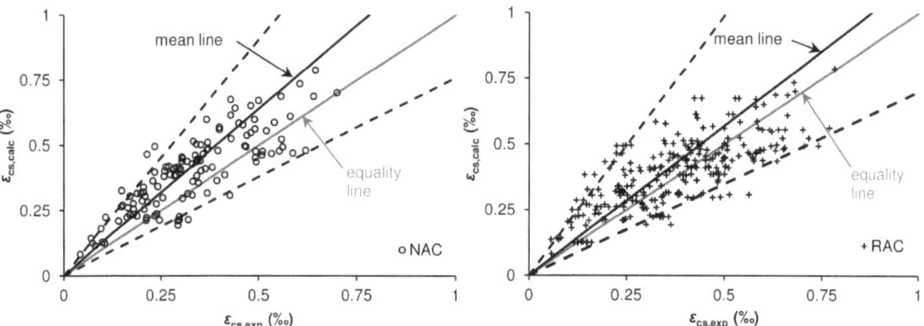

Fig. 5.5 Calculated versus experimental shrinkage strain values for all NAC and RAC data points reported in [22]

mation was found for RAC – these results are in agreement with those reported by Silva et al. [21]who also looked at the Eurocode 2 [41], ACI 209R-92 [9], and B3 [42] models. This difference could not be explained only by the differing compressive strength, and must therefore come from the differences in shrinkage [22].

Based on a regression that took into account bias in the time domain (by weighting data according to time decades), the authors proposed a global correction coefficient for RAC shrinkage:

$$\varepsilon_{cs,RAC}(t, t_s) = \xi_{cs,RAC} \cdot \varepsilon_{cs}(t, t_s) = \left(\frac{RCA\%}{f_{cm}}\right)^{0.30} \cdot \varepsilon_{cs}(t, t_s) \geq \varepsilon_{cs}(t, t_s) \qquad (5.2)$$

where $\varepsilon_{cs,RAC}(t,t_s)$ is the shrinkage strain of RAC cured until time t_s at time t, $\xi_{cs,RAC}$ is the RAC shrinkage correction coefficient, *RCA%* is the mass percentage of coarse RCA in RAC relative to the total mass of coarse aggregate, f_{cm} is the RAC compressive strength, and $\varepsilon_{cs}(t,t_s)$ is the shrinkage strain calculated according to the *fib* Model Code 2010 as for NAC.

As can be seen, the model has the benefit of using only data known at the design stage – the compressive strength of RAC and the content of coarse RCA.

In preparation of the new Eurocode 2-revision [11] and the *fib* Model Code 2020 [44], the proposal of Equation (5.2) was further adjusted. A single variable related to RA was adopted for these codes, α_{RA} = (mass of fine and coarse RA)/(mass of total fine and coarse aggregates). Considering results of the French national research project RECYBETON and the proposal for correcting RAC shrinkage strain it put forward, the following proposal for shrinkage correction was made by Tošić et al. [12]:

$$\varepsilon_{cs,RAC}(t, t_s) = (1 + 0.8 \cdot \alpha_{RA}) \cdot \varepsilon_{cs}(t, t_s) \tag{5.3}$$

Comparing the correction coefficients of Equations (5.2) and (5.3) in Fig. 5.6 for different compressive strength classes, it can be seen that the proposed expression is on the safe side. Considering that α_{RA} = 0.6 corresponds to an RAC with 100 % of coarse RA, the proposed increase in shrinkage can be seen to be around 50 %, in line with literature review results [20, 21] but also adopting a relatively conservative approach, which is justified considering the associated uncertainties.

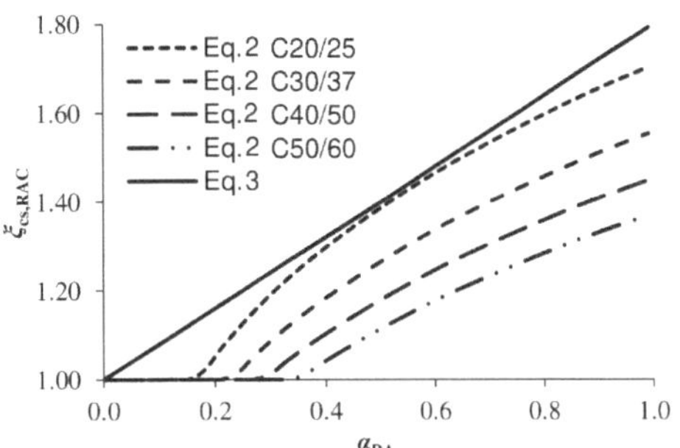

Fig. 5.6   Comparison of shrinkage correction coefficients $\xi_{cs,RAC}$ using Equations (5.2) and (5.3) [12]

5.3 Creep of Recycled Aggregate Concrete

One of the first documented records of concrete creep was made by Hatt [45] in the early 20th century. By mid-century, the phenomenon had been tied to the adsorption and desorption of water, and it was clear that analyses based on thermodynamics were needed [46]. The development of knowledge, and above all, testing equipment, enabled detailed investigations into the nanogranular origin of creep [47]: nanoindendation tests on C-S-H revealed that creep is likely caused by the rearrangement of nano-scale particles around limit packing densities following the free-volume dynamics theory of granular physics [47]. Such phenomena were also theoretically explained, for example by the "micropre-stress-solidification theory" formulated by Bažant et al. [16] in which creep is explained by changes in microstructure, modelled by the gradual deposition of non-aging layers of new hydration products on the walls of capillary pores in the hardened cement paste.

The microprestress-solidification theory formed the basis of the B3 and then later the B4 models for creep [19, 42]. While more precise than others, the B4 model requires a large number of input variables (such as the aggregate/cement ratio, water/cement ratio and cement content). Other engineering models such as the ACI 209R-92 [9] and the *fib* Model Code 2010 [6] models require less data for input, typically only compressive strength.

Similar to the development of models for shrinkage, the formulation of creep models was enabled by large databases of experimental results, and in particular, the NU-ITI database which contains approximately 1400 creep curves, divided relatively uniformly between basic and drying creep.

An important aspect when reporting experimental results on concrete creep is the form in which they are expressed. Namely, throughout literature, this can be in the form of creep strain, ε_{cc}, experimental creep coefficient $\varphi_{exp} = \varepsilon_{cc}/\varepsilon_{ci}$ (where ε_{ci} is the initial strain at loading), specific creep $\varepsilon_{cc}/\sigma_c$ (i.e., creep strain per unit stress), or creep compliance J_c. The most general and optimal way of reporting results is through the creep compliance which, in the case of linear creep (i.e., exposure of concrete to stresses below approximately 40–50 % of compressive strength) gives the following relation:

$$\varepsilon_{c\sigma}(t, t_0) = J_c(t, t_0) \cdot \sigma_c(t_0) \tag{5.4}$$

where $\varepsilon_{c\sigma}$ is the stress-dependent strain of concrete and t_0 is the age of concrete at loading.

Using such a formulation allows the subsequent derivation of any type of creep coefficient, as for example in the case of the *fib* Model Code 2010:

$$\varepsilon_{c\sigma}(t, t_0) = \left(\frac{1}{E_c(t_0)} + \frac{\varphi(t, t_0)}{E_{ci}} \right) \cdot \sigma_c(t_0) \tag{5.5}$$

where $E_c(t_0)$ and E_{ci} are the moduli of elasticity at the age of loading and 28 days, respectively. It should then be noted that the creep coefficient as formulated by the *fib* Model Code 2010 is not equal to the experimental creep coefficient $\varphi_{exp} = \varepsilon_{cc}/\varepsilon_{ci}$. This is an important aspect to be taken into account when comparing experimental results among themselves and with code predictions.

Finally, even though almost 800 creep curves in the NU-ITI database are for concretes with some type of admixture or additive, there is no information on the creep of RAC. As with shrinkage, the adhered mortar on RA particles influences creep in several ways: from an internal curing effect, to less restraint to creep as well as a certain creep of the adhered mortar itself. In the following subsections, experimental results on RAC creep are overviewed, and new and modified existing models for calculating the creep of RAC are discussed.

5.3.1 Experimental Results on RAC Creep

The number of studies that report research on RAC creep is also significant: Lye et al. [48] report a literature search that yielded 100 publications with experimental data on RAC creep. As in the case of RAC shrinkage, comprehensive literature reviews exist on RAC creep, which form the basis of this section, namely, the studies by Lye et al. [48], Silva et al. [49] and Tošić et al. [50].

In terms of the division into basic and drying creep, there are even less results than for shrinkage – Tošić et al. [50] report only one study analysing RAC basic and drying creep separately [51]; hence all of the analyses rely only on "total" creep (i.e. the sum of basic and drying).

In a comprehensive literature search, Lye et al. [48] found 100 publications on RAC creep from 27 countries published over a period of 30 years. The experimental data included RAC with RCA, RMA, MRA, fine, coarse, as well as both fine and coarse RA. The authors summarized the data graphically, Fig. 5.7, presenting the increase of RAC creep relative to a reference NAC. It should be noted that the data in the collected studies is reported and compared under different aspects: experimental creep coefficient, creep strain, or specific creep.

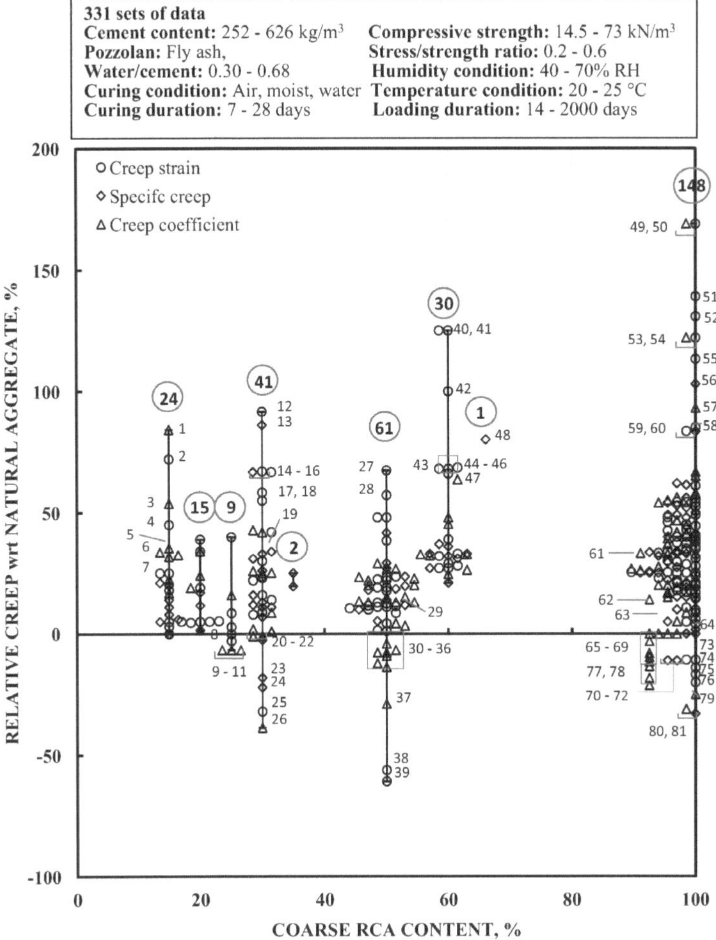

Fig. 5.7   Influence of coarse RCA content on creep of RAC [48]

The results show a similar tendency as with shrinkage, i.e., an increase in creep with a rising coarse RCA content so that the average increase for RAC with 20 % and 100 % of coarse RCA is 12 % and 32 %, respectively. The values at 100 % of coarse RCA is in agreement with other studies who suggest an increase in the range of 10–50 % [48, 50]. Beside coarse RCA content, the other important factor was identified as RAC compressive strength, so that at 100 % of coarse RCA the relative increase in creep was 35 %, 30 % and 25 % for f_{cm} in the ranges of 15–40, 41–50 and 51–70 MPa, respectively.

In terms of other influencing factors, the use of RMA or MRA leads to even higher creep compared with RCA, as does the use of only fine or both fine and coarse RA (as compared with using only coarse). However, due to scarce data, quantifying these increases is difficult. Lye et al. [48] present a useful summary of explanations proposed in the literature for increasing or decreasing RAC creep relative to reference NAC, Table 5.2.

Table 5.2 Suggested causes for differences in RAC and reference NAC creep [48]

Relative change in creep	Suggested cause	No. of studies [48]
Increase	RA adhered mortar	15
	Higher porosity/water absorption of RA	13
	More cement needed for equal strength	11
	Higher total mortar volume in RAC	4
	Higher w/c ratio for RAC	3
	Higher permeability of RAC	3
	Lower density of RAC	2
	Lower modulus of elasticity of RAC	2
	Higher stress-strength ratio	2
	Influence of RA composition	2
	Porosity of RA delays moisture loss	1
	Lower compressive strength of RAC	1
	Higher initial elastic strain of RAC	1
	No explanation	22
Decrease	Lower w/c ratio for RAC	4
	Higher compressive strength of RAC	3
	Higher density of RAC	2
	RA has more effect on elastic strain than creep strain	1
	Adhered mortar less susceptible to creep	1
	Lower water absorption of RAC	1
	No explanation	3
No change	Higher compressive strength of RAC	2
	No explanation	2

Silva et al. [49] investigated the effect of other parameters on RAC creep. In terms of mixing procedure, the use of dry RA and water absorption compensation seems preferable to the use of SSD RA, whereas specific mixing approaches such as the "two-stage mixing approach" developed by Tam et al. [52] or the "equivalent mortar volume" method developed by Fathifazl et al. [24] can bring certain reductions in creep as well.

In terms of SCMs, the use of fly ash can reduce creep strain [49], but this can mostly be attributed to a delayed strength formation over time, which reduces the stress-strength ratio over time. At the same time, more studies are critically needed on the inclusion of other SCMs such as ground granulated blast furnace slag (GGBFS) or natural pozzolans.

Finally, Tošić et al. [50] analysed the time evolution of RAC creep on 46 creep time curves (32 RAC and 14 companion NAC) from 10 studies, in order to identify whether differences are attributable to horizontal or vertical scaling (i.e., are the differences only related to the magnitude of creep or also to its time evolution), Fig. 5.8. The authors converted all experimental data to creep coefficients according to the *fib* Model Code 2010 and analysed whether, within individual time curves, the ratio $\varphi_{MC,RAC}/\varphi_{MC,NAC}$ remained stable or changed over time. The authors found no indication of horizontal scaling of RAC creep [50], i.e., only the magnitude of RAC creep is important for analysis and not its kinetics.

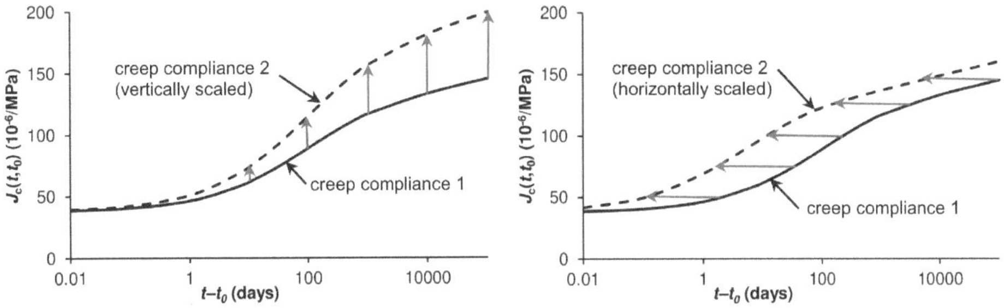

Fig. 5.8 Differences between vertical (left) and horizontal (right) scaling of creep curves [50]

5.3.2 Modelling of the Creep of Recycled Aggregate Concrete

Similar to the case of RAC shrinkage, various types of models have been proposed for RAC creep. One of the first ones was by Fathifazl and Razaqpur [53], who proposed modifying factors for the ACI 209R-92 creep model [9]. The authors proposed two coefficients, K_{RM} and K_{RC}, which take into account the effect of residual/adhered mortar and recoverable creep of RA, respectively. The coefficients were formulated as

$$K_{RM} = \frac{[1 - V_{RCA} \cdot (-RMC + 1 + R)]^{1.2 + 0.6 \cdot \frac{E_{RAC}}{E_{NAC}}}}{[1 - V_{RCA}]^{1.33}} \tag{5.6}$$

$$K_{RC} = 1 - \beta \cdot K_t \cdot \left(\frac{V_{RCA} \cdot RMC}{1 - V_{RCA} \cdot (1 + R)}\right)^{1.33} \tag{5.7}$$

where V_{RCA} is the volume of RA in RAC, RMC is the residual mortar content in the RA, R is the volumetric ratio of coarse NA to coarse RA in RAC, E_{RAC} and E_{NAC} are the moduli of elasticity of RAC and NAC, respectively, β is the percentage of recoverable creep in RCA (0 and 1 correspond to 100 % and 0 % of recoverable creep, respectively), and K_t is the coefficient for time under load ($t^{0.6}/(10 + t^{0.6})$).

As can be seen, the model requires knowledge of a lot of properties of RCA and RAC, making it difficult for routine design applications. Subsequently, similar models were proposed by other authors [39, 54].

Silva et al. [49] compared predictions of RAC creep using models such as the ACI 209R-92, *fib* Model Code 2010, B3, and GL2000 models and found all of them to be more or less inaccurate in predicting RAC creep. Analysing the Eurocode 2 creep model [41], Peña Torres et al. [55] found that the prediction for NAC should be multiplied by $(1 + 0.33\alpha_{RA})$ in order to take into account the effect of RA.

Continuing in this line, Tošić et al. [50] analysed predictions by the *fib* Model Code 2010 on 46 individual creep curves. The authors first confirmed the suitability of the model's mathematical formulation for describing the time evolution of creep, Fig. 5.9. Subsequently, the calculated versus experimental creep comparison was performed for

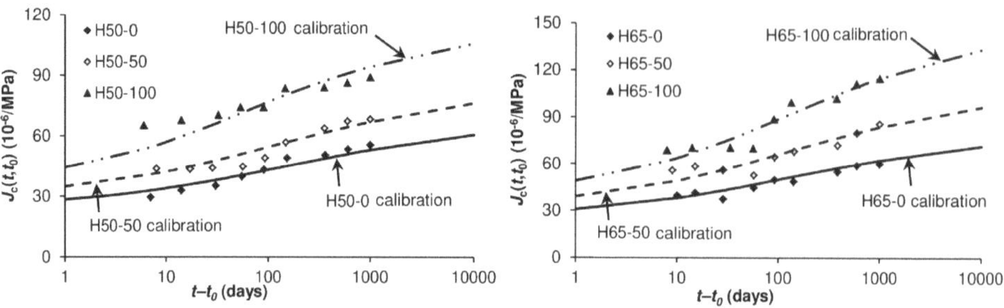

Fig. 5.9 Calibration of individual creep curves from [40], [50]

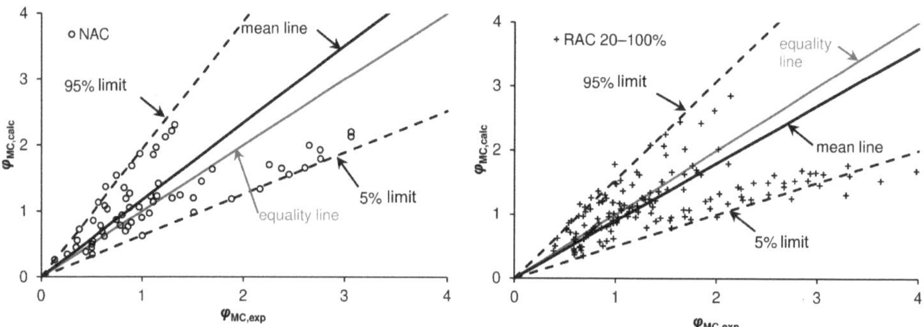

Fig. 5.10 Calculated versus experimental creep coefficient values for all NAC and RAC data points reported in [50]

RAC and reference NAC creep, Fig. 5.10. A clear underestimation of RAC creep relative to reference NAC was observed.

Based on a regression that took bias in the time domain into account (by weighting data according to time decades), the authors proposed a global correction coefficient for RAC creep:

$$\varphi_{RAC}(t, t_0) = \xi_{cc,RAC} \cdot \varphi(t, t_0) = 1.12 \cdot \left(\frac{RCA\%}{f_{cm}}\right)^{0.15} \cdot \varphi(t, t_0) \geq \varphi(t, t_0) \qquad (5.8)$$

where $\varphi_{RAC}(t,t_0)$ is the creep coefficient at time t of RAC loaded at time t_0, $\xi_{cc,RAC}$ is the RAC creep correction coefficient, $RCA\%$ is the mass percentage of coarse RCA in RAC, f_{cm} is the RAC compressive strength, and $\varphi(t,t_0)$ is the creep coefficient calculated according to the *fib* Model Code 2010 as for NAC.

As can be seen, the model has the benefit of using only data known at the design stage – the compressive strength of RAC and the content of coarse RCA.

In preparations of the new Eurocode 2-revision [11] and the *fib* Model Code 2020 [44], the proposal of Equation (5.8) was further adjusted. As stated earlier for shrinkage, a single variable related to RA was adopted for these codes, α_{RA} = (mass of fine and coarse RA)/(mass of total fine and coarse aggregates). Considering results of the French national research project RECYBETON and the RAC creep correction proposed therein , the following proposal for creep correction was made [12]:

$$\varphi_{RAC}(t, t_0) = (1 + 0.6 \cdot \alpha_{RA}) \cdot \varphi(t, t_0) \qquad (5.9)$$

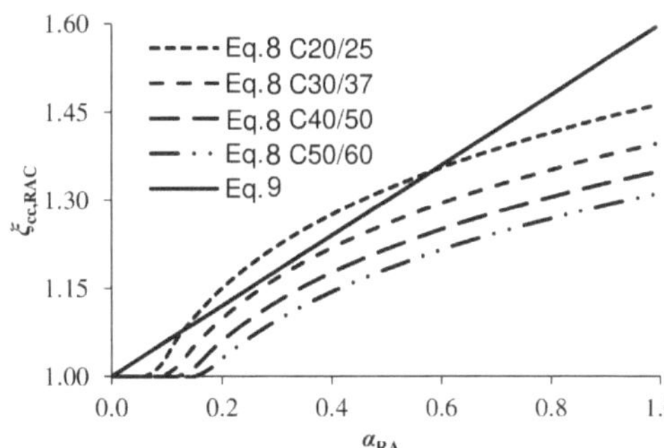

Fig. 5.11 Comparison of creep correction coefficients $\xi_{cc,RAC}$ using Equations (5.8) and (5.9) [12]

The correction coefficients of Equations (5.8) and (5.9) are compared in Fig. 5.11 for different compressive strength classes. Considering that $\alpha_{RA} = 0.6$ corresponds to an RAC with 100 % of coarse RA, the proposed increase in creep can be seen to be around 36 %, in line with literature review results [48, 50].

5.4 Long-Term Deflections of Recycled Aggregate Concrete Structures

Experimental research on long-term deflections of concrete structures is scarce when compared to research on other aspects of concrete behaviour. This is mostly attributable to the fact that such tests require a significant financial effort, good organization, and a long time. Additionally, there are a lot of factors that affect deflections of concrete structures and that need to be accounted for [3]:

- geometrical properties of the member (span, shape, and size of cross-section);
- moduli of elasticity of concrete and reinforcement;
- concrete tensile strength;
- area and distribution of reinforcement in cross-sections;
- load intensity and history;
- stiffness reduction caused by cracking and tension stiffening;
- member structural system (shape of bending moment diagram);
- moment redistribution in statically indeterminate systems caused by stiffness reduction;
- shrinkage and creep;
- moment redistribution in statically indeterminate systems caused by shrinkage and creep.

As an example of the difficulties faced when executing and later interpreting long-term deflection tests, the speed of load application can be considered. Namely, in certain tests it is reported that load was applied during several hours and initial deflections were only then recorded [56]. During this time, significant initial creep would have occurred, distorting any later interpretation of data.

In terms of NAC, in 1988 Espion [57] compiled a database of 397 long-term results from 45 different experimental campaigns. Beside this database, only a few studies performed afterwards have been well-documented [58], treating "traditional" reinforced concrete structures. Based on many of these results, different deflection calculation methods were developed. Within the framework of the *fib*, starting from the work by Jaccoud and Favre [59], the ζ-method of interpolating curvatures (or deflections) between the uncracked and fully cracked states was developed and subsequently adopted in *fib* Model Codes and the Eurocode 2. At the same time, based on the work by Branson [60] the ACI effective moment of inertia method was developed.

This summary should provide a sound basis for research on deflections of RAC members – a frame of reference and guidance in terms of test conceptualization and modelling approaches. The next sections will describe the available experimental research on deflections of RAC structures as well as analytical and numerical approaches to RAC deflection control.

5.4.1 Experimental Results on Deflections of Recycled Aggregate Concrete Structures

Deflections of RAC structural members have been the topic of a relatively small number of research campaigns [43, 61–69]. In total, close to 80 RAC and reference NAC beams have been tested over a period of 15 years. This number is not insignificant in itself, however, considering the number of parameters that can be varied, it is still challenging to draw general conclusions. An additional issue is the incomplete reporting of some of the results, e.g., in the form of conference proceedings with fewer details.

As a first step in the analysis, the results of experimental research on the ultimate flexural strength of RAC beams can also be used to comment on their short-term service load behaviour. Considering results from literature compiled in a database by Tošić et al. [70], it was noted that researchers report crack spacings/widths and deflections under service load of reinforced RAC beams as larger than those of reference NAC beams by 30–70 % and 20–100 %, respectively. Such results already demonstrate an effect of RA on deflections through a lower modulus of elasticity (higher deflections) and weaker ITZ (more pronounced cracking).

Maruyama et al. [61] compared two NAC beams ('VC') with two RAC beams made from both coarse and fine RCA ('CFRC'). The NAC and RAC (produced with a *w/c* ratio of 0.6) had different 28-day compressive strengths, 36.4 and 21.4 MPa, respectively, as well as moduli of elasticity, 24.4 and 18.1 GPa, respectively. Two beams were kept and loaded under "wet" conditions (sealed with saturated paper at room temperature) and the other two beams were under "dry" conditions (exposed to room temperature after one week of wet curing). The experiment lasted 380 days.

The beams had a 150/200 mm cross-section and a 2800 mm span. They were loaded in four-point bending with a flexural span of 800 mm. The reinforcement ratio was 1.05 % consisting of 2 $\varnothing$13 mm bars with a yield stress of approximately 340 MPa. The experimental setup is shown in Fig. 5.12. The beams were loaded so as to induce a stress of 100 MPa in the reinforcement.

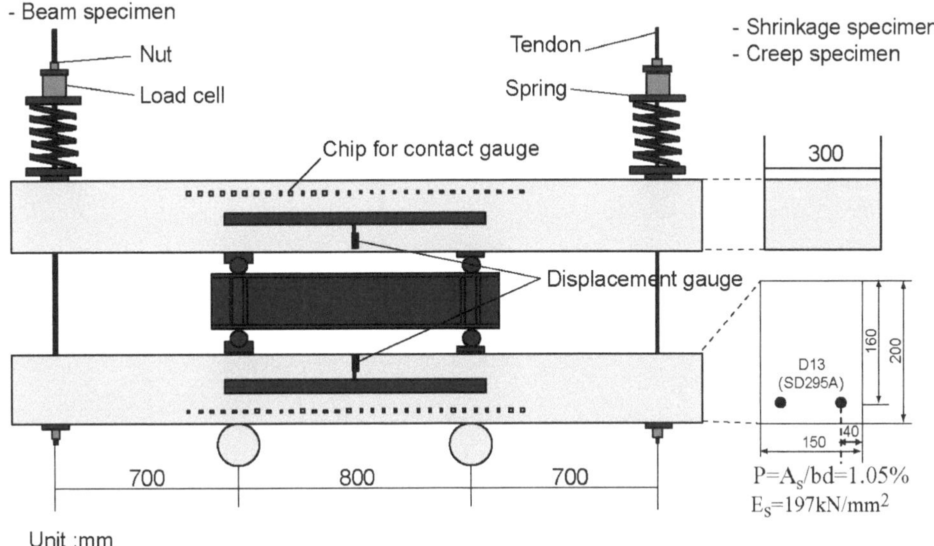

Fig. 5.12 Setup for the sustained loading test used in [61]

The authors did not report deflections but curvature at midspan. For beams under the "wet" curing regime, there was practically no increase in curvature after loading whereas the beams under "dry" conditions experienced an increase in curvature with the RAC beam's curvature being 1.8 times that of NAC.

Łapko and Grygo [68] tested two concrete beams with a 120/200 mm cross-section and a 3200 mm span under a constant load for 15 weeks in four point bending. The longitudinal tensile reinforcement consisted of 3 $\varnothing$12 mm bars, the longitudinal compressive reinforcement consisted of 2 $\varnothing$10 mm bars and stirrups were $\varnothing$6 mm bars. The study is reported in a conference paper and, hence, little information is provided beside the observation that "the differences between deflections of the two beams after 15 weeks of testing were about 20 %" [68].

Ajdukiewicz and Kliszczewicz [62] tested the long-term behaviour of both columns and beams, analysing the effect of RCA quality. The authors first produced concrete from quartzite, granite, and basalt natural aggregates; then, they crushed these concretes to produce coarse and fine RCA. With the RA, the authors produced nine concrete mixtures in which both the quality and size of RCA was varied – either only coarse RCA was used or both fine and coarse RCA. Beams with a 200/300 mm cross-section and 2400 mm span were tested for 400 days. The general conclusion for any type of RCA quality was that RAC beams had larger deflections than NAC beams, and the largest deflections were observed in beams with both fine and coarse RCA.

Choi and Yun [63] used both fine and coarse RCA of good quality from demolished concrete structures in Daejeon, South Korea: coarse RCA had a water absorption of 1.86 % and fine RCA of 3.64 %. Three beams were tested: one NAC beam (C30-0.5ω), one RAC beam with coarse RCA (RL30-0.5ω), and one with fine RCA (RH30-0.5ω). The achieved compressive strengths were 31.61, 39.66, and 36.10 MPa for C30-0.5ω, RL30-

0.5ω and RH30-0.5ω, respectively, whereas the moduli of elasticity were 26.25, 28.48, and 27.53 GPa, respectively. The tested beams had a 170/200 mm cross-section and a 2000 mm span. They were reinforced with a 0.5 % reinforcement ratio including 2 $\varnothing$10 mm bars in the tension zone. Compressive reinforcement consisted of 2 $\varnothing$6 mm bars, and $\varnothing$6 mm stirrups were used, spaced at 100 mm. The beams were loaded to 50 % of their ultimate load capacity for approximately one year (Fig. 5.13).

In terms of results, fewer cracks were observed in the NAC beam and overall smaller deflections were reached. The increase in deflection over time (expressed in terms of percentage of the initial deflection) was 108 %, 98 % and 94 % for C30-0.5ω, RL30-0.5ω and RH30-0.5ω, respectively. Nonetheless, an overall low level of cracking was observed and, therefore, the usefulness of these results for discussing deflections of cracked RAC members is diminished.

One of the largest experimental programmes was performed by Knaack and Kurama [64]: 18 reinforced beams were tested under sustained loads – 6 NAC and 12 RAC beams. Coarse RCA (with a water absorption of 6.06 %) was used to replace 0 %, 50 % and 100 % of NA (NAC, RAC50, RAC100). The compressive strength of all the concretes varied relatively widely with f_{cm} for NAC between 33 and 50 MPa, for RAC50 between 38 and 50 MPa, and for RAC100 between 36 and 49 MPa.

The beams had a 150/230 mm cross-section and a 3700 mm span and were tested in four point bending for 119 days. There were two reinforcement layouts: the UT series with only tensile reinforcement and the UC and CC series with tensile and compression reinforcement as well as stirrups. The tensile reinforcement was the same for all series and consisted of 2 $\varnothing$16 mm Grade 420 bars, while the compression and transverse reinforcement in series UC and CC consisted of 2 $\varnothing$10 mm Grade 420 bars and $\varnothing$10 mm stirrups spaced at 95 mm, respectively. Six beams were cast in each series. The distinction between series was based on cracking: series UT and UC were loaded with an imposed load that was designed not to induce immediate cracking, whereas the CC series was loaded with an imposed load such that there was immediate cracking of the beams.

The authors report that increasing RCA content leads to increasing deflections – the increase was larger for the "uncracked" series (69 % from NAC to RAC100) than for the "cracked" series (21 %), due to the larger role the concrete under creep plays in the uncracked beams.

Seara-Paz et al. [71] also performed a very comprehensive experimental investigation. The authors used coarse RA with water absorption of 5.4 % to produce two series of concretes with w/c ratios of 0.50 (H50) and 0.65 (H65) with 0 %, 20 %, 50 % and 100 % of RA. The achieved compressive strengths ranged from 60 to 43 MPa for H50 concretes (H50-0 to H50-100) and from 47 to 32 MPa for H65 concrete (H65-0 to H65-100).

Within the experimental programme, 8 beams with 200/300 mm cross-sections and 3400 mm spans were tested in four point bending for 1000 days. The beams were reinforced with 2 $\varnothing$16 mm bars in the tension zone (reinforcement ratio 0.8 %), 2 $\varnothing$8 mm bars in the compression zone, and $\varnothing$8 mm stirrups spaced at 180 mm. The applied load was determined according to the concrete compressive strength and modulus of elasticity with an aim of obtaining a maximum compressive stress of $0.4f_{cm}(t_0)$, where t_0 is the concrete age at loading.

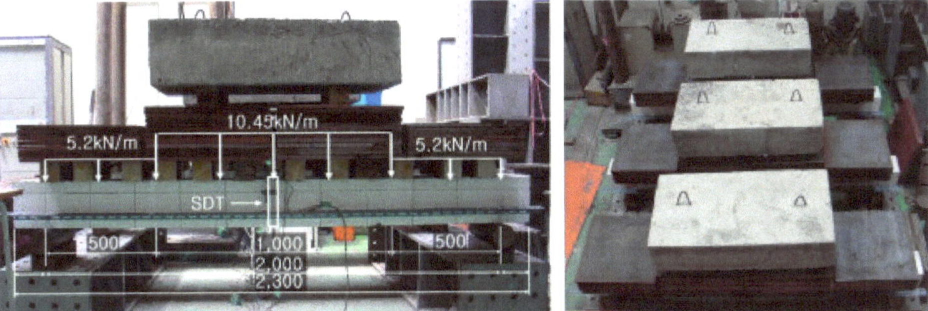

Fig. 5.13 Setup for the sustained loading test used in [63]

In terms of results, the authors discuss "normalized deflections," i.e., deflection at time t divided by the initial deflection, a_t/a_0, and find it increasing with RA content – for NAC, the increase in deflection over 1000 days was 60–70 % of the initial deflection, whereas for RAC100 it was 120–150 %. Nonetheless, it should be kept in mind that the compressive strengths, moduli of elasticity, creep, and shrinkage were all different between the concretes, so separating the material from structural effects cannot be performed easily.

Tošić et al. [43] tested two NAC and two RAC beams with 100 % of coarse RCA with a water absorption of 3.3 %. The beams had a 160/200 mm cross-section and a 3200 mm span, and were tested in four point bending over 450 days. The beams had a reinforcement ratio of 0.58 % with 2 $\varnothing$10 mm bars in tension and 2 $\varnothing$6 mm bars in compression, with $\varnothing$6 mm stirrups. One pair of NAC and RAC beams were loaded at 7 days and the other pair at 28 days. Accordingly, for the beams loaded at 7 days the $\sigma_c(t_0)/f_{cm}(t_0)$ ratio was 0.60 and for the beams loaded at 28 days it was 0.45.

The concretes had similar mechanical properties: compressive strengths of 40.7 and 37.4 MPa for NAC and RAC, respectively, with moduli of elasticity of 32.2 and 30.8 GPa, respectively, and tensile strengths of 2.4 and 2.5 MPa, respectively. In terms of result, the normalized deflections, i.e., the a_t/a_0 ratio (where a_t is the deflection at time t and a_0 is the initial deflection at loading) showed no differences between NAC and RAC beams loaded at 7 days and slightly larger values for RAC loaded at 28 days compared with its reference NAC. Additionally, the RAC beams had smaller crack spacings and crack widths than reference NAC beams.

Zhu et al. [67] tested beams with 0 %, 50 % and 100 % of coarse RCA with water absorption of 4.67 % for 3045 days under four point bending, Fig. 5.15. The beams had a 130/260 mm cross-section and 1800 mm span. The compressive strengths of the concretes were between 31.7 and 34.0 MPa and moduli of elasticity between 28 and 32 GPa.

The loading procedure was such that load was added in 2 kN increments until a principal crack width of 0.2 mm was measured. The results showed a clear increase in deflections (both initial and over time) with increasing RCA content.

Cao et al. [66] recently tested 16 RAC beams produced with both coarse and fine RCA (with water absorptions of 4.70 % and 11.32 %, respectively). The test parameters included RCA content and size, as well as concrete compressive strength and stress-

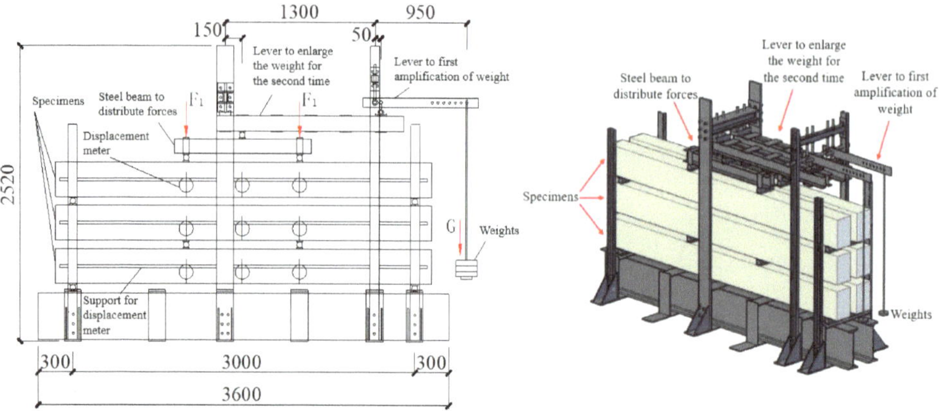

Fig. 5.14   Setup for the sustained loading test used in [66]

strength ratio of the concrete in the top compressed fibre of the tested beams. The beam specimens had a 200/300 mm cross-section and 3000 mm span, with 4 $\varnothing$14 mm bars in tension, 2 $\varnothing$12 mm bars in compression, and $\varnothing$6 mm stirrups. The beams were loaded in a stacked configuration in four point bending (Fig. 5.14) for two years.

The authors found larger deflections of RAC beams compared with reference NAC and a pronounced increase in deflections when fine RCA was used. Additionally, the increase in deflections over time was more pronounced for lower stress/load levels than for high ones. This goes in line with previous findings and can be explained by the fact that at lower stress/load levels, the concrete is less cracked so that the concrete contribution to tension stiffening is more pronounced and a larger part of the cross-section is in compression; in both instances, the higher shrinkage and creep of RAC lead to a larger deformation of RAC beams compared with reference NAC.

Finally, Brandes and Kurama [65] performed the only experimental tests on the long-term deflections of precast/prestressed RAC beams. In total, 18 beams were tested with 150/230 mm cross-sections and 5500 mm span, and with 0 %, 50 % and 100 % of coarse RCA with water absorption 3.4–4.4 %. The beams were prestressed with 2 $\varnothing$12.7 mm strands. In 12 of the beams, the strands were pretensioned to 70 % of their ultimate strength, which was 1862 MPa (series U/uncracked beams), whereas in the remaining 6 beams, the strands were pretensioned up to 50 % of their ultimate strength (series C/cracked beams). The compressive strengths of the beams varied between 25 and 45 MPa. The beams were loaded in four point bending for 80–250 days.

The authors found prestress losses to increase with increasing RCA content, attributing the result to greater shrinkage and creep strains of RAC: for full RCA incorporation, the short- and long-term prestress losses were around 1 % and 2–7 %, respectively, of the initial stress. Deflections also increased with increasing RCA content: the largest increases were found in the C-series beams (cracked under service load) – in this case the deflections of RAC beams at the end of testing were 1.25 and 1.37 times that of the corresponding NAC beam.

5.4.2 Modelling of Deflection Behaviour of Recycled Aggregate Concrete Structures

In terms of models for calculating deflections of RAC structures, both analytical and numerical work has been done so far. With respect to analytical models for deflections of RAC members, Tošić et al. [72] proposed a modification to the *fib* Model Code 2010 ζ-method of deflection or curvature interpolation between the cracked and uncracked states [6].

The authors considered the general form of the ζ-method based on curvatures (where curvatures can be changed directly for deflections a):

$$\left(\frac{1}{r}\right)_{eff} = \zeta \cdot \left(\frac{1}{r}\right)_2 + (1 - \zeta) \cdot \left(\frac{1}{r}\right)_1 \tag{5.10}$$

where $(1/r)_{\text{eff}}$ is the effective/interpolated curvature, while $(1/r)_1$ and $(1/r)_2$ are curvatures in the uncracked and fully cracked states, respectively, and ζ is a distribution coefficient taking into account tension stiffening:

$$\zeta = \begin{cases} 1 - \beta \cdot \left(\dfrac{M_{cr}}{M}\right)^2 & \text{for } M \geq \sqrt{\beta} \cdot M_{cr} \\ 0 & \text{for } M < \sqrt{\beta} \cdot M_{cr} \end{cases} \tag{5.11}$$

where β is a coefficient accounting for the influence of the duration of loading or repeated loading.

$$\beta = 1.0 \quad \textit{for single, short – term loading}$$
$$\beta = 0.5 \quad \textit{for sustained or repeated loading} \tag{5.12}$$

The authors calibrated the model against experimental results from three studies [43, 64, 71] which comprised a database of 10 NAC and 15 RAC beams providing 20 and 30 deflections, respectively (initial and "final reported", time-dependent deflection). The range of parameter values in the database is presented in Table 5.3.

Table 5.3 Range of parameters in the NAC and RAC beam database [72]

Beams	RCA (%)	b (mm)	d (mm)	f_{cm} (MPa)	L (mm)	L/d (-)	ρ_1 (%)	t_0 (d)	t (d)	M_{sus}/M_{cr} (-)
NAC	0	150–	169–	30.5–60.7	3200–	13.7–	0.58–	7–42	119–	0.81–3.35
RAC	50, 100	200	249	28.1–51.8	3700	18.9	1.32		1000	0.68–2.52

The authors calculated the deflections for the beams in the database using *fib* Model Code 2010 predictions for all properties based on compressive strength and detected an underestimation of RAC deflections by the model, relative to NAC beams, especially for time-dependent deflections. Therefore, the authors attempted to identify whether the differences originated from the misestimation of the RAC modulus of elasticity, shrinkage, and creep (i.e., only from the material level) or also from a difference in tension stiffening (structural level). After implementing the changes for creep and shrinkage predictions based on Equations (5.8) and (5.2), respectively, and a reduction of the modulus of elasticity by $(1 - 0.3 RCA\%/100)$, the authors still detected an underestimation of RAC deflections. Therefore, it was concluded that changes to tension stiffening modelling were also necessary. Hence, the following change to the β coefficient was proposed:

$$\beta_{0,\,RAC} = 0.75 \quad \textit{for single, short – term loading}$$
$$\beta_{t,\,RAC} = 0.25 \quad \textit{for sustained or repeated loading} \tag{5.13}$$

Subsequently, within the preparation of the new Eurocode 2-revision [11], considering creep and shrinkage changes according to Equations (5.9) and (5.3), respectively, and a modification of the modulus of elasticity by $(1 - 0.25\alpha_{RA})$, the following change to β was proposed [12]:

$$\beta = 1.0 \quad \textit{for single, short – term loading}$$
$$\beta_{t,\,RAC} = 0.25 \quad \textit{for sustained or repeated loading} \tag{5.14}$$

Knaack and Kurama [64] and Brandes and Kurama [65] calculated the deflections of reinforced and prestressed RAC beams, respectively, predicting their own experimental results based on ACI 318 Building Code Requirements [73]. For the case of reinforced RAC and NAC beams, the authors found the code to underestimate time-dependent deflections but with a similar precision for both RAC and NAC; similarly, in the case of prestressed RAC beams the design predictions were not significantly affected by RCA incorporation.

Knaack and Kurama [74] developed a time-dependent concrete material model for the open-source structural analysis framework of OpenSees [75], TDConcrete. The authors formulated a constitutive model for concrete linear in compression accounting for shrinkage and creep according to the ACI 209R-92 model [9] that could simulate the service load deflection behaviour of structural members using a discretization into layers and a nonlinear analysis. Using the developed model, the authors performed a successful simulation of their own experimental results as well as a Monte Carlo parametric study to determine the reliability of the model validation [74]. Subsequently, the same authors studied the effect of RAC on the service-load deflections of reinforced concrete columns in frame structures [76]. Through a parametric numerical study using the TDConcrete model, the authors investigated the time-dependent service-load deflections of RAC columns (interior and exterior) of different frame structures. The authors found that increasing coarse RCA content led to larger deflections, as did higher water absorption of RCA and a higher content of deleterious materials (used as inputs for modifications of creep and shrinkage). At the same time, deflections of columns with RAC made from high-quality coarse RCA (with low water absorption and deleterious material content) did not differ significantly from those of NAC columns.

Finally, Tošić and Kurama [77] further developed the TDConcrete model into a version that incorporated the *fib* Model Code 2010 shrinkage and creep models [6] and a nonlinear concrete behaviour in both compression and tension, TDConcreteMC10NL. The constitutive models were fully integrated into the OpenSees framework [75, 78, 79]. Using the material model, the authors performed a parametric numerical study of NAC and RAC one-way members: simply supported and continuous one-way slabs and simply supported T-section beams. The authors varied the coarse RCA content in RAC (0 %, 25 % and 50 %), concrete strength class (C25/30 and C40/50), quasi-permanent-to-design load (0.52 and 0.64), relative humidity (50 % and 70 %), and height (200 and 300 mm for simply supported and continuous slabs, and 500 and 700 mm for simply supported T-section beams). The authors then analysed different L/d ratios and deflections in terms of the a/a_{lim} ratio (where a_{lim} is the limiting deflection = $L/250$). The authors produced a series of graphs (an example for a simply supported slab with h = 200 mm is given in Fig. 5.15, and for simply supported T-section beams with h = 500 mm in Fig. 5.16).

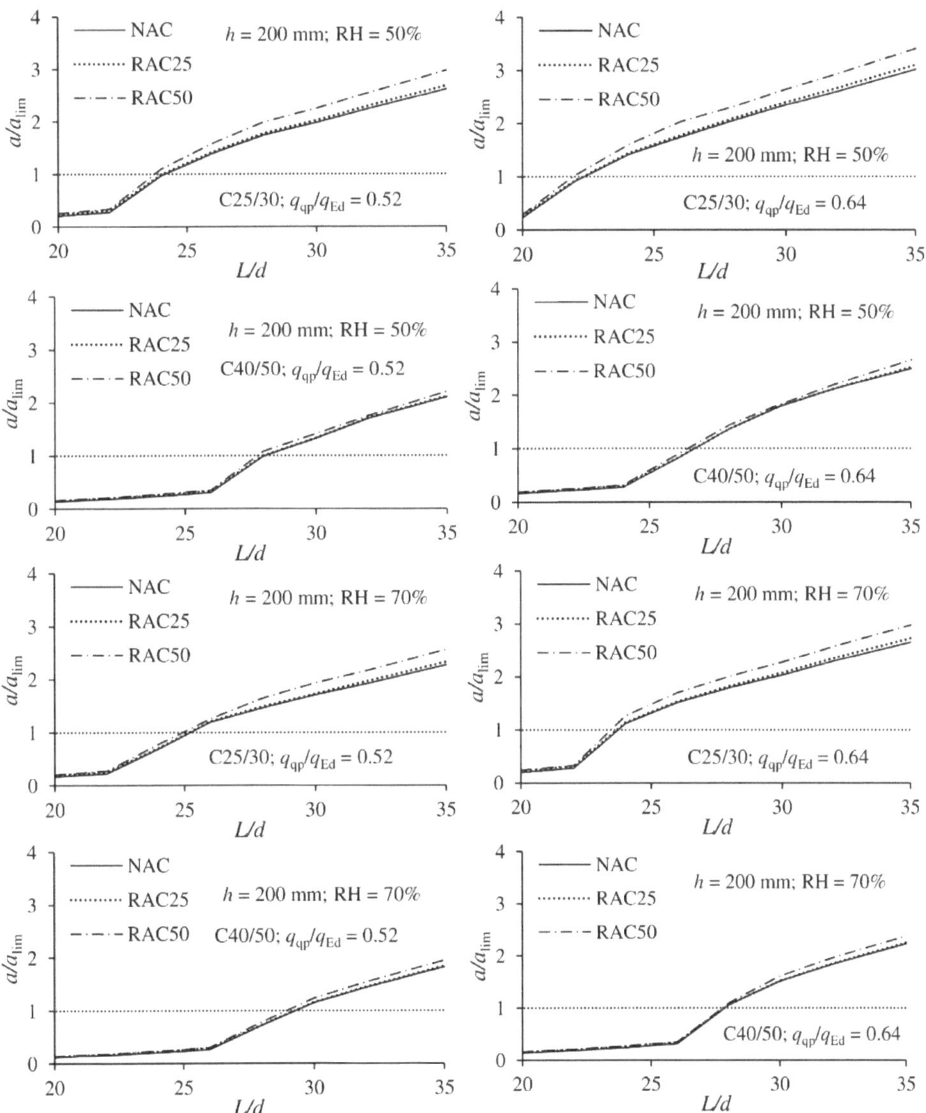

Fig. 5.15 Relationship between a/a_{lim} and L/d ratios for a simply supported one-way slab with h = 200 mm [77]

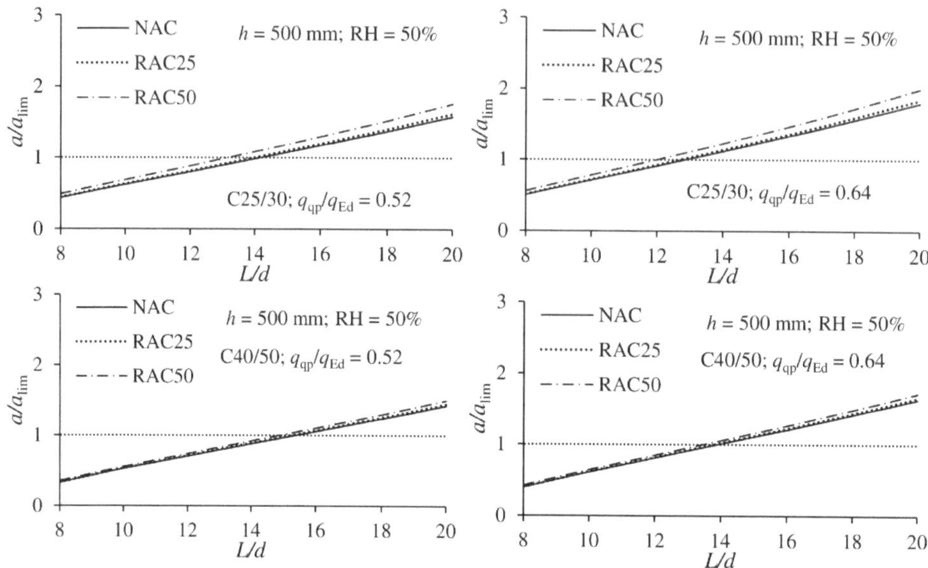

Fig. 5.16   Relationship between a/a_{lim} and L/d ratios for a simply supported T-beam with h = 500 mm [77]

The results enabled the authors to identify the effect of different parameters, in particular, revealing that for RAC with 25 % of coarse RCA (when changes in tension stiffening can be neglected [77]), there are practically no differences between NAC and RAC members, i.e. RAC25 can be used for the same L/d ratios as NAC. When RAC50 is considered, differences can be observed but they are relatively minor for higher humidity and higher compressive strength. In the end, it can be seen that RAC50 one-way slabs can be used for only slightly lower L/d ratios than NAC slabs and, furthermore, their deflection behaviour can be improved by additional tension and compression reinforcement.

In terms of simply supported T-section beams, the differences between NAC and RAC50 are even smaller and not significant in the range of typical L/d ratios used for such members [77].

5.5 Conclusions

In this chapter, a review of long-term deformation properties of structural RAC was presented. Shrinkage and creep of RAC and deflections of RAC structural members were discussed by presenting a critical review and summary of the available literature. Based on the presented information, the following conclusions can be drawn:

- The shrinkage of RAC generally increases with increasing RA content and decreases with increasing RAC compressive strength. On average, with 100 % of coarse RCA, the shrinkage of RAC increases by approximately 35 % relative to a reference NAC (within a range of 10–60 %). The increase is larger for RAC with fine RCA as well as for RAC with RMA or MRA;

- The differences between NAC and RAC shrinkage exist for both autogenous and drying shrinkage but there are not enough experimental results available to fully distinguish the effect of RA on each type of shrinkage. The change in shrinkage seems to be largely a direct scaling in magnitude, with no significant changes in its time evolution;
- Analytical models have been proposed and integrated into the new Eurocode 2-revision allowing for a simple prediction of RAC shrinkage;
- The creep of RAC generally increases with increasing RA content and decreases with increasing RAC compressive strength. On average, with 100 % of coarse RCA, the creep of RAC increases by approximately 30 % relative to a reference NAC (within a range of 10–50 %). The increase is larger for RAC with fine RCA as well as for RAC with RMA or MRA;
- The differences between NAC and RAC creep exist for both basic and drying creep, but there are not enough experimental results available to fully distinguish the effect of RA on each type of creep. The change in creep seems to be largely a direct scaling in magnitude, with no significant changes in its time evolution;
- Analytical models have been proposed and integrated into the new Eurocode 2-revision allowing for a simple prediction of RAC creep;
- Experimental results for deflections of RAC beams show a larger deformability compared with reference NAC beams. The increase in deflections is attributable to a lower modulus of elasticity and higher shrinkage and creep of RAC, as well as a weaker tension stiffening effect;
- Analytical and numerical models have been developed and integrated into the new Eurocode 2-revision and the open-source OpenSees structural analysis framework. They allow for modelling and predicting the deflection behaviour of RAC structures.

The contents of this chapter should serve only as a general introduction to the topic and the interested reader is directed to the cited references for further and more detailed information. Overall, while complex, the topic of long-term deformations of RAC and RAC structures is sufficiently well understood to enable successful modelling and predictions. Nonetheless, research should continue in this direction in order to provide more experimental results for model validation and refinement, as well as to facilitate increased practical application of structural RAC.

5.6 Acknowledgements

The authors acknowledge Prof. Jorge de Brito for the part played as Editor of this SED in providing this opportunity as well as helpful guidance in preparing this document. Permissions to use copyright images in the figures are gratefully acknowledged.

5.7 References

[1] FIB Bulletin 52, Structural Concrete – Textbook on Behaviour, Design and Performance, International Federation for Structural Concrete (fib), Lausanne, 2010. https://doi.org/10.1017/cbo9781107415324.004

[2] A.W. Beeby, S. Narayanan, Designer's guide to Eurocode 2: Design of concrete structures, Thomas Telford, London, 2005.

[3] N. Pecić, Improved method for deflection control of reinforced concrete structures, Faculty of Civil Engineering, Belgrade University, 2012.

[4] CEB-FIP, Model Code 1978, Comite Euro-International du Béton, Paris, 1978.

[5] CEB-FIP, Model Code 1990, Comite Euro-International du Béton, Paris, 1991. https://doi.org/10.1680/ceb-fipmc1990.35430.

[6] FIB, fib Model Code for Concrete Structures 2010, International Federation for Structural Concrete (fib), Lausanne, 2013. https://doi.org/10.1002/9783433604090

[7] ACI 318-14, Building code requirements for structural concrete (ACI 318-14) and commentary, Farmington Hills, MI, 2014.

[8] ACI 318-19, Building code requirements for structural concrete (ACI 318-19) and commentary, Farmington Hills, MI, 2019.

[9] ACI 209R-92, Prediction of Creep, Shrinkage, and Temperature Effects in Concrete Structures, American Concrete Institute, Farmington Hills, MI, 1992.

[10] ACI 435R-95, Control of deflection in concrete Structures, American Concrete Institute, Farmington Hills, MI, 2003.

[11] prEN1992-1-1, Eurocode 2: Design of concrete structures – Part 1-1: General rules, rules for buildings, bridges and civil engineering structures, CEN, Brussels, 2021.

[12] N. Tošić, J.M. Torrenti, T. Sedran, I. Ignjatović, Toward a codified design of recycled aggregate concrete structures : Background for the new fib Model Code 2020 and Eurocode 2, Structal Concrete (2020) 1–23. https://doi.org/10.1002/suco.202000512.

[13] V.W.Y. Tam, M. Soomro, A.C.J. Evangelista, A review of recycled aggregate in concrete applications (2000–2017), Construction and Building Materials, 172 (2018) 272–292. https://doi.org/10.1016/j.conbuildmat.2018.03.240

[14] J.-L. Gálvez-Martos, D. Styles, H. Schoenberger, B. Zeschmar-Lahl, Construction and demolition waste best management practice in Europe, Resources, Conservation and Recycling 136 (2018) 166–178. https://doi.org/https://doi.org/10.1016/j.resconrec.2018.04.016

[15] H. Le Chatelier, Experimental researches on the constitution of hydraulic mortars, McGraw Publishing Company, New York, 1887.

[16] Z.P. Bažant, A.B. Hauggaard, S. Baweja, F. Ulm, Microprestress-solidification theory for concrete creep. I: Aging and drying effects, Journal of Engineering Mechechanics 123 (1997) 1188–1194. https://doi.org/10.1061/(asce)0733-9399(1997)123:11(1188)

[17] E. Tazawa, Autogenous shrinkage of concrete, E & FN Spon, London, 1999.

[18] M.H. Hubler, R. Wendner, Z.P. Bažant, Comprehensive database for concrete creep and shrinkage: Analysis and recommendations for testing and recording, ACI Materials Journal 112 (2015) 547–558. https://doi.org/10.14359/51687453

[19] Z.P. Bažant, M.H. Hubler, R. Wendner, RILEM draft recommendation: TC-242-MDC multi-decade creep and shrinkage of concrete: material model and structural analysis, Materials and Structures, 48 (2015) 753–770. https://doi.org/10.1617/s11527-014-0485-2

[20] C.Q.C.-Q. Lye, R.K.R.K. Dhir, G.S.G.S. Ghataora, R.K.R.K. Dhir, Shrinkage of recycled aggregate concrete, in: Proceedings of the Institution of Civil Engineers – Structures and Buildings, ICE, 2016: pp. 1–25. https://doi.org/10.1680/jstbu.15.00138

[21] R.V. V. Silva, J. De Brito, R.K.K. Dhir, Prediction of the shrinkage behavior of recycled aggregate concrete: A review, Construction and Building Materials, 77 (2015) 327–339. https://doi.org/10.1016/j.conbuildmat.2014.12.102

[22] N. Tošić, A. de la Fuente, S. Marinković, Shrinkage of recycled aggregate concrete: experimental database and application of fib Model Code 2010, Materials and Structures, 51 (2018) 126. https://doi.org/10.1617/s11527-018-1258-0

[23] Y. Mao, J. Liu, C. Shi, Autogenous shrinkage and drying shrinkage of recycled aggregate concrete: A review, Journal of Cleaner Production, 295 (2021). https://doi.org/10.1016/j.jclepro.2021.126435

[24] G. Fathifazl, A. Ghani Razaqpur, O. Burkan Isgor, A. Abbas, B. Fournier, S. Foo, Creep and drying shrinkage characteristics of concrete produced with coarse recycled concrete aggregate, Cement and Concrete Composites, 33 (2011) 1026–1037. https://doi.org/10.1016/j.cemconcomp.2011.08.004

[25] V. Corinaldesi, G. Moriconi, Recycling of wastes from building demolition in low-shrinkage concretes, in: Creep, Shrinkage Durab. Mech. Concr. Concr. Struct. – Proc. 8th Int. Conf. Creep, Shrinkage Durab. Mech. Concr. Concr. Struct., 2009: pp. 811–815.

[26] A. Adessina, Caractérisation expérimentale et modélisation multi-échelle des propriétés mécaniques et de durabilité des bétons à base de granulats recyclés, Université Paris-Est, 2018.

[27] S.Y. Abate, K. Il Song, J.K. Song, B.Y. Lee, H.K. Kim, Internal curing effect of raw and carbonated recycled aggregate on the properties of high-strength slag-cement mortar, Construction and Building Materials, 165 (2018) 64–71. https://doi.org/10.1016/j.conbuildmat.2018.01.035

[28] A. Gonzalez-Corominas, M. Etxeberria, Effects of using recycled concrete aggregates on the shrinkage of high performance concrete, Construction and Building Materials, 115 (2016) 32–41. https://doi.org/10.1016/j.conbuildmat.2016.04.031

[29] Q. Wang, Y. Geng, Y. Wang, H. Zhang, Drying shrinkage model for recycled aggregate concrete accounting for the influence of parent concrete, Engineering Structures, 202 (2020) 109888. https://doi.org/10.1016/j.engstruct.2019.109888

[30] J.M. Gómez-Soberón, Shrinkage of concrete with replacement of aggregate with recycled concrete aggregate, ACI Symposium Publication, 209 (2002) 475–496. https://doi.org/10.14359/12516

[31] W. Zhang, S. Wang, P. Zhao, L. Lu, X. Cheng, Effect of the optimized triple mixing method on the ITZ microstructure and performance of recycled aggregate concrete, Construction and Building Materials, 203 (2019) 601–607. https://doi.org/10.1016/j.conbuildmat.2019.01.071

[32] H. Zhang, Y.-Y. Wang, D.E. Lehman, Y. Geng, Autogenous-shrinkage model for concrete with coarse and fine recycled aggregate, Cement and Concrete Composites, 111 (2020). https://doi.org/10.1016/j.cemconcomp.2020.103600

[33] M.H. Hubler, R. Wendner, Z.P. Bažant, Statistical justification of Model B4 for drying and autogenous shrinkage of concrete and comparisons to other models, Materials and Structures, 48 (2015) 797–814. https://doi.org/10.1617/s11527-014-0516-z

[34] Y. Fan, J. Xiao, V.W.Y.V.W.Y. Tam, Effect of old attached mortar on the creep of recycled aggregate concrete, Structural Concrete, 15 (2014) 169–178. https://doi.org/10.1002/suco.201300055

[35] Y. Guo, X. Wang, J. Qian, Physical model of drying shrinkage of recycled aggregate concrete, Journal of Wuhan University of Technology-Materials Science Edition, 30 (2015) 1260–1267. https://doi.org/10.1007/s11595-015-1305-4

[36] Z. Lv, C. Liu, C. Zhu, G. Bai, H. Qi, Experimental study on a prediction model of the shrinkage and creep of recycled aggregate concrete, Applied Sciences, 9 (2019) 1–11. https://doi.org/10.3390/app9204322

[37] Q. Wang, Y.-Y.Y.Y. Wang, Y. Geng, H. Zhang, Experimental study and prediction model for autogenous shrinkage of recycled aggregate concrete with recycled coarse aggregate, Construction and Building Materials, 268 (2021) 121197. https://doi.org/10.1016/j.conbuildmat.2020.121197

[38] Y. Guo, X. Wang, Experimental study on forecasting mathematical model of drying shrinkage of recycled aggregate concrete, Mathematical Problems in Engineering, 2012 (2012). https://doi.org/10.1155/2012/567812

[39] A.M. Knaack, Y.C. Kurama, Design of concrete mixtures with recycled concrete aggregates, ACI Materials Journal (2013). https://doi.org/10.14359/51685899

[40] S. Seara-Paz, B. González-Fonteboa, F. Martínez-Abella, I. González-Taboada, Time-dependent behaviour of structural concrete made with recycled coarse aggregates. Creep and shrinkage, Construction and Building Materials, 122 (2016) 95–109. https://doi.org/10.1016/j.conbuildmat.2016.06.050

[41] EN 1992-1-1, Eurocode 2: Design of concrete structures – Part 1-1: General rules and rules for buildings, CEN, Brussels, 2004.

[42] Z.P. Bažant, S. Baweja, Creep and shrinkage prediction model for analysis and design of concrete structures – model B3, Materials and Structures, 28 (1995) 357–365. https://doi.org/10.1007/bf02473152

[43] N. Tošić, S. Marinković, N. Pecić, I. Ignjatović, J. Dragaš, Long-term behaviour of reinforced beams made with natural or recycled aggregate concrete and high-volume fly ash concrete, Construction and Building Materials, 176 (2018) 344–358. https://doi.org/10.1016/j.conbuildmat.2018.05.002

[44] S. Matthews, A. Bigaj-van Vliet, J. Walraven, G. Mancini, G. Dieteren, fib Model Code 2020: Towards a general code for both new and existing concrete structures, Structural Concrete, 19 (2018) 969–979. https://doi.org/10.1002/suco.201700198

[45] W.K. Hatt, Notes on the effect of time element in loading reinforced concrete beams, in: ASTM Proceedings, 1907: pp. 421–433.

[46] T.C. Powers, The thermodynamics of volume change and creep, Materials and Structures, 1 (1968) 487–507. https://doi.org/10.1007/bf02473638

[47] M. Vandamme, F.-J. Ulm, Nanogranular origin of concrete creep, Proceedings of the National Academy of Sciences (PNAS), 106 (2009) 10552–10557. https://doi.org/10.1073/pnas.0901033106

[48] C.Q. Lye, R.K. Dhir, G.S. Ghataora, H. Li, Creep strain of recycled aggregate concrete, Construction and Building Materials, 102 (2016) 244–259. https://doi.org/10.1016/j.conbuildmat.2015.10.181

[49] R.V. Silva, J. de Brito, R.K. Dhir, Comparative analysis of existing prediction models on the creep behaviour of recycled aggregate concrete, Engeering Structures, 100 (2015) 31–42. https://doi.org/10.1016/j.engstruct.2015.06.004

[50] N. Tošić, A. de la Fuente, S. Marinković, Creep of recycled aggregate concrete: Experimental database and creep prediction model according to the fib Model Code 2010, Construction and Building Materials, 195 (2019) 590–599. https://doi.org/10.1016/j.conbuildmat.2018.11.048

[51] J.M. Goméz-Soberón, Creep of concrete with substitution of normal aggregate by recycled concrete aggregate, ACI Symposium Publication SP209-25. (2002) 461–474. https://doi.org/10.14359/12515 or http://upcommons.upc.edu/handle/2117/2562

[52] V.W.Y. Tam, X.F. Gao, C.M. Tam, Microstructural analysis of recycled aggregate concrete produced from two-stage mixing approach, Cement and Concrete Research, 35 (2005) 1195–1203. https://doi.org/10.1016/j.cemconres.2004.10.025

[53] G. Fathifazl, G. Razaqpur, Creep rheological models for recycled aggregate concrete, ACI Materials Journal, 2 (2013) 115–125. https://doi.org/10.14359/51685526

[54] Y. Geng, M. Zhao, H. Yang, Y. Wang, Creep model of concrete with recycled coarse and fine aggregates that accounts for creep development trend difference between recycled and natural aggregate concrete, Cement and Concrete Composites, 103 (2019) 303–317. https://doi.org/10.1016/j.cemconcomp.2019.05.013

[55] P.P. Torres, E. Ghorbel, G. Wardeh, Towards a New Analytical Creep Model for Cement-Based Concrete Using Design Standards Approach, Buildings. 11 (2021) 155. https://doi.org/10.3390/buildings11040155

[56] G.W. Washa, P.G. Fluck, Effect of compressive reinforcement on the plastic flow of reinforced concrete beams, ACI Journal Proceedings, (1952) 89–108. https://doi.org/10.14359/11806

[57] B. Espion, Long-Term Sustained Loading Test on Reinforced Concrete Beams, Université Libre de Bruxelles Service Génie Civil, Brussels, 1988.

[58] R.I. Gilbert, S. Nejadi, An experimental study of flexural cracking in reinforced concrete members under sustained loads, Kensington, 2004.

[59] J.-P. Jaccoud, R. Favre, Flèche des structures en béton armé: Vérification experimentale d'une méthode de calcul, Lausanne, 1982.

[60] D.E. Branson, G.A. Metz, Instantaneous and time-dependent deflections of simple and continuous reinforced concrete beams – Alabama Highway Research Report, 1963.

[61] I. Maruyama, Y. Oka, R. Sato, Time-dependent Behavior of Reinforced Recycled Concrete Beams, in: CONCREEP 7, 2005: pp. 1–6.

[62] A. Ajdukiewicz, A. Kliszczewicz, Long-term behaviour of reinforced-concrete beams and columns made of recycled aggregate, in: fib Symposium Concrete Engineering for Excellence and Efficiency. June 8–10, 2011: pp. 479–482.

[63] W.-C. Choi, H.-D. Yun, Long-term deflection and flexural behavior of reinforced concrete beams with recycled aggregate, Materials & Design, 51 (2013) 742–750. https://doi.org/10.1016/j.matdes.2013.04.044

[64] A.M. Knaack, Y.C. Kurama, Sustained service load behavior of concrete beams with recycled concrete aggregates, ACI Structural Journal, 112 (2015) 565–578. https://doi.org/10.14359/51687799

[65] M.R. Brandes, Y.C. Kurama, Service-load behavior of precast/prestressed concrete beams with recycled concrete aggregates, ACI Structural Journal, 115 (2018) 861–873. https://doi.org/10.14359/51702133

[66] W. Cao, Y. Liu, Q. Qiao, Y. Feng, S. Peng, Time-dependent behavior of full-scale recycled aggregate concrete beams under long-term loading, Materials (Basel), 13 (2020) 1–22. https://doi.org/10.3390/ma13214862

[67] C. Zhu, C. Liu, G. Bai, J. Fan, Study on long-term performance and flexural stiffness of recycled aggregate concrete beams, Construction and Building Materials, 262 (2020) 120503. https://doi.org/10.1016/j.conbuildmat.2020.120503

[68] A. Łapko, R. Grygo, Long term deformations of recycled aggregate concrete (RAC) beams made of recycled concrete, in: Modern Building Materials, Structures and Techniques, Vilnius Gediminas Technical University, Vilnius, 2010: pp. 709–712.

[69] C. Liu, Z. Lv, C. Zhu, G. Bai, Y. Zhang, Study on calculation method of long term deformation of RAC beam based on creep adjustment coefficient, KSCE Journal of Civil Engineering, (2018) 1–8. https://doi.org/10.1007/s12205-018-0131-6

[70] N. Tošić, S. Marinković, I. Ignjatović, A database on flexural and shear strength of reinforced recycled aggregate concrete beams and comparison to Eurocode 2 predictions, Construction and Building Materials, 127 (2016) 932–944. https://doi.org/10.1016/j.conbuildmat.2016.10.058.

[71] S. Seara-Paz, B. González-Fonteboa, F. Martínez-Abella, D. Carro-Lopez, D. Carro-López, Long-term flexural performance of reinforced concrete beams with recycled coarse aggregates, Construction and Building Materials, 176 (2018) 593–607. https://doi.org/10.1016/j.conbuildmat.2018.05.069

[72] N. Tošić, S. Marinković, J. de Brito, Deflection control for reinforced recycled aggregate concrete beams: Experimental database and extension of the fib Model Code 2010 model, Structural Concrete, 20 (2019) 2015–2029. https://doi.org/10.1002/suco.201900035

[73] ACI 318-11, Building code requirements for structural concrete (ACI 318-11) and commentary, American Concrete Institute, Farmington Hills, MI, 2011.

[74] A.M.A.M. Knaack, Y.C.Y.C. Kurama, Modeling time-dependent deformations: Application for reinforced concrete beams with recycled concrete aggregates, ACI Structural Journal, 115 (2018) 175–190. https://doi.org/10.14359/51701153

[75] F. McKenna, OpenSees: A framework for earthquake engineering simulation, Computing in Science & Engineering, 13(4) (2011). https://doi.org/10.1109/mcse.2011.66

[76] A.M. Knaack, Y.C. Kurama, Effect of recycled concrete coarse aggregates on service-load deflections of reinforced concrete columns, Engineering Structures, 204 (2020) 109955. https://doi.org/10.1016/j.engstruct.2019.109955

[77] N. Tošić, Y. Kurama, Parametric numerical study on service-load deflections of reinforced recycled aggregate concrete slabs and beams based on fib Model Code 2010, Structural Concrete, 21 (2020) 2854–2868. https://doi.org/10.1002/suco.202000015

[78] N. Tošić, A. Knaack, Y. Kurama, Source code for OpenSees time-dependent concrete material models, GitHub. (2019). https://github.com/ntosic87/OpenSees (accessed December 23, 2019).

[79] N. Tošić, A. Knaack, Y. Kurama, Supporting documentation for time-dependent concrete material models in OpenSees, Mendeley Data, V3. (2019). https://doi.org/10.17632/z4gxnhchky.3

Chapter 6

Durability of Recycled Aggregate Concrete

Miren Etxeberria

Department of Civil and Environmental Engineering, Universitat Politècnica de Catalunya. Barcelona TECH, Spain.

6.1 Introduction

Recycled aggregates (RA) are now well-known to be suitable for structural concrete production [1]. The employment of RA (obtained from concrete or ceramic waste) in recycled aggregate concrete (RAC) production as a structural material has been widely analysed and validated in many applications [2–4]. However, extensive and precise knowledge of its long-term performance under extreme environmental conditions is essential for this material to be accepted and adopted in real-world applications. Durability is a crucial parameter for the structural applications of RAC. The relationship between the properties of RA (and its content used in concrete production) and the durability of RAC has often been discussed. In this chapter, the durability of RAC is evaluated through various properties such as carbonation resistance, chloride ion penetration, permeability, freeze-thaw resistance, alkali-silica reaction, and sulphate attack. The type of RA (concerning its origin and size) and the amount employed have been considered to validate the RAC's durability for its structural application.

6.2 Carbonation Resistance

Carbonation is responsible for lowering the concrete pH, which in turn initiates corrosion in reinforced concrete. The carbonation rate depends on the permeability and moisture content of concrete, the environment's CO_2 content, and relative humidity. The highest carbonation rate occurs at 50–70 % relative humidity in ambient air; however, the carbonation rate is minimal when concrete is very dry or nearly saturated. Because of carbonation, the pH of the pore solution is reduced from 13.0–13.5 to below 10. If the carbonation depth is the same as that of the cover's depth, the passive layer on the steel surface (of high pH) will dissolve, and the corrosion of reinforcement will start [5]. The carbonation resistance of RAC depends on many factors such as recycled concrete aggregate (RCA) replacement ratio, cement content, quality and crushing process of RCA, and properties of the parent concrete of RCA.

According to Le and Bui [6], in most cases the carbonation depth of RAC is reported to be higher than natural aggregate concrete (NAC), and it increases with the percentage rise of RCA in use [7–9]. Guo et al. [10], Adessina et al. [11], and Silva et al. [9], among other researchers, also stated that RAC produced with a high percentage of coarse RCA achieved higher carbonation depth than NAC. Zeng [12] stated that the best replacement percentage of natural aggregates (NA) with RCA was 50 %, in order to prevent a decrease in carbonation resistance. Pedro et al. [13] documented that when the quality of the cement paste in the new concrete was found to be considerably better than that adhered to the RCA, an increase of the carbonation coefficient with increasing levels of NA replacement was inevitable. In addition, it was found that RAC employing 100 % of fine RCA achieved a 8.7 times higher carbonation depth than that of NAC [8, 14].

However, some researchers [15, 16] have stated that the RAC mixes (produced with coarse RCA) had similar or higher resistance to carbonation than NAC due to the old adhered mortar. Etxeberria [17] also achieved a similar conclusion when 50 % of uncarbonated RCA were employed. According to Leemann and Loser [18], there was no systematic difference in the carbonation coefficient independent of the replacement levels of RCA. However, RCA particles that are either porous or already carbonated at the time of concrete production can lead to a local increase of carbonation depth. Nevertheless, the impact on the carbonation coefficient is not significant, as there is only an increase of 10 % at a given compressive strength. Levy and Helene [19] suggest that carbonation depth depends strongly on the chemical composition of concrete and not merely on its physical aspects. They found that the carbonation depth of concrete produced with coarse RCA was lower when compared with the reference concrete due to the increase of alkalinity of the mix. In addition, fine RCA can contain calcium hydroxide, which may increase the material's alkaline reserve. The newly formed phases from the hydration of unhydrated cement (if any) in the fine RCA could cause an increase in carbonation resistance [20].

It is known that concrete (NAC and RAC) produced using fly ash (FA) as a binder have a higher carbonation depth than the concretes produced using ordinary Portland cement [14]. Moreover, according to Tam et al. [14], the addition of blast furnace slag did not increase the carbonation depth of RAC. Concrete made with a combination of FA (up to 50 %) and 50 % RCA (uncarbonated) (CRAFA) using a w/c ratio of 0.50 could improve carbonation resistance and eco-efficiency required for new sustainable concrete in comparison to that of concrete produced with FA and NA [17]. This fact is probably due to the effect of fly ash-based RAC, which can be emphasized in terms of higher consumption of Calcium Hydroxide (CH) and greater extent of secondary reactions [21]. Faella et al. [22] also determined that adding FA increases the carbonation rate in concrete, with a higher increase in concrete produced with NA than that with RCA. Faella et al. [22] and Kurda et al. [23] determined that a considerable carbonation rise occurred when 100 % of RCA was used with 50 % of FA. However, Sim and Park [24] concluded that the RAC produced using FA, with 100 % of coarse RCA and increasing the percentages of fine RCA decreased carbonation in samples. It was claimed that this resulted from the increased amount of $Ca(OH)_2$ in the mix from the fine RCA's adhered mortar. A similar conclusion was obtained by Levy and Helene [19].

Corinaldesi and Moriconi [25] concluded that the use of FA binders in RAC increased the carbonation rate. However, in their study, the CRAFA mixes were prepared with a lower water/binder ratio [25]; as a result of the refinement of the pore system, carbonation did not present risks of reinforcement corrosion. The use of FA in concrete production improves the internal pore structure of concrete [5, 26]. Moreover, Tian et al. [27] concluded that the most important influence on the depth of carbonation of the CRAFA was that of the water/binder ratio, signifying that the carbonation age and the FA content had a minor effect. However, Xiao et al. [16] found that, although the use of FA in RAC production improved the internal pore structure by decreasing the porosity of the concrete, it also resulted in the reduction of the total alkaline content that can be carbonated, resulting in greater carbonation depths.

6.3 Chloride Ion Penetration

The penetration of chloride ions with or without carbonation is a principal cause of steel reinforcement corrosion. Of the many variables that control chloride ingress in concrete, porosity is the main factor and is primarily defined by the water-cement ratio, degree of hydration and the coarse-to-fine aggregate ratio [28].

Research has shown that RAC generally possesses a lower chloride ion diffusion resistance than NAC when using ordinary Portland cement (OPC) [10, 11, 15, 29]. There is also an apparent increase in the migration coefficient with an increasing replacement ratio (up to 100%) of the use of fine RCA due to the resulting porosity increase [8].

Silva et al. [30] carried out a systematic literature review of a significant number of studies on the effect of ion penetration resistance in concrete using RA of different types, sizes, and qualities. They concluded that the chloride ion penetration tends to increase with the increasing replacement ratio of coarse RCA. However, they also concluded that the increased penetration could be mitigated if supplementary cementitious materials are employed or if the w/c ratio is reduced.

The chloride-ion penetration is found to be significantly attenuated by adding FA in concrete produced with NA [26, 31, 32], and a similar effect happens to RAC mixes [22, 25]. According to Faella et al. [22], RAC samples presented a higher chloride ion penetration than the NAC when both were produced with the same amount of FA binder. However, the RAC showed adequate resistance to chloride ions for structural concrete use. In addition, when more FA was employed, the difference between RAC and NAC behaviour was lower. Besides, Etxeberria and Castillo [33] described that the RAC could achieve a higher chloride ion resistance than the NAC when the RCA is produced from a parent concrete made using FA. Kurda et al. [23], Kim et al. [34], Kou and Poon [35], among other researchers, determined that the total charge passed through concrete (measured in Coulombs) rose with increasing levels of coarse RCA replacement and the opposite occurred when increasing the incorporation levels of FA. Sim and Park [24] evaluated chloride ion penetration in fine RAC with 15 % and 30 % of FA and determined that the use of FA improved their chloride resistance. They concluded that RAC could achieve sufficient chloride ion penetration resistance for structural concrete application. Kou and Poon [36] also found that as the fine RCA content increased, the resistance to chloride-ion penetration increased when that concrete was produced with a high amount of FA. It was concluded that the decrease in total charge passed may have been due to the pozzolanic reactions between that addition and the old mortar of the fine RCA.

Wang et al. [37] described that the addition of mineral admixture could significantly reduce the chloride ion migration coefficient of the concrete. Several researchers [38, 39] found that 50 %–70 % blast-furnace slag (by weight) substitution in cement can reduce chloride ion penetration by 18 %–49.8 %, increasing the chloride resistance considerably. Etxeberria and Gonzalez-Corominas [40] concluded that the concrete produced employing up to 100 % of coarse RA, CEM III blast furnace slag cement, and seawater achieved more adequate durability properties than those of conventional concrete produced employing Portland cement, freshwater, and raw aggregates. In addition, they concluded that the influence of water/binder ratio and binder type on the rate of chloride penetration in a marine environment is higher than that of aggregate porosity. Berndt [39] determined that mixes containing pozzolanic material exhibited an about 50 % lower chloride diffusion coefficient when compared with the corresponding NAC and RAC mixes produced with OPC.

The use of fine RCA negatively affects the resistance to chloride ion penetration when OPC is used [41–43]. However, as mentioned above [35], the concrete produced with fine RCA using FA achieved high resistance. In addition, several researchers [44, 45] reported that fully replacing the fine NA fraction with fine RA from crushed ceramic bricks led to enhanced resistance to chloride ion penetration (i.e., the chloride migration coefficient decreased by 57 % when compared to the control concrete). Various researchers [46, 47] concluded that the chloride resistance of concrete made with up to 50 % of ceramic or mixed fine RAs at 1 year was similar or better than that of conventional concrete. The presence of ceramic fine grains caused a reduction of chloride ion penetrability of concrete. The explanation for this was that this was prompted by the pozzolanic reactions between the fine ceramic RA and the cement. However, the use of 100 % fine RA from crushed sanitary ware increased the migration coefficient by 38 %.

6.4 Permeability

Permeability is related to the microstructure of RAC and depends on the porosity, interconnectivity of pores, presence of cracks, and water content. The reduction in permeability is closely related to the ongoing hydration process. However, porosity is critical, as the RA are porous and sometimes present micro-cracking. Consequently, the incorporation of RA increases the permeability [6, 15, 48]. As a norm, the porosity of RAC rises with the increase in the replacement ratio of RCA, and it increases more strongly when fine RCA is employed [49]. In the same manner, the absorption capacity of RAC with a higher coarse RA content rises even further with increasing w/c ratios [15]. Furthermore, it was shown that the negative effect of RCA is significantly lower for low w/c ratios. In this case, both the porosity and absorption of the RCA are partially isolated from the surrounding less absorbent cement paste.

According to Limbachiya [50], there is only a slight effect when the replacement ratio of NA by RCA is less than 30 %. However, the use of coarse and fine RCA increased the absorption capacity of concrete [8].

On the other hand, Kou et al. [51] concluded that RAC produced with 100 % RCA after 5 years of curing achieved the lowest porosity. Thus, the RCA significantly improved the long-term interfacial properties of the new concrete, probably due to the long-term self-cementing effects of the old cement mortar and the interaction between the new cement paste and the old cement mortar. In addition, the influence of steam curing was especially beneficial in reducing the average pore size of RAC compared to those concrete mixes that only underwent air curing [52]. This reduction was higher in mixes produced with lower quality RCA. Wang et al. [53] described that RAC submitted to carbonation achieved a higher pore tortuosity, indicating that the pore structure was more complex and RAC had better durability. The shape of RCA also influenced RAC's permeability. Matias et al. [54] concluded that mixes containing more elongated RCA led to increases close to 19 % relative to the control concrete, whilst, when rounder RCA was used, it only increased by 12 %.

Vieira et al. [32] showed that the use of fine recycled aggregates, obtained from crushed brick aggregates, achieved 31 % lower water absorption than the NAC concrete. The authors explained that this may have resulted from the pozzolanic reactions between the high SiO_2 and Al_2O_3 contents of the fine crushed ceramic brick aggregates and the cement's hydration products. However, the permeability increased by 38 % when 100 % of fine RA from crushed sanitary ware was used due to the extensive agglomeration of these particles, which avoided adequate dispersion in the mix.

According to Correia et al. [55], the water absorption by capillary action of RAC was also greater than that of the NAC, and it was significantly higher when coarse and fine RCA aggregates were replaced [56]. Kappor et al. [57] concluded that the use of silica fume (SF) and metakaolin (MK) reduced the capillary absorption of RAC when 50 % of coarse RCA was employed in concrete production. However, they were not so effective in reducing the capillary capacity when 100 % of RCA were used. Moreover, they achieved a similar tendency in water permeability depth. According to Levy and Helene [19], capillary absorption can also be reduced by using FA to replace cement.

According to Dodds et al. [58], CEM III/A (B: Blast furnace slag) concrete mixes produced with up to 100 % RCA content had a lower 24 h sorption coefficient by a factor 1.1 to 2.2 than the control CEM I concrete mixes at 91 days (Fig. 6.1). Consequently, the concrete produced with CEM III/A achieved a better durability performance than that of a control CEM I concrete and a positive finding for the wider implementation of coarse RCA to produce sustainable structural concrete. Tam et al. [14] described that the mineral admixtures effectively reduced deformation and permeability in RAC. They have proved very beneficial in providing long-term durability performance and cost savings in RAC production.

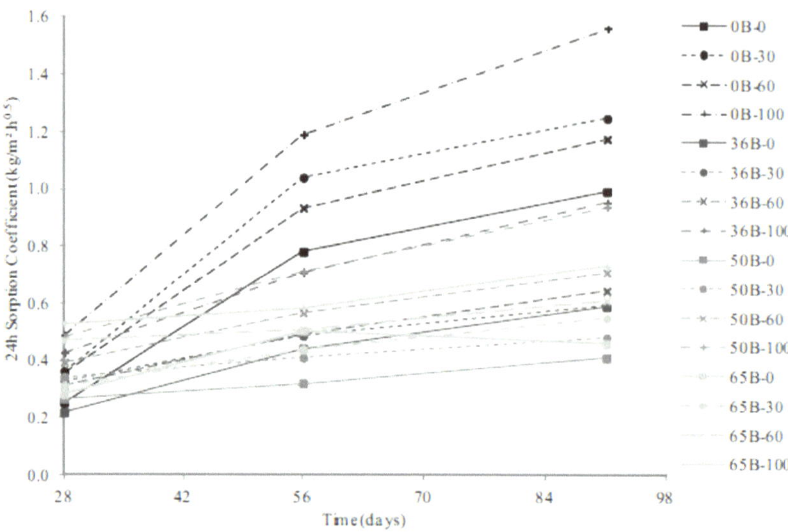

Fig. 6.1 24 h sorptivity coefficient for RAC (produced with blast furnace slag) [58]

6.5 Alkali-Silica Reaction

Alkali-silica reaction (ASR) is considered one of the most nefarious concrete degrading agents causing expansion due to a gel formation that swells in contact with water. Generally, the RCA is formed by 65–70 % (in volume) of original aggregates (OA), and the rest part is a residual mortar (RM) [59]. In principle, the extent of alkali-aggregate reaction (AAR) in RCA concrete is mainly controlled by the content of alkali components available to react. Moreover, certain studies [60, 61] described that the AAR performance of RCA concrete is closely related to the OA of RCA. Johnson and Shehata [61] used four different OA (Alberta, Bernier, Potsdam, and Springhill) for concrete production, and later the four concrete mixes were crushed for RCA production. The potential alkali reactivity of four OA aggregates and the produced RCA were determined following ASTM C 1260 specification. The recycled mortars were produced using different percentages of fine RCA. Table 6.1 shows that the AAR performance of RCA concrete is closely related to the OA of RCA. In addition, it was observed that a continuous increase in AAR expansion happened with the rise of proportions of RCA from 25 to 100 % in the concrete mix [61–63].

Table 6.1 Expansion of mortars produced with different aggregates [61]

Location/Mortar	OA-mortar	100 %RCA-mortar	50 %RCA-mortar	25 %RCA-mortar
Alberta	0.360	0.338	0.316	0.231
Bernier	0.170	0.132	0.083	0.090
Potsdam	0.090	0.073	0.066	0.065
Springhill	0.460	0.357	0.300	0.220

McCarthy et al. [64] also reported that RCA concrete exhibited higher AAR expansion when the OA had a higher alkali reactivity. This study clearly indicated that both the reactivity of OA and its reaction extent in parent concrete could have an effect on the AAR expansion. Other researchers [62, 65] found that the AAR expansion could be lower when a significant fraction of the reactive silica had been consumed.

The influence of the RM on the ASR expansion of RCA concrete is more complicated since it is affected by both the reactive silica and alkali content in RM. Etxeberria and Vazquez [66] showed that the fine aggregates of RM could possess high alkali reactivity. They found that the expansion of RM was higher than that of RCA. In addition, the enrichment of alkali in the RM can also accelerate the AAR in RAC. Therefore, the alkali leaching from old cement paste in RM should also be considered for the AAR in recycled concrete. Some studies [61, 67] reported that the expansion of specimens cast with unwashed RCA was higher than those using washed RCA. The low expansion of the concrete with washed RCA was attributed to the loss of alkalis during washing. However, the RM surrounding the reactive particles could also limit the exposure of the OA, thereby reducing the extent of AAR of OA in recycled concrete, which was confirmed by Beauchemin et al. [62]. In addition, Barreto et al. [63] stated that RCA composed of nonreactive NA and RM are likely not to cause the alkali-silica reactivity of RAC.

Other researchers [63, 65, 67, 68] described that the RCA subjected to a secondary crushing stage has a significantly higher expansion at 14 days than the RCA produced with only a primary crushing stage. They described that with the secondary crushing, more RM was separated, and a large amount of reactive OA was crushed. Consequently, more new surfaces of OA were exposed to AAR.

The aggregates' size and porosity could also influence the concrete's expansion. Johnson and Shehata [62] concluded that concrete suffered a higher expansion when a smaller RCA aggregate was employed in RAC production. Moreover, Delobel et al. [69] described that concrete suffered a higher expansion when the RM of RCA had lower porosity.

Delobel et al. [69] also reported that the recycled concrete produced with water-saturated RCA showed lower AAR expansion than dried RCA, especially in the early age expansion.

Barreto Santos et al. [68] described that the American Society for Testing and Materials ASTM C1260 is known to be effective in detecting the potential alkali reactivity of some aggregate types. However, when RA are tested, the ASTM C1260 limits should be revised by taking into consideration the correlation with the results of the RILEM AAR-3 CPT test. In addition, Locati et al. [70] proposed a petrographic method that seemed to be a useful tool to semi-quantify the RM content in fine recycled concrete aggregate (FRCA) and the content of particles with reactive components in FRCA prior to evaluating their potential reactivity by mortar bars.

In order to mitigate the AAR in RAC, the carbonation treatment of RCA was found to be effective [71]. In addition, Li and Gress [72] described that the minimum use of supplementary cementitious materials in 25 % of FA was determined to be possible to mitigate the ASR. Shehata et al. [67] also described that the ternary blend of Portland cement with 25 % class F fly ash and 10 % metakaolin was the most effective way of mitigating AAR.

6.6 Freeze-Thaw Resistance

Several studies noted that the freezing-thawing (F-T) resistance of RAC is less than that of NAC concrete [73–75]. Zhu et al. [62] concluded that, although the performance of RAC is inferior to the NAC, RCA can still be used to prepare structural concrete for the complex environment. Yildirim et al. [76] concluded that the concrete produced with 50 % of coarse RCA at a 50 % saturation level achieved comparable performance to that of NAC concrete.

According to Liu et al. [77], the F-T resistance is directly related to the properties of the parent concrete of RCA. RAC having RCA derived from parent concrete with high frost resistance, such as a high-strength concrete or an air-entrained concrete, showed F-T resistance almost the same as NAC. However, the RAC sample prepared using RCA obtained from a non-air-entrained concrete exhibited poor frost resistance. Gokce et al. [78] also reported that the concrete produced with RCA belonging to air-entrained source concrete was highly durable against freezing and thawing and even slightly superior to the reference NAC concrete. In contrast, they described an inferior performance of concrete incorporating RCA obtained from non-air-entrained concrete, although those new RAC mixes were satisfactorily air-entrained. Moreover, they stated that while the frost soundness test distinguished the non-air-entrained type RCA and the others (air-entrained type or original one), the sulphate soundness test results were controlled mainly by the adhered mortar content. Consequently, they defined that the sulphate soundness test fails to predict the frost susceptibility of the RCA. Therefore, they concluded that the direct unbound frost soundness test offers more realistic testing conditions to judge the soundness of RCA than the sulphate soundness test.

According to Richardson et al. [79], the incorporation of RCA can provide, due to its porosity, a greater intrinsic system of voids to the cementitious matrix. This fact may lead to a reduction in the compressive strength of the concrete. However, this material provides greater durability to the freeze-thaw cycles due to the greater ease of dissipation of hydraulic pressures.

Amorim et al. [80] determined that the 15 % replacement of the coarse natural aggregate by the coarse RCA was efficient in withstanding the expansion stresses of the water, presenting a greater durability factor and not suffering substantial loss of stiffness or physical wear. It was also observed that samples containing 50 % RCA had a durability factor superior to the ones obtained from the reference mixes. In addition, although the reference specimens presented the lowest performances due to their low water/cement ratio and consequent lower capillary porosity, they obtained a high average durability factor. Therefore, they concluded that using the RCA as an alternative way to air incorporation was effective, presenting a durability factor at the end of the cycle higher than that of the reference mixes.

Medina et al. [81] also concluded that concrete with up to 25 % coarse ceramic aggregates was more freeze-thaw resistant than conventional concrete. As a result, the scaling rate was lower, and the cracks were narrower in RAC. Bogas et al. [82] produced RAC using fine RCA. They reported that the freeze-thaw resistance was more affected by the w/c ratio rather than the type of aggregate used. In addition, in general, the internal freeze-thaw resistance of concrete did not decrease with the incorporation of FRCA; the higher porosity of FRCA may better contribute to hydraulic pressure dissipation.

6.7 Sulphate Attack

The high soluble sulphate content in RA may cause the internal sulphate attack of concrete [6]. Therefore, in order to avoid the formation of ettringite or thaumasite, some authors proposed limiting the maximum content of acid-soluble sulphates of RCA to 0.3 % or 0.8 % [6].

According to Feng et al. [83], RAC produced with 100 % replacement exhibited poor sulphate resistance, and the recommended RCA replacement ratio should be less than 50 %. In addition, Corral-Higuera et al. [84] evaluated the durability of reinforced RAC structures exposed to a high-sulphate solution. They found that increasing the supplementary cementitious materials contributes to the increased resistance to RAC sulphate attack.

Bulatovic et al. [85] described that concrete with RCA are more prone to sulphate attack than NAC. However, the sulphate resistance of concrete with RCA can be achieved if RCA fulfils the quality requirements given in EN 206 and if the proper cement type is used (i.e., CEM III LH/SR). They described that the type of cement and water to cement ratio had more significant influence on the concrete sulphate resistance than the origin of aggregates. Concrete made with CEM III LH/SR shows higher sulphate resistance despite the lower compressive strength values of concrete with this type of cement compared to the concrete with CEM I. They concluded that sulphate resistant concrete with RCA and CEM I could be obtained if a low water-cement ratio w/c=0.4 was employed for concrete production.

Zega et al. [86] analysed the sulphate resistance of RAC by partly submerging the samples in sulphate-rich soil for 10 years. They concluded that RCA content has no significant impact on sulphate resistance. Santillan et al. [87] also subjected RAC samples to 10 years of sulphate attack. They concluded that the increased porosity provided by RCA plays an ambiguous role in the external sulphate attack mechanism. On the one hand, it increases the sulphate penetration rate, mostly observed near the surface of specimens, and, on the other hand, damage can only be possible due to the generation of cracking due to internal stresses, and stresses are not produced when there is sufficient space for precipitation of reaction products or a lower stiffness of the whole material. The balance of both effects suggests a certain optimal content of RCA for improved external sulphate attack (ESA) resistance. Consequently, they concluded that the influence of RCA content on ESA resistance is indiscernible in laboratory conditions.

Boudali et al. [88] described that the mixes incorporating RCA and recycled fine concrete (as the binder) showed better performance than a mix with natural aggregate and pozzolana under cyclic environmental conditions combined with sodium sulphate exposure. The authors explained that this effect was possible because the recycled fine concrete binder achieved a higher sulphate resistance due to a high alumina oxide content in the used pozzolana, which led to lower sulphate resistance. In addition, due to a higher space available in RCA, this concrete could absorb the expansive stresses caused by gypsum and ettringite formation. However, the employed RCA had a water absorption capacity of 2.5 %, which was categorized as excellent quality RCA.

Lee et al. [89] suggested that the use of around 50 % replacement level of fine recycled aggregates showed a beneficial effect on the sulphate resistance of mortar specimens to both sodium and magnesium sulphate attacks. However, it was found that the resistance of the mortar specimens with 100 % recycled fine aggregates was not satisfactory. The incorporation of RA fines with higher water absorption led to the more pronounced deterioration of the mortar specimen, irrespective of the types of sulphate solutions.

6.8 Conclusions

After analysing the different durability properties, the general conclusion is that RAC produced with up to 50 % of coarse RCA (RAC50) using OPC achieves adequate properties and, sometimes, even better than NAC in terms of carbonation, freeze-thaw resistance, and sulphate attack. However, supplementary cementitious materials are required in the RCA50 to achieve satisfactory or better durability properties than NAC for chloride penetration resistance, permeability, and alkaline aggregate properties. The use of fine RCA increases the porosity of concrete, but the alkaline amount of the mix also increases, which may mitigate the porosity increase by the reaction between RCA and supplementary cementitious materials, in addition to increasing the carbonation resistance. In contrast, the use of fine recycled ceramic aggregates achieved lower porosity than the NAC produced with OPC due to pozzolanic reactions between the high SiO_2 and Al_2O_3 contents of the fine crushed ceramic brick aggregates and the cement hydration products.

The durability of these recycled aggregate concretes will improve and obtain very good properties for structural applications, reducing the water-cement ratio used in concrete production compared to that in NAC

The specific conclusion obtained for each of the properties studied:

Carbonation Resistance

The carbonation resistance of RAC is usually lower than that of NAC when OPC is employed for concrete production and decreases with increasing RCA content. However, RAC's carbonation resistance can be higher than that of NAC when concretes are produced with fly ash and uncarbonated RCA.

Chloride Ion Penetration

RAC generally shows a lower chloride ion diffusion resistance than NAC when using ordinary Portland cement (OPC).

Although the concretes' chloride ion penetration resistance increases with the use of FA, the RAC achieved a higher chloride ion penetration than the NAC when both were produced with the same amount of FA binder and water-cement ratio. However, the use of coarse and fine RCA along with supplementary cementitious materials in concrete production could achieve adequate resistance for application in structural concrete. This may be due to the pozzolanic reactions between the additions and the old mortar of the fine RCA.

Permeability

The porosity and permeability of RAC rise with the increase in the replacement ratio of RCA, and it increases more strongly when fine RCA is employed. In order to achieve a similar permeability value to NAC, RAC should be produced with a lower water/cement ratio than the NAC, or a low replacement percentage of coarse natural aggregates by recycled aggregates (30 %) should be employed. However, concrete produced with 100 % of RCA could achieve a lower porosity than conventional concrete after 5 years of curing. RCA has proved to be very beneficial in long-term durability performance.

Alkali-Silica Reaction

The extent of alkali-aggregate reaction (AAR) in RCA concrete is mainly controlled by the content of alkali components available to react and by the alkali reactivity of RCA's original aggregate (OA).

The washed RCA (attributed to the loss of alkalis during washing), has lower reactivity, and in general, old mortar (RM) surrounding the OA also reduced the reactivity of RCA.

The ASTM C1260 limits should be revised by taking into consideration the correlation with the results of the AAR assessment using concrete prism test (CPT), RILEM AAR-3 CPT test..

Freeze-Thaw Resistance

RAC produced using RCA obtained from non-air entrained concrete usually exhibits poor frost resistance. However, concrete with RCA derived from parent concrete with high frost resistance shows a freeze-thaw resistance almost similar to NAC.

It seems that the direct unbound frost soundness test offers more realistic testing conditions to judge the soundness of RCA than the sulphate soundness test.

Sulphate Attack

It appears that the porosity of RCA can reduce the internal stresses in RAC caused by an external sulphate attack mechanism, having acceptable behaviour when 50 % of RCA is employed for concrete production. In addition, the type of cement and the water-to-cement ratio have a more significant influence on the concrete sulphate resistance than the origin of aggregates.

Currently, the structural applications of RAC are quite limited. Based on the above discussion, further research is needed to develop confidence in the utilization of RCA. Detailed guidelines and standards are still to be developed to use the RAC for structural applications [90].

6.9 References

[1] W. Chen *et al.*, "Adopting recycled aggregates as sustainable construction materials: A review of the scientific literature," *Construction and Building Materials*, vol. 218, pp. 483–496, 2019. https://doi.org/10.1016/j.conbuildmat.2019.05.130

[2] D. Pedro, J. De Brito, and L. Evangelista, "Influence of the use of recycled concrete aggregates from different sources on structural concrete," *Construction and Building Materials*, vol. 71, pp. 141–151, Nov. 2014. https://doi.org/10.1016/j.conbuildmat.2014.08.030

[3] M. Etxeberria, Marí R. Antonio, and E. Vázquez, "Recycled aggregate concrete as structural material," *Materials and Structures*, vol. 40, pp. 529–541, 2007. https://doi.org/10.1617/s11527-006-9161-5

[4] J. Xiao, *Recycled Aggregate Concrete Structures*. Springer-Verlag GmbH, 2018. https://doi.org/10.1007/978-3-662-53987-3

[5] A. Neville, *Properties of concrete*, 5 ed. London, UK: Longman, 2011.

[6] H. B. Le and Q. B. Bui, "Recycled aggregate concretes – A state-of-the-art from the microstructure to the structural performance," *Construction and Building Materials*, vol. 257, p. 119522, 2020. https://doi.org/10.1016/j.conbuildmat.2020.119522

[7] T. C. Hansen, *Recycling of demolished concrete and masonry*. London (UK): E&FN Spon, 1992. https://doi.org/10.1201/9781482267075

[8] L. Evangelista and J. De Brito, "Durability performance of concrete made with fine recycled concrete aggregates," *Cement and Concrete Composites*, vol. 32, no. 1, pp. 9–14, Jan. 2010. https://doi.org/10.1016/j.cemconcomp.2009.09.005

[9] R. V. Silva, R. Neves, J. De Brito, and R. K. Dhir, "Carbonation behaviour of recycled aggregate concrete," *Cement and Concrete Composites*, vol. 62, pp. 22–32, 2015. https://doi.org/10.1016/j.cemconcomp.2015.04.017

[10] H. Guo *et al.*, "Durability of recycled aggregate concrete – A review," *Cement and Concrete Composites*, vol. 89, pp. 251–259, 2018. https://doi.org/10.1016/j.cemconcomp.2018.03.008

[11] A. Adessina, A. Ben Fraj, J. F. Barthélémy, C. Chateau, and D. Garnier, "Experimental and micromechanical investigation on the mechanical and durability properties of recycled aggregates concrete," *Cement and Concrete Research*, vol. 126, no. September, 2019. https://doi.org/10.1016/j.cemconres.2019.105900

[12] X. Zeng, "Progress in the research of carbonation resistance of RAC," *Construction and Building Materials*, vol. 230, p. 116976, 2020. https://doi.org/10.1016/j.conbuildmat.2019.116976

[13] D. Pedro, J. de Brito, and L. Evangelista, "Performance of concrete made with aggregates recycled from precasting industry waste: influence of the crushing process," Materials and Structures, vol. 48, no. 12, pp. 3965–3978, 2015. https://doi.org/10.1617/s11527-014-0456-7

[14] V. W. Y. Tam, M. Soomro, A. C. J. Evangelista, and A. Haddad, "Deformation and permeability of recycled aggregate concrete – A comprehensive review," *Journal of Building Engineering*, vol. 44, no. December, p. 103393, 2021. https://doi.org/10.1016/j.jobe.2021.103393

[15] C. Thomas, J. Setién, J. A. Polanco, P. Alaejos, and M. Sánchez De Juan, "Durability of recycled aggregate concrete," *Construction and Building Materials*, vol. 40, pp. 1054–1065, 2013. https://doi.org/10.1016/j.conbuildmat.2012.11.106

[16] J. Z. Xiao, B. Lei, and C. Z. Zhang, "On carbonation behavior of recycled aggregate concrete," *Science China Technological Sciences*, vol. 55, no. 9, pp. 2609–2616, 2012. https://doi.org/10.1007/s11431-012-4798-5

[17] M. Etxeberria, "Evaluation of eco-efficient concretes produced with fly ash and uncarbonated recycled aggregates," *Materials (Basel)*, vol. 14, no. 24, 2021. https://doi.org/10.3390/ma14247499

[18] A. Leemann and R. Loser, "Carbonation resistance of recycled aggregate concrete," *Construction and Building Materials*, vol. 204, pp. 335–341, 2019. https://doi.org/10.1016/j.conbuildmat.2019.01.162

[19] S. M. Levy and P. Helene, "Durability of recycled aggregates concrete: a safe way to sustainable development," *Cement and Concrete Research*, vol. 34, no. 11, pp. 1975–1980, Nov. 2004. https://doi.org/10.1016/j.cemconres.2004.02.009

[20] M. Nedeljković, J. Visser, B. Šavija, S. Valcke, and E. Schlangen, "Use of fine recycled concrete aggregates in concrete: A critical review," *Journal of Building Engineering*, vol. 38, no. January, 2021. https://doi.org/10.1016/j.jobe.2021.102196

[21] S. Sunayana and S. V. Barai, "Partially fly ash incorporated recycled coarse aggregate based concrete: Microstructure perspectives and critical analysis," *Construction and Building Materials*, vol. 278, p. 122322, 2021. https://doi.org/10.1016/j.conbuildmat.2021.122322

[22] C. Faella, C. Lima, E. Martinelli, M. Pepe, and R. Realfonzo, "Mechanical and durability performance of sustainable structural concretes: An experimental study," *Cement and Concrete Composites*, vol. 71, pp. 85–96, 2016. https://doi.org/10.1016/j.cemconcomp.2016.05.009

[23] R. Kurda, J. de Brito, and J. D. Silvestre, "Combined influence of recycled concrete aggregates and high contents of fly ash on concrete properties," *Construction and Building Materials*, vol. 157, pp. 554–572, 2017. https://doi.org/10.1016/j.conbuildmat.2017.09.128

[24] J. Sim and C. Park, "Compressive strength and resistance to chloride ion penetration and carbonation of recycled aggregate concrete with varying amount of fly ash and fine recycled aggregate," *Waste Management*, vol. 31, no. 11, pp. 2352–2360, 2011. https://doi.org/10.1016/j.wasman.2011.06.014

[25] V. Corinaldesi and G. Moriconi, "Influence of mineral additions on the performance of 100 % recycled aggregate concrete," *Construction and Building Materials*, vol. 23, no. 8, pp. 2869–2876, Aug. 2009. https://doi.org/10.1016/j.conbuildmat.2009.02.004

[26] B. Lothenbach, K. Scrivener, and R. D. Hooton, "Supplementary cementitious materials," *Cement and Concrete Research*, vol. 41, no. 12, pp. 1244–1256, 2011. https://doi.org/10.1016/j.cemconres.2010.12.001

[27] Y. L. S. Fang Tian, Wei Xin Hu, He Ming Cheng, "Carbonation Depth of Recycled Aggregate Concrete Incorporating Fly Ash," *Advanced Material Resdearch*, vol. 261–263, pp. 217–222, 2011. https://doi.org/10.4028/www.scientific.net/amr.261-263.217

[28] K. H. Obla, C. L. Lobo, and H. Kim, "Tests and Criteria for Concrete Resistant to Chloride Ion Penetration," ACI Materials Journal, vol. 113, no. 5, pp. 621–631, 2016. https://doi.org/10.14359/51689107

[29] N. Otsuki, S. Miyazato, and W. Yodsudjai, "Influence of recycled aggregate on interfacial transition zone, strength, chloride penetration and carbonation of concrete," *Journal of Materials in Civil Engineering*, vol. 15, no. 5, pp. 443–451, 2003. https://doi.org/10.1061/(asce)0899-1561(2003)15:5(443)

[30] R. V. Silva, J. De Brito, R. Neves, and R. Dhir, "Prediction of chloride ion penetration of Recycled Aggregate Concrete," *Materials Research*, vol. 18, no. 2, pp. 427–440, 2015. https://doi.org/10.1590/1516-1439.000214

[31] G. Xu and X. Shi, "Characteristics and applications of fly ash as a sustainable construction material: A state-of-the-art review," *Resources, Conservation and Recycling*, vol. 136, no. August 2017, pp. 95–109, 2018. https://doi.org/10.1016/j.resconrec.2018.04.010

[32] M. VM, "Durability of concrete incorporating high-volume of low-calcium (ASTM Class F) fly ash," *Cement and Concrete Composites*, vol. 12, no. 4, pp. 271–7, 1990. https://doi.org/10.1016/0958-9465(90)90006-j

[33] M. Etxeberria and S. Castillo, "How the Carbonation Treatment of Different Types of Recycled Aggregates Affects the Properties of Concrete," *Sustainability*, vol. 15, 2023. https://doi.org/10.3390/su15043169

[34] K. Kim, M. Shin, and S. Cha, "Combined effects of recycled aggregate and fly ash towards concrete sustainability," *Construction and Building Materials*, vol. 48, pp. 499–507, 2013. https://doi.org/10.1016/j.conbuildmat.2013.07.014

[35] S. C. Kou and C. S. Poon, "Long-term mechanical and durability properties of recycled aggregate concrete prepared with the incorporation of fly ash," *Cement and Concrete Composites*, vol. 37, no. 1, pp. 12–19, 2013. https://doi.org/10.1016/j.cemconcomp.2012.12.011

[36] S. C. Kou and C. S. Poon, "Properties of self-compacting concrete prepared with coarse and fine recycled concrete aggregates," *Cement and Concrete Composites*, vol. 31, no. 9, pp. 622–627, Oct. 2009. https://doi.org/10.1016/j.cemconcomp.2009.06.005

[37] J. Wang, J. Zhang, and D. Cao, "Pore characteristics of recycled aggregate concrete and its relationship with durability under complex environmental factors," *Construction and Building Materials*, vol. 272, p. 121642, 2021. https://doi.org/10.1016/j.conbuildmat.2020.121642

[38] K. Y. Ann, H. Y. Moon, Y. B. Kim, and J. Ryou, "Durability of recycled aggregate concrete using pozzolanic materials," *Waste Management*, vol. 28, no. 6, pp. 993–999, 2008. https://doi.org/10.1016/j.wasman.2007.03.003

[39] M. L. Berndt, "Properties of sustainable concrete containing fly ash, slag and recycled concrete aggregate," *Construction and Building Materials*, vol. 23, no. 7, pp. 2606–2613, 2009. https://doi.org/10.1016/j.conbuildmat.2009.02.011

[40] M. Etxeberria and A. Gonzalez-Corominas, "Properties of Plain Concrete Produced Employing Recycled Aggregates and Sea Water," *International Journal of Civil Engineering*, vol. 16, no. 9, pp. 993–1003, 2018. https://doi.org/10.1007/s40999-017-0229-0

[41] L. Evangelista and J. De Brito, "Durability of crushed fine recycled aggregate concrete assessed by permeability-related properties," *Magazine of Concrete Research*, vol. 71, no. 21, pp. 1142–1150, 2019. https://doi.org/10.1680/jmacr.18.00093

[42] A. Mardani-Aghabaglou, M. Tuyan, and K. Ramyar, "Mechanical and durability performance of concrete incorporating fine recycled concrete and glass aggregates," Materials and Structures, vol. 48, no. 8, pp. 2629–2640, 2015. https://doi.org/10.1617/s11527-014-0342-3

[43] A. Z. Bendimerad, B. Delsaute, E. Rozière, S. Staquet, and A. Loukili, "Advanced techniques for the study of shrinkage-induced cracking of concrete with recycled aggregates at early age," *Construction and Building Materials*, vol. 233, p. 117340, 2020. https://doi.org/10.1016/j.conbuildmat.2019.117340

[44] F. Pacheco-Torgal and S. Jalali, "Reusing ceramic wastes in concrete," *Construction and Building Materials*, vol. 24, no. 5, pp. 832–838, 2010. https://doi.org/10.1016/j.conbuildmat.2009.10.023

[45] T. Vieira, A. Alves, J. de Brito, J. R. Correia, and R. V. Silva, "Durability-related performance of concrete containing fine recycled aggregates from crushed bricks and sanitary ware," *Materials & Design*, vol. 90, pp. 767–776, 2016. https://doi.org/10.1016/j.matdes.2015.11.023

[46] M. Etxeberria and I. Vegas, "Effect of fine ceramic recycled aggregate (RA) and mixed fine RA on hardened properties of concrete," *Magazine of Concrete Research*, vol. 67, no. 12, pp. 645–655, 2014. https://doi.org/10.1680/macr.14.00208

[47] M. Etxeberria and A. Gonzalez-Corominas, "The assessment of ceramic and mixed recycled aggregates for high strength and low shrinkage concretes," *Materials and Structures*, vol. 51, no. 5, pp. 1–21, 2018. https://doi.org/10.1617/s11527-018-1244-6

[48] W. H. Kwan, M. Ramli, K. J. Kam, and M. Z. Sulieman, "Influence of the amount of recycled coarse aggregate in concrete design and durability properties," *Construction and Building Materials*, vol. 26, pp. 565–573, Jul. 2011. https://doi.org/10.1016/j.conbuildmat.2011.06.059

[49] J. M. Gómez-Soberón, "Porosity of recycled concrete with substitution of recycled concrete aggregate," *Cement and Concrete Research,* vol. 32, no. 8, pp. 1301–1311, 2002. https://doi.org/10.1016/s0008-8846(02)00795-0

[50] M. C. Limbachiya, T. Leelawat, and R. K. Dhir, "Use of recycled concrete aggregate in high-strength concrete," *Materials and Structures*, vol. 33. pp. 574–580, 2000. https://doi.org/10.1007/bf02480538

[51] S. C. Kou, C. S. Poon, and M. Etxeberria, "Influence of recycled aggregates on long term mechanical properties and pore size distribution of concrete," *Cement and Concrete Composites*, vol. 33, no. 2, pp. 286–291, 2011. https://doi.org/10.1016/j.cemconcomp.2010.10.003

[52] A. Gonzalez-Corominas, M. Etxeberria, and C. S. Poon, "Influence of steam curing on the pore structures and mechanical properties of fly-ash high performance concrete prepared with recycled aggregates," *Cement and Concrete Composites*, vol. 71, 2016. https://doi.org/10.1016/j.cemconcomp.2016.05.010

[53] J. Wang, J. Zhang, and D. Cao, "Pore characteristics of recycled aggregate concrete and its relationship with durability under complex environmental factors," *Con-

struction and Building Materials, vol. 272, p. 121642, 2021. https://doi.org/10.1016/j.conbuildmat.2020.121642

[54] D. Matias, J. de Brito, A. Rosa, and D. Pedro, "Durability of Concrete with Recycled Coarse Aggregates: Influence of Superplasticizers," *Journal of Materials in Civil Engineering*, vol. 26, no. 7, p. 06014011, 2014. https://doi.org/10.1061/(asce)mt.1943-5533.0000961

[55] J. R. Correia, J. De Brito, and A. S. Pereira, "Effects on concrete durability of using recycled ceramic aggregates," Materials and Structures, vol. 39, no. 2, pp. 169–177, 2006. https://doi.org/10.1617/s11527-005-9014-7[57]

[56] R. Zaharieva, F. Buyle-Bodin, F. Skoczylas, and E. Wirquin, "Assessment of the surface permeation properties of recycled aggregate concrete," *Cement and Concrete Composites*, vol. 25, no. 2, pp. 223–232, 2003. https://doi.org/10.1016/s0958-9465(02)00010-0

[57] K. Kapoor, S. P. Singh, and B. Singh, "Durability of self-compacting concrete made with Recycled Concrete Aggregates and mineral admixtures," *Construction and Building Materials*, vol. 128, pp. 67–76, 2016. https://doi.org/10.1016/j.conbuildmat.2016.10.026

[58] W. Dodds, C. Goodier, C. Christodoulou, S. Austin, and D. Dunne, "Durability performance of sustainable structural concrete: Effect of coarse crushed concrete aggregate on microstructure and water ingress," *Construction and Building Materials*, vol. 145, pp. 183–195, 2017. https://doi.org/10.1016/j.conbuildmat.2017.03.232

[59] J. Zhang, C. Shi, Y. Li, X. Pan, C. S. Poon, and Z. Xie, "Influence of carbonated recycled concrete aggregate on properties of cement mortar," *Construction and Building Materials*, vol. 98, pp. 1–7, 2015. https://doi.org/10.1016/j.conbuildmat.2015.08.087

[60] Z. Peng *et al.*, "Alkali-aggregate reaction in recycled aggregate concrete," *Journal of Cleaner Production*, vol. 255, p. 120238, 2020. https://doi.org/10.1016/j.jclepro.2020.120238

[61] R. Johnson and M. H. Shehata, "The efficacy of accelerated test methods to evaluate Alkali Silica Reactivity of Recycled Concrete Aggregates," *Construction and Building Materials*, vol. 112, pp. 518–528, 2016. https://doi.org/10.1016/j.conbuildmat.2016.02.155

[62] S. Beauchemin, B. Fournier, and J. Duchesne, "Evaluation of the concrete prisms test method for assessing the potential alkali-aggregate reactivity of recycled concrete aggregates," *Cement and Concrete Research*, vol. 104, February 2018, pp. 25–36, 2018. https://doi.org/10.1016/j.cemconres.2017.10.008

[63] M. Barreto Santos, J. de Brito, A. Santos Silva, and H. Hasan Ahmed, "Study of ASR in concrete with recycled aggregates: Influence of aggregate reactivity potential and cement type," *Construction and Building Materials*, vol. 265, p. 120743, 2020. https://doi.org/10.1016/j.conbuildmat.2020.120743

[64] M. J. McCarthy, L. J. Csetenyi, J. E. Halliday, and R. K. Dhir, "Evaluating the effect of recycled aggregate on damaging AAR in concrete," *Magazine of Concrete Research*, vol. 67, no. 11, pp. 598–610, 2015. https://doi.org/10.1680/macr.14.00260

[65] M. P. Adams *et al.*, "Applicability of Standard Alkali-Silica Reactivity Testing Methods for Recycled Concrete Aggregate," *14th International Conference on Alkali-Ag-*

gregate Reaction, May, 2012, Austin Texas. https://icaarconcrete.org/wp-content/uploads/2022/10/14ICAAR-AdamsMP-1.pdf

[66] M. Etxeberria and E. Vázquez, "Reacción álcali sílice en el hormigón debido al mortero adherido del árido reciclado," *Materiales de Construcción*, vol. 60, no. 297, pp. 47–58, 2010. https://doi.org/10.3989/mc.2010.46508

[67] M. H. Shehata, C. Christidis, W. Mikhaiel, C. Rogers, and M. Lachemi, "Reactivity of reclaimed concrete aggregate produced from concrete affected by alkali-silica reaction," *Cement and Concrete Research*, vol. 40, no. 4, pp. 575–582, 2010. https://doi.org/10.1016/j.cemconres.2009.08.008

[68] M. Barreto Santos, J. de Brito, A. Santos Silva, and A. Hawreen, "Evaluation of alkali-silica reaction in recycled aggregates: The applicability of the mortar bar test," *Construction and Building Materials*, vol. 299, no. August 2020, 2021. https://doi.org/10.1016/j.conbuildmat.2021.124250

[69] F. Delobel, D. Bulteel, J. M. Mechling, A. Lecomte, M. Cyr, and S. Rémond, "Application of ASR tests to recycled concrete aggregates: Influence of water absorption," *Construction and Building Materials*, vol. 124, pp. 714–721, 2016. https://doi.org/10.1016/j.conbuildmat.2016.08.004

[70] F. Locati, C. Zega, G. Coelho dos Santos, S. Marfil, and D. Falcone, "Petrographic method to semi-quantify the content of particles with reactive components and residual mortar in ASR-affected fine recycled concrete aggregates," *Cement and Concrete Composites*, vol. 119, no. August 2020, p. 104003, 2021. https://doi.org/10.1016/j.cemconcomp.2021.104003

[71] C. Shi, Y. Li, J. Zhang, W. Li, L. Chong, and Z. Xie, "Performance enhancement of recycled concrete aggregate – A review," *Journal of Cleaner Production*, vol. 112, pp. 466–472, 2016. https://doi.org/10.1016/j.jclepro.2015.08.057

[72] H. C. Scott and D. L. Gress, "Mitigating alkali silica reaction in recycled concrete," *American Concrete Institute, ACI Special Publications*, vol. SP-219, pp. 61–76, 2004. https://doi.org/10.14359/13139

[73] R. Zaharieva, F. Buyle-Bodin, and E. Wirquin, "Frost resistance of recycled aggregate concrete," *Cement and Concrete Research*, vol. 34, no. 10, pp. 1927–1932, 2004. https://doi.org/10.1016/j.cemconres.2004.02.025

[74] S. M. S. Kazmi, M. J. Munir, Y. F. Wu, I. Patnaikuni, Y. Zhou, and F. Xing, "Effect of recycled aggregate treatment techniques on the durability of concrete: A comparative evaluation," *Construction and Building Materials*, vol. 264, p. 120284, 2020. https://doi.org/10.1016/j.conbuildmat.2020.120284

[75] P. Zhu, Y. Hao, H. Liu, X. Wang, and L. Gu, "Durability evaluation of recycled aggregate concrete in a complex environment," *Journal of Cleaner Production*, vol. 273, p. 122569, 2020. https://doi.org/10.1016/j.jclepro.2020.122569

[76] S. T. Yildirim, C. Meyer, and S. Herfellner, "Effects of internal curing on the strength, drying shrinkage and freeze-thaw resistance of concrete containing recycled concrete aggregates," *Construction and Building Materials*, vol. 91, pp. 288–296, 2015. https://doi.org/10.1016/j.conbuildmat.2015.05.045

[77] K. Liu, J. Yan, Q. Hu, Y. Sun, and C. Zou, "Effects of parent concrete and mixing method on the resistance to freezing and thawing of air-entrained recycled aggre-

gate concrete," *Construction and Building Materials*, vol. 106, pp. 264–273, 2016. https://doi.org/10.1016/j.conbuildmat.2015.12.074

[78] A. Gokce, S. Nagataki, T. Saeki, and M. Hisada, "Identification of frost-susceptible recycled concrete aggregates for durability of concrete," *Construction and Building Materials*, vol. 25, no. 5, pp. 2426–2431, 2011. https://doi.org/10.1016/j.conbuildmat.2010.11.054

[79] A. Richardson, K. Coventry, and J. Bacon, "Freeze/thaw durability of concrete with recycled demolition aggregate compared to virgin aggregate concrete," *Journal of Cleaner Production*, vol. 19, no. 2–3, pp. 272–277, 2011. https://doi.org/10.1016/j.jclepro.2010.09.014

[80] N. S. Amorim Júnior, G. A. O. Silva, and D. V. Ribeiro, "Effects of the incorporation of recycled aggregate in the durability of the concrete submitted to freeze-thaw cycles," *Construction and Building Materials*, vol. 161, pp. 723–730, 2018. https://doi.org/10.1016/j.conbuildmat.2017.12.076

[81] C. Medina, M. I. Sánchez De Rojas, and M. Frías, "Freeze-thaw durability of recycled concrete containing ceramic aggregate," *Journal of Cleaner Production*, vol. 40, pp. 151–160, 2013. https://doi.org/10.1016/j.jclepro.2012.08.042

[82] J. A. Bogas, J. De Brito, and D. Ramos, "Freeze-thaw resistance of concrete produced with fine recycled concrete aggregates," *Journal of Cleaner Production*, vol. 115, pp. 294–306, 2016. https://doi.org/10.1016/j.jclepro.2015.12.065

[83] F. Xie, J. Li, G. Zhao, C. Wang, Y. Wang, and P. Zhou, "Experimental investigations on the durability and degradation mechanism of cast-in-situ recycled aggregate concrete under chemical sulfate attack," *Construction and Building Materials*, vol. 297, p. 123771, 2021. https://doi.org/10.1016/j.conbuildmat.2021.123771

[84] R. Corral-Higuera *et al.*, "Sulfate attack and reinforcement corrosion in concrete with recycled concrete aggregates and supplementary cementing materials," *International Journal of Electrochemical Science*, vol. 6, no. 3, pp. 613–621, 2011. https://doi.org/10.1016/s1452-3981(23)15020-6

[85] V. Bulatović, M. Melešev, M. Radeka, V. Radonjanin, and I. Lukić, "Evaluation of sulfate resistance of concrete with recycled and natural aggregates," *Construction and Building Materials*, vol. 152, pp. 614–631, 2017. https://doi.org/10.1016/j.conbuildmat.2017.06.161

[86] C. J. Zega, G. S. Coelho Dos Santos, Y. A. Villagrán-Zaccardi, and A. A. Di Maio, "Performance of recycled concretes exposed to sulphate soil for 10 years," *Construction and Building Materials*, vol. 102, pp. 714–721, 2016. https://doi.org/10.1016/j.conbuildmat.2015.11.025

[87] L. R. Santillán, F. Locati, Y. A. Villagrán-Zaccardi, and C. J. Zega, "Long-term sulfate attack on recycled aggregate concrete immersed in sodium sulfate solution for 10 years," *Materiales de Construcción*, vol. 70, no. 337, pp. 1–14, 2020. https://doi.org/10.3989/mc.2020.06319

[88] S. Boudali, D. E. Kerdal, K. Ayed, B. Abdulsalam, and A. M. Soliman, "Performance of self-compacting concrete incorporating recycled concrete fines and aggregate exposed to sulphate attack," *Construction and Building Materials*, vol. 124, pp. 705–713, 2016. https://doi.org/10.1016/j.conbuildmat.2016.06.058

[89] S.-T. Lee, R. N. Swamy, S.-S. Kim, and Y.-G. Park, "Durability of Mortars Made with Recycled Fine Aggregates Exposed to Sulfate Solutions," *Journal of Materials in Civil Engineering*, vol. 20, no. 1, pp. 63–70, 2008. https://doi.org/10.1061/(asce)0899-1561(2008)20:1(63)

[90] F. Colangelo, R. Cioffi, and I. Farina, *Handbook of sustainable concrete and industrial waste management*. Elsevier. Woodhead Publishing, 2022. https://doi.org/10.1016/c2019-0-04591-8

Chapter 7

Fatigue of Recycled Aggregate Concrete

José Sainz-Aja, Carlos Thomas

LADICIM (Laboratory of Materials Science and Engineering), University of Cantabria. E.T.S. de Ingenieros de Caminos, Canales y Puertos, Av./Los Castros 44, 39005 Santander, Spain.

7.1 Introduction

The construction sector, which is a significant source of global solid waste, is attempting to minimize waste by recycling construction and demolition debris into recycled aggregates (RA) for use in recycled aggregate concrete (RAC) [1]. Initial experiments with RAC began after WWII [2–5], and extensive developments have been made in lab research and practical application, demonstrating that RAC can perform well mechanically and in terms of durability [6, 7, 8–12]. However, RA can vary due to different sources [13, 14] and crushing processes [15, 16]. Despite technical regulations allowing RA usage in various countries [17, 18], see Table 7.1, limitations exist, including potential strength loss from increased water absorption [7, 19–21]. This is mitigated by pre-saturating the aggregates or using superplasticiser additives to decrease water content [22–24], though understanding the fatigue effect on RAC remains a challenge.

Although it is unusual for concrete elements to be exposed to variable loads that can lead to fatigue failure, there are a number of typical concrete elements that may be exposed to such loads, such as: railway superstructure elements, railway superstructures, sleepers or slab tracks [25], rail and road bridges [26], offshore structures subject to variable wind and tidal loads [26, 27], and/or wind generators [28]. As Skarżyński et al. state, knowledge about the effect of cyclic loads on concrete is currently very limited [28].

As a starting point in the study of the fatigue behaviour of RAC, the research began with bending fatigue, since this is the most common type of stress that appears, and it is, for example, the type of fatigue damage that appears in the case of prestressed concrete bridges. There are several publications that analyse this effect on concretes with 100 % RA [27, 29]. These studies conclude that the presence of RA reduces the fatigue life of the concrete.

Concrete is a material that is barely capable of resisting tensile stresses, therefore, in general, bending fatigue studies must be carried out on reinforced or fibre-reinforced concrete. To be able to evaluate more accurately the behaviour of RAC under cyclic loading, other authors have chosen to characterize plain concrete under compressive fatigue.

Fatigue is a subcritical process that makes the cracks grow, reducing the strength capacity of the components, so that component failure can occur under loads lower than expected. Concrete, which is a material that inherently has micro-cracks, can be damaged due to the repeated action of loads far below its ultimate strength. These micro-cracks are mainly found either in the cement paste or at the paste-aggregate interface. In the specific case of RAC, as there is generally a greater amount of mortar, this means that a priori, fatigue has a greater influence than in the case of conventional concretes.

Table 7.1 Summary of the percentage of substitution of natural aggregate by recycled aggregate allowed in each country

Country	Type of aggregate	Percentage of substitution allowed*	
		Type 1[b] (%)	Type 2[b] (%)
Germany	X0 (coarse)	< 45	< 35
	XC1 a XC4 (coarse)	< 45	< 35
	XF1 y XF3 (coarse)	< 35	< 25
	XA1 (coarse)	< 25	< 25
	Fine	0	0
		Percentage of substitution allowed	
Netherlands	Coarse	< 20	
	Fine	0	
Belgium	Coarse	< 100	
	Fine	< 100 (with some restrictions)	
Denmark	Coarse	< 100	
	Fine	< 20	
UK	Coarse	< 20	
	Fine	0	
Spain	Coarse	< 20	
	Fine	0	

*Where Type 1[b] is concrete waste while Type 2[b] is material from construction and demolition.

As these micro-cracks grow, this has an impact on the mechanical behaviour of the concrete. In particular, the increasing cracks make the concrete more deformable [30–32]. Fig. 7.1(a) shows an example of how the strain of a specimen evolves during a fatigue test, and several authors have classified this evolution in 3 phases [33, 34]. The first of these zones, which can be seen in the first stage of the test, is characterised by a decreasing strain rate. This is a stabilisation stage of the fatigue process. The second stage of the process is characterised by an approximately constant strain rate. Finally, the third stage is characterised by an increasing strain rate, i.e., the deformations grow progressively until the specimen breaks. Similarly, this increase in the flexibility of the specimens has an inversely proportional effect on the elastic modulus of the concrete, see Fig. 7.1(b).

(a)   (b)

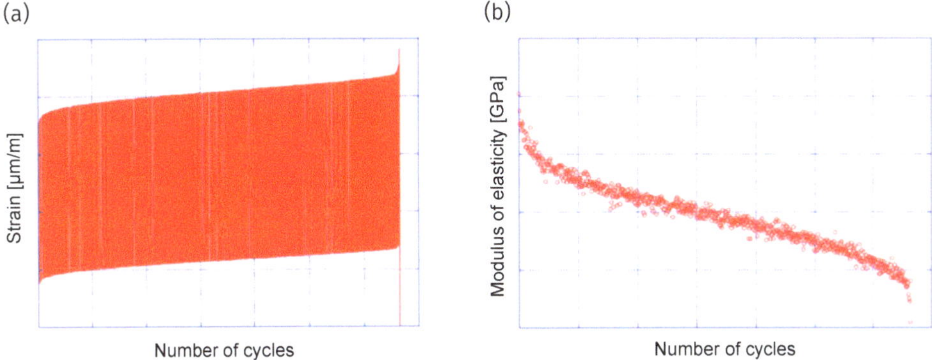

Fig. 7.1 Evolution of the strain (a) and the young modulus (b) as function of cycles during a
fatigue test.

In the case of the elastic modulus, it is also possible to identify the 3 stages described in
the case of the deformations.

Regarding the mechanisms culminating in specimen failure, it has been shown that
these are like those obtained in the case of conventional concretes. This is because, as
previously indicated, fatigue failure is caused by the growth of interconnected cracks that
weaken the concrete [28], a process that occurs similarly whether RA is present or not.
It should be noted that, as indicated above, since fatigue is a subcritical mechanism, the
failure of concrete specimens during fatigue tests results in the specimen failing in compression, as it has become so weakened that it cannot withstand the applied compressive
load. For this reason, it is possible to find different failure mechanisms depending on
the quality of the aggregates: for aggregates with poor tribological properties, failure is
likely to be a consequence of aggregate pulverisation, whereas for aggregates with good
tribological properties, failure is likely to occur through the aggregate paste interface. In
other words, failure will always occur through the weakest fraction of the concrete.

7.2 Fatigue Test Methodologies

7.2.1 Fatigue Limit and Fatigue Life

The "fatigue limit" or "fatigue resistance" of a material is defined as the stress below which
collapse does not take place due to the accumulation of damage, so that the material
presents "infinite life" in the considered test conditions. The "infinite life" will depend
on the circumstances to which the material is subjected, such as the range of loads, the
accumulated damage, the frequency of application of the cycles, the properties of the
concrete, and the eccentricity.

The mechanisms of fatigue damage in plain concrete are essentially the same in cases
of tensile strength, bending, or compression, so the considerations regarding static loads
will essentially be the same. Concrete is a material that primarily withstands compression
loads, so from now on the effects that these exert on it will be considered.

The fatigue limit, due to the impossibility of reaching an infinite number of cycles in a test, will correspond to a minimum number of cycles established by agreement. Once this value is exceeded, the concrete will be considered to have "infinite life" under these conditions. Depending on the type of material, common values for this criterion are $1{\cdot}10^6$, $5{\cdot}10^6$ and $1{\cdot}10^7$ cycles. In the case of concrete, $1{\cdot}10^7$ cycles is a conservative value, although in practice durability is not usually observed below this number of cycles. Experimentally, overcoming $2{\cdot}10^6$ cycles without apparent damage may be a sufficient reference to consider infinite life [35].

7.2.2 Strength Range

In a fatigue test, especially for RAC, the fundamental parameter to consider is the level or range of stresses applied on the specimen. In the case that the position of the machine can be controlled, the amplitude of fatigue will be defined by the lower limit and upper limit of the position of the interval. In some cases, such as high frequency (> 10–15 Hz), strain control can be considered (if this is possible) to prevent the material's strain-recovery capacity from being exceeded during load-discharges. However, this type of test on plain concrete is not recommended since it is a material with little deformation and, consequently, the control of the position or the strain would provide a greater uncertainty in the stress-strain correlation than in the load control case.

In general, fatigue tests on concrete will be carried out by controlling the upper and lower load limits. The test will be carried out between lower limit and upper limit of stress of the interval to which the material will be subjected.

The design of the fatigue test can be made from the Goodman diagram, essentially the same for tensile, bending, and compressive tests. Conservatively, it is assumed that with 50 % of the static resistance a concrete could withstand $1{\cdot}10^6$ cycles with a minimum level of zero stress. From this point on, the increase in the value of the minimum stress causes the range of stress that the concrete can withstand to decrease [36]. For example, if the minimum stress value is approximately 20 % of the static strength, the maximum stress that would allow the concrete to withstand $1{\cdot}10^6$ cycles is approximately 60 % of the same static strength. It is known that the influence of recycled aggregate on the fatigue resistance of concrete depends on its strength [37]. When the compressive strength is high, the percentage loss by incorporation of RA is higher. Therefore, when trying to characterize high-strength recycled concretes to fatigue, lower load ranges will be considered than in the case of low strengths.

7.2.3 Wöhler Curve or S-N Curve

It is usual to represent the fatigue behaviour of concrete in a diagram that relates, in ordinate, the maximum stress or the stress interval, referred to the value of the static property, the compressive strength, and in abscissa the number of cycles in logarithmic scale or the logarithm of the number of cycles on a linear scale. The representation of the maximum stress of the range in relation to the compressive strength versus the number of cycles on a logarithmic scale is called a Wöhler or S-N diagram [38].

When the logarithm of the number of cycles is zero ($N = 1$), the maximum load corresponds to the static value of the property studied. It is observed that as the maximum stress level decreases, the number of cycles that the concrete can withstand increases. When the same concrete is characterized in different stress ranges, as the width of the interval becomes larger, the number of cycles that the test specimen can withstand decreases. The Wöhler curve, in general, presents an asymptote that defines the "fatigue limit" sought. However, as mentioned above, in the case of concrete the number of cycles for which "infinite life" will be agreed is below $1\cdot10^7$, before which no asymptotic behaviour or endurance of this material is observable. To obtain the S-N curve, it is necessary to have many specimens to be tested for the different compression load ranges. As concrete is a very heterogeneous material, the number of specimens necessary for fatigue must be in accordance with the number of samples that have been required to determine the compressive strength. Thus, for example, it is common for the S-N curve to represent a 50 % probability of failure and use no less than 10 specimens in each of the stress ranges considered.

Once the compressive strength of concrete is known, the test campaign to determine the S-N curve can start from a maximum value of compressive strength, for example 70 % of the static resistance, testing 10 specimens to failure. From this range, depending on the desired precision in obtaining the S-N curve, as many ranges of test strength as required will be considered, for example the maximum values of 50 % and 90 % of the stress. It is essential to define a limit on the number of cycles, or "run-out", of around $1\cdot10^7$ cycles. If the specimen reaches the run-out without collapsing, it will provide very useful information if it is tested under compression, and it is possible to determine the damage caused, based on the residual resistance.

7.2.4 Wave Type and Frequency

Other variables to consider are the type of wave (sine, square, trapezoidal, triangular, etc.) and the frequency of cycles of loading. In general, the type of wave most used in fatigue tests is the sine type. The influence of the type of wave is not significant since load application speed is not a relevant parameter, although running very large load intervals may require an application of a high velocity of displacement of the test machine actuator.

Regarding the test frequency, it will generally remain constant from the first cycle. However, resonance fatigue equipment requires time to reach the natural frequency of vibration and decreases at the same time as the concrete loses its rigidity due to the accumulation of cycles. In general, fatigue life decreases with frequency because creep is more relevant.

7.2.5 Test Specimens

The shape and slenderness of the specimens used in concrete fatigue tests, just as in static compressive tests, influences the results. It is recommended to use specimens with dimensions and shapes like those used in static tests. Cylindrical specimens should be of a height twice the diameter and with dimensions that are adequate to the maximum size of the aggregate they contain. For this reason, tests on cement paste or mortar samples allow the use of smaller specimens.

To avoid the effect of eccentricity as far as possible, it will always be necessary to face, polish, or carve the open faces of the mould with special care since the small imperfections will have a greater influence on dynamic than static results. To avoid uncertainty due to facing materials, it is advisable to polish the specimens or use prisms obtained from moulds with minimum tolerances.

Characterization tests of fatigued concrete can take a long time. If one test reaches $1 \cdot 10^7$ cycles at a frequency of 10 Hz, the test will last slightly less than 12 days. If the specimen had a young age, its properties would have evolved from the beginning to the end of the test. Furthermore, to obtain an S-N curve, a considerable number of specimens has to be be tested for each stress range. It is not possible to obtain reliable and comparable values of concretes that come from the same batch if the tested specimens are of such an age that their mechanical properties are still evolving. However, it can be assumed that after 365 days of curing the concrete reduces its evolution speed sufficiently [30] and, therefore, it is recommended to carry out fatigue characterizations on concrete that have cured for more than one year..

7.3 Fatigue Behaviour of Recycled Aggregate Concrete Under Compression and Bending Cyclic Loadings

Concrete elements subjected to cyclic loading are in many cases subjected to bending stresses. For this reason, it is possible to find studies on the effect of bending fatigue on concrete elements. However, since concrete is a material that is not able to withstand tensile stresses, such elements are usually reinforced.

Xiao et al. [27] analysed the effect of 4-point bending fatigue on RAC with 100 % RA using $150 \times 150 \times 550$ mm prismatic specimens. They found that the presence of RA reduced the fatigue life compared to concretes with natural aggregates. In this work, they found that the failure mechanisms of specimens with RA subjected to bending were like those obtained in static tests. This is because, as already mentioned, fatigue is a subcritical process.

Arora and Singh [29] also analysed the effect of bending fatigue on concretes with 100 % replacement of natural aggregates by RA. In their work they used a total of 64 prismatic specimens of $100 \times 100 \times 500$ mm. Similar to what Xiao et al. [27] observed, they obtained a strength loss in the concretes with RA of between 7 % to 8 % compared to the reference concrete.

In the case of the bending fatigue studies, in general, these are conducted with well reinforced or fibre-reinforced concretes. In this type of concrete, the material used as reinforcement is mainly responsible for withstanding the tensile stresses. For this reason, the effect of the presence of RA is diluted. To appreciate more clearly the effect of the presence of RA on the fatigue behaviour of concrete, some authors have characterised the compressive fatigue with tests on plain concrete.

Xiao et al. [27], analysed the effect of uniaxial compressive fatigue on concrete with 100 % replacement of natural aggregates by RA. For this purpose, they used 100 × 100 × 300 mm prismatic specimens. This work concludes that the presence of RA does not reduce the fatigue life of the concrete.

Thomas et al. [30, 32] analysed the effect of different percentages of substitution of natural aggregates by RA (0, 20, 50 and 100 % respectively) using different water/cement ratios, resulting in a total of 24 mixes. They used different fatigue characterisation methods on these mixes and concluded that if the replacement of natural aggregates by RA is limited to 20 %, the RAC will not lose fatigue strength.

7.3.1 Strain Versus Cycles

Thomas et al. [30, 32] demonstrated that if the specimen reaches the run-out, as shown in Fig. 7.2, a moderate quasi-logarithmic increase in the maximum and minimum strain from the first to the last cycle is observed. This same phenomenon occurs in the case of conventional concretes. In this type of behaviour, the difference between the maximum and minimum strain slightly increases throughout the test.

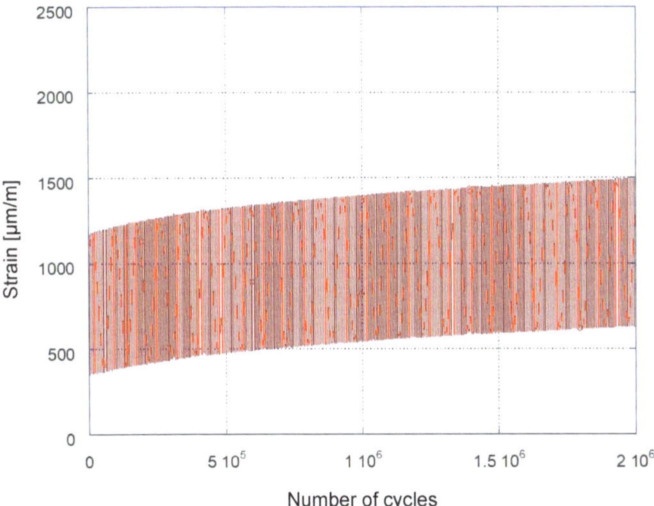

Fig. 7.2   Strain versus cycles of a concrete specimen that exceeds run-out cycles.

If the specimen fails before run-out, Fig. 7.3, three different stages (A, B and C) are observed. In this case, the three stages representing approximately 15 %, 70 %, and 15 % of the total lifetime respectively. The first stage A is like that observed in conventional concretes exceeding run-out cycles. This is characterized by a quasi-logarithmic increase in the maximum and minimum strain, the increase rate of the maximum strain being slightly higher than the minimum. This stage corresponds to the initial formation of micro-cracks. These same phenomena occur in the case of conventional concretes.

To observe the micro-cracks of RAC, some samples have been analysed by scanning electron microscopy (SEM). Fig. 7.4 shows the generation of micro-cracks in a pore of the cement paste of concrete. The second stage B shows an almost linear increase in strain, which is caused by the accumulation of damage due to fatigue cycles. In this second stage, the crack growth results in a loss of stiffness evidenced by the slight increase of the gap between the maximum and minimum strain. The stress limits are also constant so the stiffness in this stage decreases linearly.

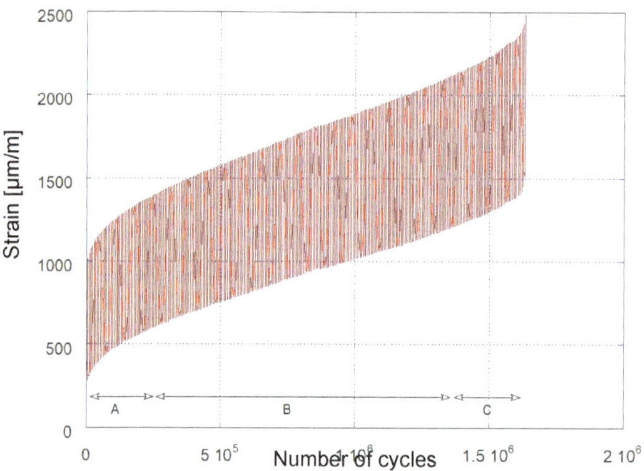

Fig. 7.3   Strain versus cycles of a concrete specimen with failure before the run-out cycles.

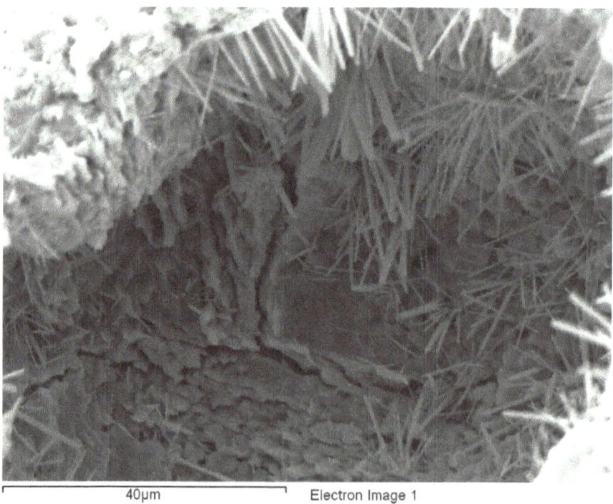

Fig. 7.4   Initial formation of micro-cracks.

The stage is associated with the growth and stabilization of the formed cracks. Fig. 7.5 shows the growth of cracks through the cement paste and the cement-paste interface. A third and final stage C is observed. In this stage the interconnections of cracks cause the collapse of the concrete under fatigue loading.

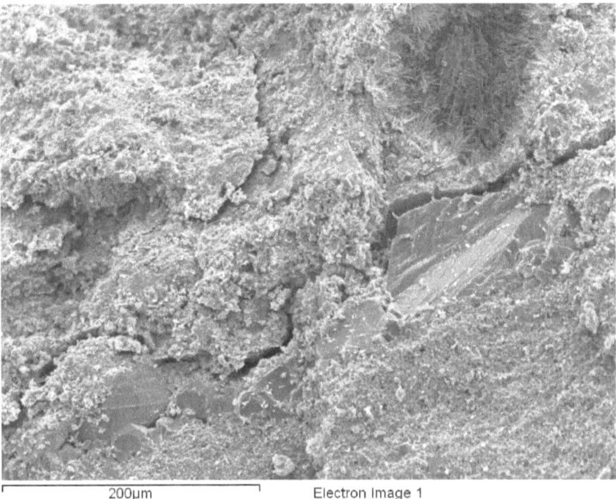

Fig. 7.5 Crack growth.

Fig. 7.6 Interconnections of cracks.

The increase in deformation, unlike the previous stages A and B, is exponential: an accelerated loss of stiffness is accompanied by a greater gap between the upper and lower envelopes of maximum and minimum strain with the cycles. Fig. 7.6 shows the interconnections of cracks causing the collapse of the concrete.

7.3.2 Fatigue Limit Versus Compressive Strength

Fig. 7.7 shows the results obtained by Thomas et al. [30, 32] for fatigue limit versus the compressive strength of the corresponding concretes for the different substitution degrees (0 %, 20 %, 50 % and 100 %).

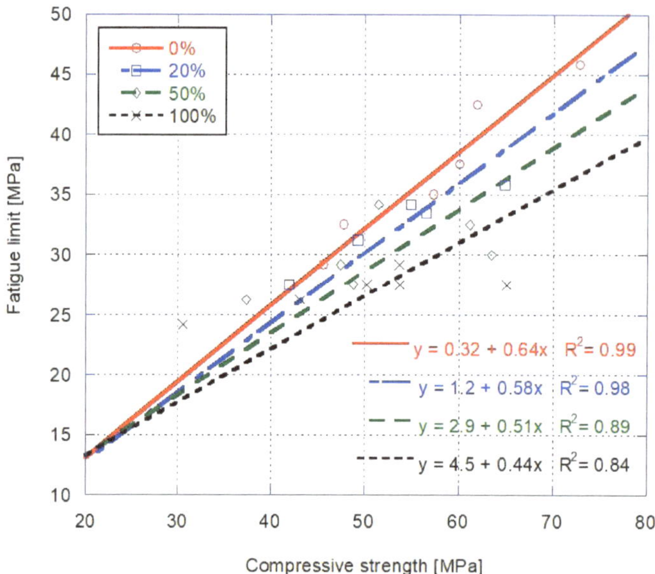

Fig. 7.7   Fatigue limit versus the compressive strength of recycled concretes.

A linear behaviour of the fatigue limit versus the compressive strength is observed for the different degrees of substitution. The curves tend to converge when the compressive strengths are lower than 30 MPa. Below this value of compressive strength, the fatigue damage is ruled by the characteristics of the cement paste, the presence of recycled aggregate having no influence. Due to the characteristics of the RA, the compressive strength of the original concrete should be slightly higher than 30 MPa. This suggests that RAC with a compressive strength of the cement paste less than the RA compressive strength fail by processes originating in the cement paste. However, when the cement paste compressive strength is higher than 30 MPa the failure initiation takes place in the RA.

Taking a 60 MPa concrete, the incorporation of 20 % RA reduces the fatigue limit of the concrete by around 10 %. This difference is greater when higher compressive strengths are considered. When the substitution is total, the slope of the fatigue limit is approximately 25 % between control and recycled concretes with the same compressive strength.

7.4 Fatigue Failure Micro-Mechanisms by Micro-CT Analysis

Although fatigue damage is of great interest in concrete, there is still a great lack of knowledge regarding the mechanisms of fatigue failure in this type of material. In order to deepen the knowledge of the effect of fatigue in these materials and to know how it affects the application of load cycles on concrete specimens, several authors have used a novel technique, the micro-computed axial tomography (micro-CT), which has great potential to discover the effect of load cycles on concrete.

Computed tomography makes it possible to generate 3D models of the specimen being analysed, allowing the different phases that compose the sample to be identified. The main advantage of this technique compared to others is that it allows to distinguish not only at a topological level, but also to distinguish what the sample looks like inside. This technique makes it possible to identify the crack growth process qualitatively and quantitatively in a concrete or mortar specimen.

Regarding the use of micro-CT to analyse the effect of fatigue on concrete, Skarżyński et al. [28] used this powerful technique to analyse fracture evolution in concrete compressive fatigue, analysing the evolution process of open and closed porosity and fissures. In this study, they observed that as the number of cycles increased, the volume of pores and fissures in the specimen increased with a hyperbolic tendency. Within this evolution he distinguished three zones, micro-crack, transient and crack extension part. Vicente et al. [39] analysed the damage suffered by a concrete with metallic fibres being able to obtain information on the failure micro-mechanisms in the interaction between the metallic fibres and the concrete.

Regarding the use of micro-CT to analyse the effect of fatigue on RAC, only one case study has been found [40]. In this work, Sainz-Aja et al. isolated the most critical fraction of RAC, fine aggregate and manufactured mortar. In this work, three different materials were used: firstly, a mortar with natural siliceous aggregate that was used as a reference, secondly, a recycled mortar with RA from crushed railway ballast, and lastly, a second recycled mortar with aggregate from crushed railway sleepers that were out of service. Locati fatigue tests were performed on these three materials. Before the start of each step of the Locati test, a micro-CT scan of the recycled mortar sample was performed, and the steps were performed until failure of the specimens occurred. In this way, it was possible to see the evolution of the specimens throughout the fatigue tests. In each of the scans, the internal cracking state of the specimens was visually analysed, and the volume of accessible pores, closed pores, and cracks was quantified.

In this work, it was concluded that similar to what was observed by Skarżyński et al. [28], as the number of cycles increases, the volume of pores and cracks increases, so this information can be used as a fatigue damage parameter. However, it is indicated that the evolution of the pore volume is lighter, while the evolution of the crack volume is a significant parameter and, therefore, more suitable to be taken as a reference. It was also noted that in mortars with very low load values no cracks were found, so it was possible to estimate a fatigue limit for these mortars. On the other hand, from the physical observation of the cracking process, it was concluded that the geometry of the aggregates plays a fundamental role in the fatigue resistance of both recycled and conventional concretes. This is because the presence of sharp edges produces a stress concentrating effect that facilitates the appearance of cracks and therefore reduces the fatigue life of the concrete.

7.5 Resonance Fatigue Behaviour of Recycled Concrete

One of the major problems with fatigue characterisation is that it generally requires many tests of long duration. For this reason, defining a methodology that reduces the characterisation time would not only facilitate the study of fatigue, but it would also reduce the costs necessary to perform this characterisation. To reduce these test times, one of the options is to increase the test frequency. Increasing the test frequency has two main possible consequences , as already mentioned. On the one hand, it has a direct influence on the test time and, on the other hand, it can have a certain influence on the mechanical response of the concrete.

Regarding the recommended test frequency for fatigue testing in compression, good practice recommends fatigue testing between 1 and 15 Hz [41–43]. It is well known that using frequencies lower than 1 Hz reduces the fatigue limit due to an increase in the creep damage [28, 43, 44]. On the other hand, the upper limit of 15 Hz has not a clear reason. Recent technical machines allow to use frequencies hat can exceed 100 Hz, referred to as high-frequency fatigue [45]. As is expected, high-frequency testing achieves a significant reduction in the time to reach the limit of fatigue of a material. However, doubts arise regarding the validity of comparisons with low-frequency tests [30, 32]. The main way to achieve such high frequencies in compression fatigue is by means of resonance machines, which perform the fatigue test at the resonance frequency of the assembly, which can vary throughout the test if the stiffness of the specimen is modified.

There are very few studies that analyse the effect of fatigue testing at very high frequency. Thomas et al [33] compared the performance of a high performance self-compacting concrete tested at moderate and very high frequencies. From this comparison they found that testing at a moderate frequency resulted in a higher fatigue life than that obtained at very high frequency. Despite obtaining different fatigue results, they found that the effect of increasing the number of cycles was similar in both cases and that the fracture surfaces were similar in both cases. Furthermore, from this analysis it was found that the resonance frequency was a parameter to estimate the damage suffered by the specimen. In addition, based on the results obtained, it was concluded that the test frequency recommended by good practice gives the maximum fatigue strength values, which in principle goes against good practice.

In relation to the effect of resonant fatigue on RAC, three papers have been found [34, 46, 47]. Cantero et al. [34] performed resonant fatigue tests with different percentages of RA and cement. Sainz-Aja et al. [46] compared two resonant frequency fatigue characterisation methods, namely the Locati method and the Staircase method. Sainz-Aja et al. [47] compared the fatigue performance of three types of RAC tested at moderate and resonant frequency using the Locati method.

Cantero et al. [34] carried out Locati tests on 6 different concrete mixes. Three of them used ordinary Portland cement and three used recycled cement. In addition, they used three replacement percentages of natural aggregate by mixed recycled aggregates (MRA), 0, 20, and 50 respectively. On these 6 types of concrete, they analysed the effect of fatigue on the fatigue limit, specimen strain, resonance frequency, elastic modulus, and fracture surfaces. In this work, it was concluded that the presence of MRA increased the specimen strain during resonant fatigue tests, which in turn resulted in a reduction of the resonance frequency of the test. In the compressive resonance fatigue tests, the MRA specimens exhibited fracture planes across the paste-aggregate interface, while the specimens with MRA showed cracks through the RAC aggregates and pulverisation of the ceramic fractions.

Sainz-Aja et al. [46] tested three types of RAC under resonant compressive fatigue, one with 100 % aggregates from out of use railway ballast, another with 100 % aggregates from out of use railway sleepers and a third type combining both types of RA. Three types of fatigue tests were used on these three types of concrete: a Locati test with $2 \cdot 10^5$ cycles per step, a Locati test with $5 \cdot 10^5$ cycles per step, and a Staircase test with $2 \cdot 10^6$ cycles per step. In this work they analysed the fatigue limit obtained based on the three types of tests proposed, finding that the number of cycles per step of the Locati test had no influence and observing a correlation between the Locati and Staircase methods. They also obtained similar results to the study by Cantero et al. [34] regarding the evolution of strain and resonance frequency throughout the test. Furthermore, they found that during the resonant fatigue tests there was an increase in the temperature of the specimens tested.

Sainz-Aja et al [47] analysed the influence of frequency when testing RAC. For this purpose, they performed fatigue tests at moderate frequency (10 Hz) and resonant fatigue ($\approx$90 Hz) on the same 3 mixes used in [46]. In this study they found that the very high frequency tests resulted in lower fatigue limits, which corresponds with the rest of the literature. Furthermore, this work showed that this effect was greater in cases where RA from crushed sleepers were used. It was found that in the specimens tested under resonant fatigue, there was a significant increase in the specimens' temperature, ranging between 60 and 100 °C. This thermal increase experienced by the specimens accelerates the damage caused by creep, justifying the reduction in fatigue life observed in the specimens tested under resonant fatigue. Furthermore, the fact that this reduction is caused by an increase in creep damage also justifies the higher declines of fatigue life in the case of RAC under resonant fatigue.

7.6 Effect of Temperature on Recycled Concrete

Sainz-Aja et al [46] found that during resonant fatigue tests there is a reduction in the fatigue limit and, simultaneously, an increase in the temperature of the specimens, with recorded temperatures reaching values between 60 and 100 °C. These same authors, seeking to verify whether there is a relationship between the two effects, carried out another study [48]. In this study [48] they hypothesise that compressive fatigue damage is a combination of fatigue damage and creep damage. The explanation for the premature failure of specimens tested at very high frequency is that the increase in temperature registered during the test increases the creep damage of the concrete. In this publication, the authors demonstrate that temperature plays an important role in the creep damage of concrete, something never found before. They also provide explanations to the two questions that were unanswered until now: Why does temperature increase during fatigue tests at very high frequencies? Why is this effect greater in the case of specimens with RA from concrete?

Concrete is a material that always has cracks in its interior. For every crack, two crack faces are in contact with each other. During the application of loads, a relative movement occurs between the two faces resulting in friction which leads to a certain thermal increase. The reason why the temperature rise is only noticeable in very high frequency tests is because the time between cycles is reduced, so the specimen does not have enough time to dissipate the energy generated by friction in the time between cycles, which is the case at moderate frequency.

When the aggregates used come from concrete samples, the new concrete has a higher mortar content. It is known that mortar is the material that is affected by the creep phenomenon. For this reason, the effect of increasing the frequency has a greater impact on concretes with RA from RAC.

7.7 Fatigue for Railway Superstructure Applications

There are several elements of the railway superstructure that are typically made of concrete, which are mainly the sleepers and/or slab track. Several authors have raised the possibility of manufacturing railway superstructure elements with RAC. However, as already mentioned, railway superstructure elements are one of these components where fatigue failures can occur. For this reason, it is possible to find several research works in which authors have designed railway superstructure elements in RAC and subsequently evaluated their fatigue behaviour.

The PhD Thesis "Slab track manufacture using out of service railway wastes" [49] proposes the valorisation of the waste generated in track removal as RA and its use to manufacture the new track, thus encouraging a recycling loop. In this PhD Thesis, three types of RAC were designed using the different aggregates obtained from the crushing of the railway superstructure elements. These three types of concrete were characterised in fatigue tests, obtaining the result that their fatigue limit was above the requirements of the concretes used in various commercial types of slab track. In addition, this PhD Thesis analysed the influence of replacing the conventional concrete with the designed RA concretes on the mechanical response of the track, concluding that the influence was minimal.

Apart from the possibility of manufacturing slab track with RAC, there are also some cases that propose manufacturing prestressed sleepers with RAC. González-Corominas et al. [50] proposed the use of waste from track removal to manufacture new elements. Specifically, they manufactured and tested sleepers with RAC made from RA from railway sleepers. In this work, they used a reference concrete and two RAC. The difference between the two RAC was in the percentage of natural aggregates replaced, 50 % and 100 %, respectively. The three types of sleepers produced complied with the requirements for railway sleepers and achieved similar results.

Although not manufactured with RAC, Koh et al. [51, 52] manufactured an eco-friendly prestressed concrete sleeper by replacing conventional aggregates with two types of slag by designing three mixes, a reference mix and a mix with each of the slag types. In all cases they obtained better results in the eco-friendly sleeper. Koh et al. justified this improvement by the mechanical response of the sleepers due to the better tribological properties of slag compared to natural aggregates.

7.8 Conclusions

After an extensive literature review on the state of the art of recycled concrete subjected to fatigue, the following conclusions have been obtained:

- Recycled aggregate concrete is more affected by fatigue compared to natural aggregate concrete.
- A linear relationship between fatigue limit and compressive strength is observed for different degrees of substitution.
- The micro-damage mechanisms that cause fatigue are similar to those observed in conventional concrete. The increased effect of fatigue in recycled aggregate concrete can be attributed to a higher presence of initial defects in the concrete (pre-existing cracks in the recycled aggregates).
- It has been found that for very high-frequency tests ($\approx$100 Hz), the fatigue life of concrete decreases, particularly in the case of recycled aggregate concrete. This is because high-frequency fatigue tests generate an increase in the temperature of the concrete, which accelerates creep damage in the concrete. The effect is more pronounced in the case of recycled aggregate concrete than in conventional concrete.

These findings show the importance of considering the presence of recycled aggregates and the influence of high-frequency testing on the fatigue performance of concrete structures.

7.9 References

[1]   A. Rao, K.N. Jha, S. Misra, Use of aggregates from recycled construction and demolition waste in concrete, Resources, Conservation and Recycling, 50 (2007) 71–81. https://doi.org/10.1016/j.resconrec.2006.05.010

[2]   A.M. Wagih, H.Z. El-Karmoty, M. Ebid, S.H. Okba, Recycled construction and demolition concrete waste as aggregate for structural concrete, HBRC Journal, 9 (2013) 193–200. https://doi.org/10.1016/j.hbrcj.2013.08.007

[3]   S. Kou, Reusing recycled aggregates in structural concrete, Hong Kong Polytechnic Univerisity, Hong Kong, 2006, 278 p.

[4]   M. Leite, Evaluation of the mechanical properties of concrete made with recycled aggregates from construction and demolition waste, University of Rio Grande do Sul, Brazil (2001), 290 p.

[5]   M. Solyman, Classification of recycled sands and their applications as fine aggregates for concrete and bituminous mixtures, University of Kassel, Germany (2005), 194 p.

[6]   C.S. Poon, S.C. Kou, L. Lam, Use of recycled aggregates in molded concrete bricks and blocks, Construction and Building Materials, 16 (2002) 281–289.

[7]   C. Thomas, J. Setién, J.A. Polanco, Structural recycled aggregate concrete made with precast wastes, Construction and Building Materials, 114 (2016) 536–546. https://doi.org/10.1016/j.conbuildmat.2016.03.203

[8]   C. Thomas, J. Setién, J.A. Polanco, P. Alaejos, M. Sánchez de Juan, Durability of recycled aggregate concrete, Construction and Building Materials, 40 (2013) 1054–1065. https://doi.org/10.1016/j.conbuildmat.2012.11.106

[9]   D. Soares, J. de Brito, J. Ferreira, J. Pacheco, Use of coarse recycled aggregates from precast concrete rejects: Mechanical and durability performance, Construction and Building Materials, 71 (2014) 263–272. https://doi.org/10.1016/j.conbuildmat.2014.08.034

[10] S.C. Kou, C.S. Poon, Enhancing the durability properties of concrete prepared with coarse recycled aggregate, Construction and Building Materials, 35 (2012) 69–76. https://doi.org/10.1016/j.conbuildmat.2012.02.032

[11] J. Sainz-Aja, I. Carrascal, J.A. Polanco, C. Thomas, I. Sosa, J. Casado, S. Diego, Self-compacting recycled aggregate concrete using out-of-service railway superstructure wastes, Journal of Cleaner Production, 230 (2019) 945–955. https://doi.org/10.1016/j.jclepro.2019.04.386

[12] J. Thomas, N.N. Thaickavil, P.M. Wilson, Strength and durability of concrete containing recycled concrete aggregates, Journal of Building Engineering, 19 (2018) 349–365. https://doi.org/10.1016/j.jobe.2018.05.007

[13] A. Akbarnezhad, K.C.G. Ong, C.T. Tam, M.H. Zhang, Effects of the parent concrete properties and crushing procedure on the properties of coarse recycled concrete aggregates, Journal of Materials in Civil Engineering, 25 (2013) 1795–1802. https://doi.org/10.1061/(asce)mt.1943-5533.0000789

[14] A.K. Padmini, K. Ramamurthy, M.S. Mathews, Influence of parent concrete on the properties of recycled aggregate concrete, Construction and Building Materials, 23 (2009) 829–836. https://doi.org/10.1016/j.conbuildmat.2008.03.006

[15] F. Rodrigues, M.T. Carvalho, L. Evangelista, J. de Brito, Physical-chemical and mineralogical characterization of fine aggregates from construction and demolition waste recycling plants, Journal of Cleaner Production, 52 (2013) 438–445. https://doi.org/10.1016/j.jclepro.2013.02.023

[16] S.C. Angulo, Variabilidade de agregados graúdos de resíduos de construção e demolição reciclados, MSc Thesis – University of São Paulo, Brazil (2001), 172 p.

[17] K. McNeil, T.H.-K. Kang, Recycled Concrete Aggregates: A Review, International Journal of Concrete Structures and Materials, 7 (2013) 61–69. https://doi.org/10.1007/s40069-013-0032-5

[18] L. Evangelista, J. De Brito, Concrete with fine recycled aggregates: A review, European Journal of Environmental and Civil Engineering, 18 (2014) 129–172. https://doi.org/10.1080/19648189.2013.851038

[19] F. Debieb, L. Courard, S. Kenai, R. Degeimbre, Mechanical and durability properties of concrete using contaminated recycled aggregates, Cement and Concrete Composites 32 (2010) 421–426. https://doi.org/10.1016/j.cemconcomp.2010.03.004

[20] M. Martín-Morales, M. Zamorano, A. Ruiz-Moyano, I. Valverde-Espinosa, Characterization of recycled aggregates construction and demolition waste for concrete production following the Spanish Structural Concrete Code EHE-08, Construction and Building Materials, 25 (2011) 742–748. https://doi.org/10.1016/j.conbuildmat.2010.07.012

[21] M.S. de Juan, P.A. Gutiérrez, Study on the influence of attached mortar content on the properties of recycled concrete aggregate, Construction and Building Materials, 23 (2009) 872–877. https://doi.org/10.1016/j.conbuildmat.2008.04.012

[22] C. Medina, W. Zhu, T. Howind, M.I. Sánchez de Rojas, M. Frías, Influence of mixed recycled aggregate on the physical-mechanical properties of recycled concrete, Journal of Cleaner Production, 68 (2014) 216–225. https://doi.org/10.1016/J.JCLEPRO.2014.01.002

[23] P. Saravana Kumar, G. Dhinakaran, Effect of Admixed Recycled Aggregate Concrete on Properties of Fresh and Hardened Concrete, Journal of Materials in Civil Engineering, 24 (2012) 494–498. https://doi.org/10.1061/(asce)mt.1943-5533.0000393

[24] D. Matias, J. de Brito, A. Rosa, D. Pedro, Durability of concrete with recycled coarse aggregates: Influence of superplasticizers, Journal of Materials in Civil Engineering, 26 (2014). https://doi.org/10.1061/(asce)mt.1943-5533.0000961

[25] J. Sainz-Aja, J. Pombo, D. Tholken, I. Carrascal, J. Polanco, D. Ferreño, J. Casado, S. Diego, A. Perez, J.E.A. Filho, A. Esen, T.M. Cebasek, O. Laghrouche, P. Woodward, Dynamic calibration of slab track models for railway applications using full-scale testing, Computers & Structures, 228 (2020) 106180. https://doi.org/10.1016/j.compstruc.2019.106180

[26] A. Alliche, Damage model for fatigue loading of concrete, International Journal of Fatigue. 26 (2004) 915–921. https://doi.org/10.1016/j.ijfatigue.2004.02.006

[27] J. Xiao, H. Li, Z. Yang, Fatigue behavior of recycled aggregate concrete under compression and bending cyclic loadings, Construction and Building Materials, 38 (2013) 681–688. https://doi.org/10.1016/j.conbuildmat.2012.09.024

[28] Ł. Skarżyński, I. Marzec, J. Tejchman, Fracture evolution in concrete compressive fatigue experiments based on X-ray micro-CT images, International Journal of Fatigue, 122 (2019) 256–272. https://doi.org/10.1016/j.ijfatigue.2019.02.002

[29] S. Arora, S.P. Singh, Analysis of flexural fatigue failure of concrete made with 100 % Coarse Recycled Concrete Aggregates, Construction and Building Materials, 102, Part (2016) 782–791. https://doi.org/doi.https://doi.org/10.1016/j.conbuildmat.2015.10.098

[30] C. Thomas, J. Setién, J.A. Polanco, I. Lombillo, A. Cimentada, Fatigue limit of recycled aggregate concrete, Construction and Building Materials, 52 (2014) 146–154. https://doi.org/https://doi.org/10.1016/j.conbuildmat.2013.11.032

[31] C. Thomas, I. Carrascal, J. Setién, J.A. Polanco, Determinación del límite a fatiga en hormigones reciclados de aplicación estructural – Determining the fatigue limit recycled concrete structural application (In Spanish), Anales de Mecánica de la Fractur 1 (2009) 283–289.

[32] C. Thomas, I. Sosa, J. Setién, J. a. Polanco, A.I.A.I. Cimentada, Evaluation of the fatigue behavior of recycled aggregate concrete, Journal of Cleaner Production, 65 (2014) 397–405. https://doi.org/10.1016/j.jclepro.2013.09.036

[33] C. Thomas, J. Sainz-Aja, J. Setien, A. Cimentada, J.A. Polanco, Resonance fatigue testing on high-strength self-compacting concrete, Journal of Building Engineering, 35 (2021) 102057. https://doi.org/10.1016/j.jobe.2020.102057

[34] B. Cantero, J. Sainz-Aja, A. Yoris, C. Medina, C. Thomas, Resonance fatigue behaviour of concretes with recycled cement and aggregate, Applied Sciences, 11 (2021). https://doi.org/10.3390/app11115045

[35] C. Thomas, J. Setién, J.A. a. Polanco, I. Lombillo, A. Cimentada, Fatigue limit of recycled aggregate concrete, Construction and Building Materials, 52 (2014) 146–154. https://doi.org/10.1016/j.conbuildmat.2013.11.032

[36] M.E. Awad, H.K. Hilsdorf, K.D. Raithby, J.W. Galloway, S. Soretz, S.S. Takhar, I.J. Jordaan, B.R. Gamble, J.M. Hanson, N.F. Somes, T. Helgason, I.C. Jhamb, J.G. MacGregor, N.M. Hawkins, L.W. Heaton, Fatigue of Concrete: Abeles Symposium, Proceedings, 1972., (n.d.).

[37] C. Thomas, J. Setién, J.A. Polanco, P. Alaejos, M. Sánchez De Juan, Durability of recycled aggregate concrete, Construction and Building Materials, 40 (2013) 1054–1065. https://doi.org/10.1016/j.conbuildmat.2012.11.106

[38] A. Wöhler, Wöhler's experiments on the strength of metals, Engineering. 4 (1867) 160–161.

[39] M.A. Vicente, G. Ruiz, D.C. González, J. Mínguez, M. Tarifa, X. Zhang, CT-Scan study of crack patterns of fiber-reinforced concrete loaded monotonically and under low-cycle fatigue, International Journal of Fatigue, 114 (2018) 138–147. https://doi.org/10.1016/j.ijfatigue.2018.05.011

[40] J. Sainz-Aja, I. Carrascal, J.A. Polanco, C. Thomas, Fatigue failure micromechanisms in recycled aggregate mortar by μCT analysis, Journal of Building Engineering, 28 (2020). https://doi.org/10.1016/j.jobe.2019.101027

[41] A. Medeiros, X. Zhang, G. Ruiz, R.C. Yu, M. de S.L. Velasco, Effect of the loading frequency on the compressive fatigue behavior of plain and fiber reinforced concrete, International Journal of Fatigue, 70 (2015) 342–350. https://doi.org/ 10.1016/j.ijfatigue.2014.08.005

[42] J.W. Murdock, A critical review of research on fatigue of plain concrete, Engineering Experiment Station Bulletin 475. University of Illinois at Urbana Champaign, College of Engineering, 1965, 25 p..

[43] C. Rep, Considerations for Design of Concrete Structures Subjected to Fatigue Loading, Journal of the American Concrete Institute. 71 (1974) 97–121. https://doi.org/10.14359/11172

[44] European Concrete Committee, Ceb Committee GTG 15. Fatigue of concrete structures. CEB Bulletin d'information, N° 188 (1988).

[45] J. Sainz-Aja, C. Thomas, J.A. Polanco, I. Carrascal, High-Frequency Fatigue Testing of Recycled Aggregate Concrete, Applied Sciences, 10 (2019) 10. https://doi.org/10.3390/app10010010

[46] J. Sainz-Aja, C. Thomas, I. Carrascal, J.A. Polanco, J. de Brito, Fast fatigue method for self-compacting recycled aggregate concrete characterization, Journal of Cleaner Production, 277 (2020) 123263. https://doi.org/10.1016/j.jclepro.2020.123263

[47] J. Sainz-Aja, C. Thomas, J.A. Polanco, I. Carrascal, High-frequency fatigue testing of recycled aggregate concrete, Applied Sciences, 10 (2020). https://doi.org/10.3390/app10010010

[48] J.A. Sainz-Aja, I.A. Carrascal, J.A. Polanco, C. Thomas, Effect of temperature on fatigue behaviour of self-compacting recycled aggregate concrete, Cement and Concrete Composites, (2021) 104309. https://doi.org/10.1016/j.cemconcomp.2021.104309

[49] J.A. Sainz-Aja, Slab track manufacture using out of service railways wastes, PhD Thesis. University of Cantabria, Spain (2019), 325 p.

[50] A. Gonzalez-Corominas, M. Etxeberria, I. Fernandez, Structural behaviour of prestressed concrete sleepers produced with high performance recycled aggregate concrete, Materials and Structures, 50 (2017). https://doi.org/10.1617/s11527-016-0966-6

[51] T. Koh, M. Shin, Y. Bae, S. Hwang, Structural performances of an eco-friendly prestressed concrete sleeper, Construction and Building Materials, 102 (2016) 445–454. https://doi.org/10.1016/j.conbuildmat.2015.10.189

[52] T. Koh, S. Hwang, Field evaluation and durability analysis of an eco-friendly prestressed concrete sleeper, Journal of Materials in Civil Engineering, 27 (2014) B4014009. https://doi.org/10.1061/(asce)mt.1943-5533.0001109

Chapter 8

Structural Design of Recycled Aggregate Concrete

Rui Vasco Silva[1], Jorge De Brito[2]

1 PhD Civil Engineering, CERIS, Instituto Superior Técnico, Av. Rovisco Pais, 1049-001 Lisbon, Portugal
2 Full Professor, Department of Civil Engineering, Instituto Superior Técnico, Av. Rovisco Pais, 1049-001 Lisbon, Portugal

8.1 Introduction

In spite of the success of implementing recycled aggregate concrete (RAC) in building structures in a handful of countries for quite some time, it is widely regarded with distrust by most individuals in the construction industry. This widespread lack of acceptance is generally associated with the alleged lack of quality of recycled aggregates (RA) and unpredictable structural behaviour. RA from construction and demolition waste (CDW) do present a relatively lower quality when compared with natural aggregates (NA) and are likely to lead to a decline in mechanical and durability performance [1]. However, the quality of RA can be measured quantitively based on their physical properties and composition and their negative effects on concrete properties can be easily countered.

Advances in structural RAC behaviour and design have been made recently to facilitate the path towards standard practice. Studies on this have generally reported that, for equivalent cross-section geometries, the load-bearing capacity may be lower and the deflection significantly increases [2–7]. Apart from the similar failure mechanism seen in the two types of concrete [8–10], some have observed lower instant and total deflections of structural RAC elements when compared with control natural aggregate concrete (NAC) elements, after increasing the strength class of the former [11, 12]. This concept, which can be achieved by means of decreasing the effective water content [13] (possible with a combination of water reducing admixtures and the use of the water compensation method [14], which is known to result in minimum losses in slump and strength [15–19]), rather than increasing the cement content [20, 21], is probably the most effective approach to produce a material with a predictable performance both in the material and structural level [22], and sticking to the initial goal of reducing the environmental impacts. In fact, increasing the amount of cement to compensate the strength loss of incorporating RA is a practice that should be avoided in the light of the considerable environmental impacts that it represents [23].

According to EN standardization [24], RA may be composed of a mixture of several types of construction materials, but mostly of crushed concrete, natural aggregate waste, and masonry fragments, with significant amounts of other components looked at as contaminants. Considerable variation in the amount of the aforementioned components is typically observed, which significantly affects the properties of the combined material. Therefore, rather than relying on a characterization based on composition only, research has been focusing on the use of performance-based approaches to classify RA [25], which has been found to be a more comprehensive approach to estimate the performance of concrete [25, 26]. In previous studies, the authors have developed several practical rules for the prediction of the mechanical and durability related performance of RAC [7, 22, 25–32], based on the results of up to date meta-analyses and a systematic literature review on the matter. Working on the assumptions of those previous studies, this study presents a proposal for a performance-based design of reinforced RAC slabs and beams in conformity with Eurocode 2 [33]. This approach allows understanding the implications concerning the cross-section's geometry of a reinforced RAC element and adjusting it so that the same service life, load bearing capacity, and long-term deformation are guaranteed as those of a corresponding NAC element, all within the limits established by Eurocode 2 (EC2).

8.2 Methodology

8.2.1 Certified Recycled Aggregates

In recycling facilities, the beneficiation treatments may result in aggregates with varying characteristics [25]. To understand the impact of using such RA on the properties of concrete and its structural behaviour, minimum requirements must be established and standardized. Table 8.1 presents the authors' proposal for the requirements of coarse RA [35] with a particle size over 4 mm, following the test methods of Comité Européen de Normalisation (CEN) standards. It is a corrected version of that present in EN 206:2013+A1 [36] for the use of coarse RA.

The reason for this correction lies in the existence of lenient limits (and even the absence of some), which are likely to result in the certification of low-quality RA. Thus, three classes of RA are considered – A, B, and C – subdivided according to their updated physical properties [25], though their composition should also be discriminated in accordance with EN 933-11 [24]. The main changes to this characterization include:

1. An additional class of RA (i.e., class C), which corresponds to a lower-grade product obtained from the beneficiation of CDW. This RA is meant for unreinforced concrete (e.g., blinding concrete);

2. New limits for the minimum oven-dry particle density and maximum water absorption were proposed (existing limits for A and B are 2100 kg/m³ and 1700 kg/m³, respectively, and there are no limits imposed for water absorption). The proposed

Table 8.1 Proposed recommendations for the use of coarse RA according to EN 12620 [35]

Property	Clause in EN 12620:2002+A1:2008	Aggregate class	Category according to EN 12620
Fines content	4.6	A+B+C	Category or value to be declared
Flakiness index	4.4	A+B	$\leq FI_{50}$ or $\leq SI_{55}$
Resistance to fragmentation	5.2	A+B+C	$\leq LA_{50}$
Oven dried particle density (ρ_{rd})	5.5	A	$\geq 2400\ kg/m^3$
		B	$\geq 2100\ kg/m^3$
		C	$\geq 1800\ kg/m^3$
Water absorption	5.5	A	$\leq 3.5\,\%$
		B	$\leq 8.5\,\%$
		C	$\leq 15\,\%$
Constituents	5.8	A	Rc_{90}, Rcu_{95}, Rb_{10-}, Ra_{1-}, FL_{2-}, XRg_{1-}
		B	Rc_{50}, Rcu_{70}, Rb_{30-}, Ra_{5-}, FL_{2-}, XRg_{2-}
		C	Rc_{NR}, Rcu_{NR}, Rb_{NR}, Ra_{10-}, FL_{5-}, XRg_{2-}
Acid-soluble sulphate	6.3.1	A+B+C	$\leq AS_{0.8}$
Water soluble sulphate content	6.3.3	A+B+C	$\leq SS_{0.2}$
Acid-soluble chloride ion content	6.2	A+B	$0.04\,\%$
		C	NR
Effect on initial setting time	6.4.1	A+B	$\leq A_{10}$
		C	$\leq A_{40}$

Notes: FI – flakiness index (EN 933-3 [39]); SI – shape index (EN 933-4 [40]); LA – Los Angeles (EN 1097-2 [41]); AS – acid-soluble sulphate (EN 1744-1 [42]); SS – water-soluble sulphate (EN 1744-1 [42]); NR – no requirement; A – change in initial setting time (EN 1744-6 [43]). Rc – content by mass of concrete-based fragments; Rcu – concrete-based fragments + unbound aggregates; Rb – masonry-based fragments; Ra – bituminous materials; Rg – glass; X – other materials (EN 933-11 [24])

values corresponding to classes A, B, and C RA are based on the authors' previous studies [25, 26]. This performance-based classification was found to present strong correlations with the performance of the resulting concrete;

3. Additional limits for the acid-soluble chloride ion content and sulphates, as well as water-soluble sulphates for all RA classes (presently, there are no limits for acid-soluble chloride ion and sulphate content);

4. Decrease of the class corresponding to the influence of the initial setting time for classes A and B (current class is A_{40}).

In the light of the compositional variability of coarse RA, these should be considered as high-reactivity aggregates in terms of alkali-silica reactions, unless additional testing proves otherwise [37]. The alkali content of RA should be determined according to the regulations applying at the place of use and follow the guidelines of CEN/TR-16349 [38].

8.2.2 Structural Design Methodology

The same structural design principles of conventional concrete may be applied in RAC. Naturally, this must take the RA's varying physical properties into consideration, which are likely to result in some loss in performance. In the absence of experimental trials to ascertain the actual effect of a given RA on the properties of concrete, maximum replacement levels and conservative correction coefficients (Tables 8.2 and 8.3, respectively) may be applied for the design of structural concrete as per EN 206:2013+A1 [36] and EC2 [34].

Table 8.2 Proposed maximum percentage of replacement of coarse aggregates (% by volume) [35]

Aggregate class	Exposure classes		
	X0	XC1-XC4, XA1, XS1, XD1	All other exposure classes
A[a]	70 %	50 %	0 %
B[b]	40 %	30 %	0 %
C[b]	30 %	30 %	0 %

For each class of aggregate, the maximum replacement level allowed, as presented in Table 8.2, is expected to cause a maximum compressive strength loss of around 15 % in comparison with that of conventional concrete with an equivalent mix design [26]. This loss can be easily offset with a slight reduction of the effective water to cement ratio (w/c_{eff}) (Table 8.3) according to current methods for target strength correction [44]. This reduction in effective water content for the same amount of cement should be carried out additionally to the limits imposed on the maximum w/c ratio depending on the exposure class as per Annex F in EN 206:2013+A1 [36]. Guaranteeing the mechanical performance (i.e., target strength class) within the aforementioned boundaries is the role of the concrete producer. The designer should merely understand the implications of using a given amount of RA on the properties of concrete and thus on its structural behaviour (Table 8.3).

Although coarse RA leads to a decline in durability-related properties, the magnitude of which depends on the replacement level and porosity of the RA, which is represented by the aggregate class, a correction of the w/c_{eff} ratio can offset the carbonation and chloride ingress rates [29, 31]. In this proposal, only classes A and B may be used in reinforced concrete, whereas class C aggregates should only be used for unreinforced concrete. According to this approach, the cover of structural RAC elements may follow the same guidelines as those of EC2 [34].

Table 8.3 Correction factors for RAC subjected to a X0 exposure class [35]

Aggregate class	A	B	C
w/c_{eff} ratio reduction		0.05	
Tensile strength (f_{ctm})		1.00	
Modulus of elasticity (E_{cm})	0.80	0.85	0.90

For the same compressive strength class, little variation can be expected in the tensile strength (f_{ctm}) of RAC and that of conventional concrete with a similar mix design. Therefore, the corresponding formulae presented in Table 3.1 in EC2 may be used for estimating this property during the design of structural concrete:

$$f_{ctm} = 0.30 \cdot f_{ck}^{2/3} \text{ , for concrete classes} \leq \text{C50/60} \tag{8.1}$$

$$f_{ctm} = 2.12 \cdot ln(1 + \tfrac{f_{cm}}{10}) \text{ , for concrete classes} > \text{C50/60}$$

Where f_{ck} is the characteristic 28-day compressive strength of cylinders and f_{cm} the mean 28-day compressive strength of cylinders.

The higher porosity of RA in comparison with that of NA leads to greater deformability of concrete, and thus, a lower modulus of elasticity (E_{cm}) is expected for the same compressive strength. The correction factors in Table 8.3 should be multiplied by the formula in Table 3.1 in EC2 for the estimation of E_{cm} of RAC with the maximum replacement levels presented in Table 8.2:

$$E_{cm} = \alpha \cdot \beta \cdot 22 \cdot (f_{cm}/10)^{0.3} \tag{8.3}$$

Where α is the correction factor in Table 8.3 and β is the factor corresponding to the nature of the NA (1.2 for basalt aggregates, 1.0 for quartzite aggregates, 0.9 for limestone aggregates, and 0.7 for sandstone aggregates).

The aforementioned deformability can also affect the creep and shrinkage phenomena, and thus, it would be preferable for the moment to limit the use of RAC to applications less likely to present high dimensional variation related to those properties or especially sensitive to them. Nevertheless, whenever applicable, the impacts of the long-term length change of RAC due to creep may be accounted for using equation (8.4) for the effective modulus of elasticity ($E_{c.eff}$):

$$E_{c.eff} = \frac{E_{cm}}{1 + \varphi(\infty, t_0)} \tag{8.4}$$

Where E_{cm} is the modulus of elasticity of concrete (GPa), α is the creep coefficient correction factor and $\varphi(\infty, t_0)$ is the creep coefficient for a specific time period and loading. The $\varphi(\infty, t_0)$ considered is the one from section 3.1.4 of EC2 [33] for a relative humidity of 50 % (i.e., 2.5). The lower E_{cm} previously calculated will translate into a correspondingly lower $E_{c.eff}$ and thus greater deflections in the structural elements. In this study, the correction factors for the E_{cm} in equation (8.3) are shown in Table 8.4. They result from a statistical analysis of the effect of RA with different quality classes according to Table 8.1 on the modulus of elasticity of concrete [1]. The correction factors correspond to those of the 5th and 50th percentiles (P_5 and P_{50}, respectively). The former corresponds to a conservative approach with a probability of 95 % of falling within the range. The latter is an average decline of the property that should be applied with further design experience.

Table 8.4 Correction factors for E_{cm} estimation based on coarse RA content

RA class	Scenario	E_{cm} correction factor for coarse RA content		
		20 %	**50 %**	**100 %**
A	P_5	0.91	0.88	0.87
	P_{50}	0.96	0.95	0.89
B	P_5	0.86	0.78	0.65
	P_{50}	0.92	0.88	0.80

Previous studies have shown that the coefficient for accelerated carbonation and chloride ion migration are correlated with the relative compressive strength, and thus, the cover of structural elements may be calculated to guarantee a pre-established service life [29, 31]. The relationship between the cover of RAC and NAC, with the same mix design, may be calculated using the following equations:

$$\frac{X_{RAC}}{X_{NAC}} = \left(\frac{f_{cm,NAC}}{f_{cm,RAC}}\right)^{2,7} \tag{8.5}$$

$$\frac{X_{RAC}}{X_{NAC}} = \sqrt{e^{-0,023(f_{cm,RAC}-f_{cm,NAC})}} \tag{8.6}$$

Where X_{RAC} and X_{NAC} (mm) are the affected depths of RAC and NAC, respectively, and $f_{cm,RAC}$ and $f_{cm,NAC}$ are the corresponding 28-day compressive strengths. Until further research is carried out, these equations can only be applied for concrete products produced with type CEM I cement, in accordance with EN 197 [45]. Nevertheless, such an approach is only valid in cases wherein the loss in performance is accepted. The more likely approach to be used in practice, via an equivalent target strength class with a w/c_{eff} correction, usually offsets the aforementioned degradation phenomena. This translates to an equivalent service life for the same cover.

There is little information on the application of RAC in exposure classes other than those expressed in Table 8.2, and thus, the use of RA should be limited to concrete exposed to environments wherein there is sufficient evidence on the material's performance. In the specific case of concrete containing RA from non-air-entrained concrete, it is likely to present a significant decline in performance in terms of freeze-thaw resistance [46]. Without the existence of additional testing capable of establishing the technical viability of a given RA for certain conditions, the use of RA should be avoided in the production of concrete subjected to freezing and thawing, unless the components are from air-entrained concrete products, as expressed in Table 8.2. In this case, the maximum replacement level may be the same as that of the other exposure classes presented in the same table.

8.2.3 Building Design

This study covers the design of slabs and beams for a framed concrete structure (Fig. 8.1(a)). These include square corner slabs in two different possible support condi-

(a)   (b)

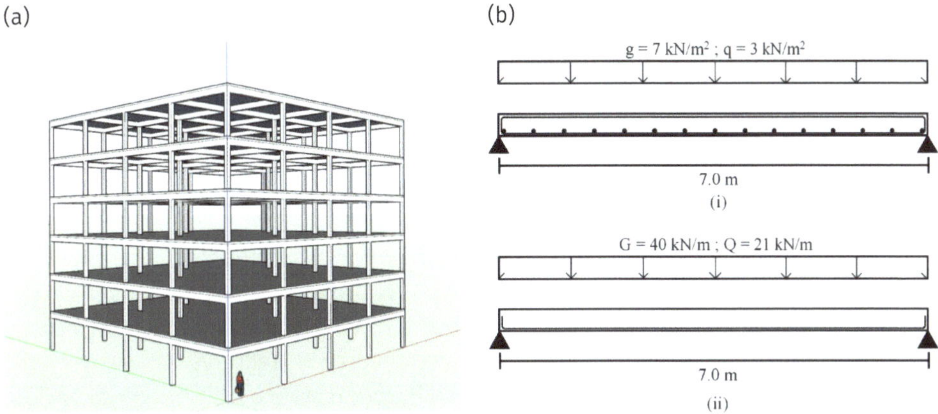

Fig. 8.1 Concrete frame structure (a) and loading conditions (b)

tions – a slab with two adjacent continuous ends and two other discontinuous as well as a fully discontinuous slab. The support conditions of the beams include the extremity of a continuously supported beam (discontinuous-continuous) and a simply supported one (discontinuous-discontinuous). A span of 7 m was considered to understand the effects of RA in relatively extreme circumstances involving high midspan deflection, which is typically the main drawback of using structural RAC. The concrete was considered to have a strength class of C30/37 for an exposure class XS1+XC3 [34] (environment where the concrete structure is exposed to air containing dissolved sea salts, but without direct contact with the seawater and is moderately wet and exposed to atmospheric CO_2). Such class is not usually applied for concrete inside buildings where reasonably impervious finishings can be expected. However, it was considered here to demonstrate the applicability of the approach in conditions susceptible to corrosion due to carbonation and chloride ingress.

To demonstrate the flexibility and practicality of the guidelines for the design of reinforced RAC elements, and their adaptability with EC2, four variables were considered in this study:

1. Incorporation ratio of coarse RA (0 %, 20 %, 50 % and 100 %);

2. RA with different quality classes (classes A and B [25]);

3. Different support conditions (previously described);

4. Structural design using the fifth and fiftieth percentiles (P_5 and P_{50}, respectively) of a 95 % confidence interval.

Starting permanent and live loads of 7 kN/m² and 3 kN/m², respectively, were considered for the slabs and 40 kN/m and 21 kN/m, respectively, for the beams (Fig. 8.1(b)). The permanent loads correspond to the slab's self-weight including floor construction. Since the design methodology involves the increase of the elements' height, the self-weight was adjusted accordingly. S500 steel reinforcement was considered. The remainder of the ULS and SLS design methodology followed that presented in previous studies [22, 47]. The iterative process of the design can be summarized with the following process:

(i) establish the expected performance of concrete and minimum nominal cover; (ii) calculate the minimum area of steel reinforcement for the pre-established cross-section and concrete properties; (iii) calculate the crack width (if above maximum permissible, a sequential increase of the height of the cross-section by steps of 5 mm will be made until compliance); (iv) calculate midspan deflection (if above midspan's length/250, a sequential increase of 5 mm will be made until compliance).

Concerning the bond strength of RAC, a recent study has shown it to be less reliable in comparison with that of NAC, and a 25 % increase in the anchorage length was proposed for concrete with 100 % RA [48]. Nevertheless, it was also ascertained that existing information is scarce and unrepresentative. Therefore, further testing is required to adequately model this behaviour, as it is safe to assume a certain degree of discrepancy between the anchorage lengths of RAC and NAC given the likely difference in Poisson's coefficient between the two even for the same strength class. Shear reinforcement design is also likely to be affected by the incorporation of RA [49] and specific correction factors may need to be applied to compensate the loss in performance [50]. However, since information in the literature is not enough to model the behaviour for different scenarios, shear reinforcement design was not considered in this study.

8.3 Results and Discussion

Table 8.5 presents the results of the E_{cm} and $E_{c,eff}$ after having applied the correction factors of Table 8.4 in equations (8.3) and (8.4). Even though this calculation was carried out on mixes with the same target strength class (i.e., C30/37), there is a decrease in modulus of elasticity with increasing RA content due to their greater deformability and more so when the quality class declines. This leads to higher deformation due to creep over a long period of time that needs to be accounted for when designing structural RAC.

Table 8.5 E_{cm} and $E_{cm,eff}$ (GPa) for mixes with varying coarse RA content of different classes

Property	Scenario	NA	A			B		
			20 %	50 %	100 %	20 %	50 %	100 %
E_{cm}	P_5	32.8	29.9	28.9	28.6	28.2	25.6	21.3
	P_{50}		31.5	31.2	29.2	30.2	28.9	26.3
$E_{cm,eff}$	P_5	9.38	8.54	8.26	8.16	8.07	7.32	6.10
	P_{50}		9.01	8.91	8.35	8.63	8.26	7.51

Table 8.6 to Table 8.9 present the solutions for the main lower steel reinforcements, the cross-sections' final height for slabs and beams, and estimated crack width (w_k). Fig. 8.2 presents the cross-sections of the NAC beam-slab connection. In the case of slabs (Tables 8.6 and 8.7), for the aforementioned loading conditions, the optimized solutions for the thickness of NAC slabs were of 160 mm and 220 mm for corner slabs (two adjacent continuous + two adjacent discontinuous) and for discontinuous slabs, respectively.

Table 8.6   Solution for the corner slab

Scenario		NA	A			B		
		20	50	100	20	50	100	
P_5	Slab thickness (mm)	160	160	160	160	160	160	165
	Min. area reinforcement (cm²/m)	3.26	3.26	3.26	3.26	3.26	3.26	3.16
	Adopted solution	φ8//0.15						
	Area (cm²/m)	3.35						
	w_k (mm) < 0.3 mm	0.244	0.244	0.244	0.244	0.244	0.243	0.241
	$L/250$ (mm) < 28 mm	16.0	18.2	19.1	19.4	19.7	22.4	14.7
P_{50}	Slab thickness (mm)	160	160	160	160	160	160	160
	Min. area reinforcement (cm²/m)	3.26	3.26	3.26	3.26	3.26	3.26	3.26
	Adopted solution	φ8//0.15						
	Area (cm²/m)	3.35						
	w_k (mm) < 0.3 mm	0.244	0.244	0.244	0.244	0.244	0.241	0.243
	$L/250$ (mm) < 28 mm	16.0	16.9	17.2	18.8	18.0	19.1	21.7

Crack width was not an issue and was always below the limit (i.e., 0.3 mm). Considering an approach where no strength loss is observed (i.e., all RAC mixes present a strength class of C30/37), it is obvious that the required minimum area of reinforcement is the same, regardless of the replacement level. Therefore, it is reasonably straightforward to guarantee the EC2's Ultimate Limit State (ULS) of RAC with the assumption of equivalent strength classes. However, in terms of Serviceability Limit State (SLS), to comply with the maximum deflection (L/250 < 28 mm, in this case), a slight increase to the slab thickness was required to compensate for the materials' greater deformability (noticeable in the lower $E_{c,eff}$ – Table 8.5) when compared to a conventional NAC slab subjected to the same loading conditions.

Table 8.7   Solution for the discontinuous slab

Scenario		NA	A			B		
		20	50	100	20	50	100	
P_5	Slab thickness (mm)	220	220	220	220	220	225	225
	Min. area reinforcement (cm²/m)	3.90	3.90	3.90	3.90	3.90	3.90	3.83
	Adopted solution	φ8//0.125						
	Area (cm²/m)	4.02						
	w_k (mm) < 0.3 mm	0.275	0.275	0.275	0.274	0.274	0.273	0.272
	$L/250$ < 0.028	22.4	24.7	25.6	25.9	26.2	19.8	25.3
P_{50}	Slab thickness (mm)	220	220	220	220	220	220	225
	Min. area reinforcement (cm²/m)	3.90	3.90	3.90	3.90	3.90	3.90	3.83
	Adopted solution	φ8//0.125						
	Area (cm²/m)	4.02						
	w_k (mm) < 0.3 mm	0.275	0.275	0.275	0.274	0.274	0.274	0.273
	$L/250$ < 28 mm	22.4	23.4	23.6	25.3	24.4	25.6	19.1

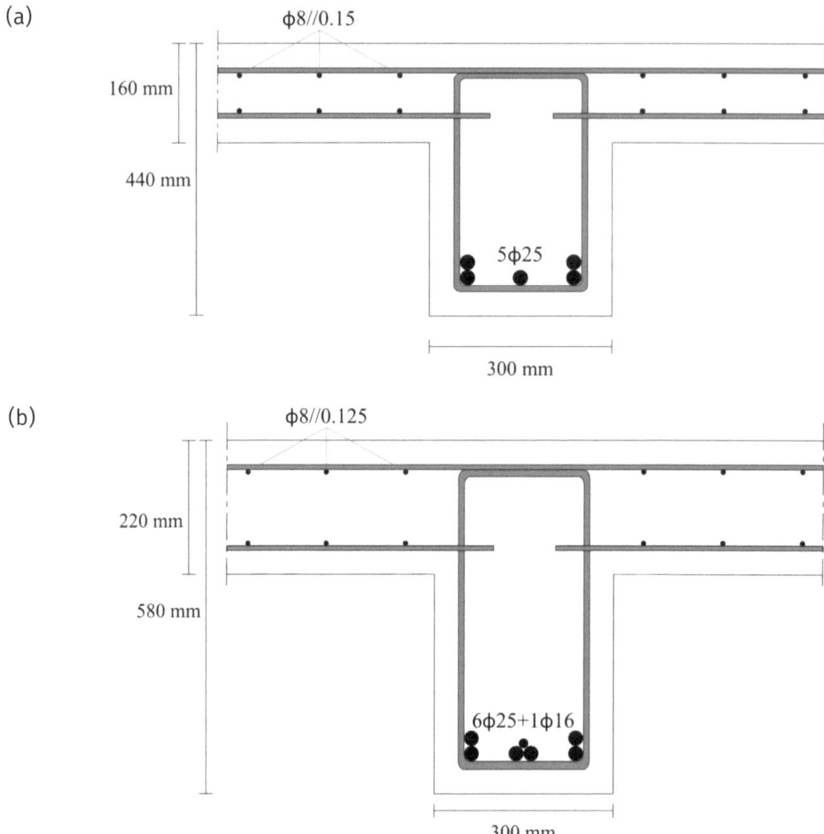

Fig. 8.2 Cross-sections of the beam-slab connection

Even though the adopted initial thickness of 160 mm for NAC is optimized in terms of SLS midspan deflection (any value lower than that would result in non-compliance), it led to a calculated deflection of 16 mm (limit of 28 mm; 12 mm difference). This leeway permitted a greater deflection within the limit by most RAC mixes without the need of increasing the slab's thickness. The only case for corner slabs where this was not possible (Table 8.6) was for a RAC with 100 % coarse RA of class B using P_5 correction coefficients. Nevertheless, in this case, a mere 5 mm increase was enough to comply with the approach. Furthermore, using the P_{50} approach, no alterations were needed.

In the case of the discontinuous slab (Table 8.7), less leeway existed in the case of NAC (22.4 mm vs. 28 mm; 6 mm difference). Consequently, P_5-designed RAC elements with 50 % and 100 % coarse RA of class B required a 5 mm increase to comply with the SLS limits. In the P_{50} approach, this increase was only necessary for 100 % coarse RA of class B.

Table 8.8 and Table 8.9 present the design solution of the continuous and simply supported beams, respectively. Naturally, a much higher depth was required for the latter given the higher bending moment (440 mm vs. 580 mm, respectively). Beams, in comparison with slabs, require a greater relative increase in cross-section height in order to comply with SLS limits.

Table 8.8 Solution for the continuous beam

Scenario		NA	A			B		
		20	50	100	20	50	100	
P_5	Beam height (mm)	440	450	450	450	450	460	480
	Min. area reinforcement (cm²)	23.55	22.53	22.53	22.53	22.53	21.62	20.07
	Adopted solution	5φ25						
	Area (cm²)	24.54						
	w_k (mm) < 0.3 mm	0.195	0.188	0.189	0.189	0.189	0.185	0.178
	$L/250$ (mm) < 28 mm	27.0	26.6	27.1	27.2	27.4	27.1	26.6
P_{50}	Beam height (mm)	440	440	450	450	450	450	460
	Min. area reinforcement (cm²)	23.55	23.55	22.53	22.53	22.53	22.53	20.07
	Adopted solution	5φ25						
	Area (cm²)	24.54						
	w_k (mm) < 0.3 mm	0.195	0.193	0.193	0.189	0.188	0.189	0.178
	$L/250$ (mm) < 28 mm	27.0	27.6	27.7	27.2	26.5	27.1	26.6

Table 8.9 Solution for the simply supported beam

Scenario		NA	A			B		
		20	50	100	20	50	100	
P_5	Beam height (mm)	580	590	590	590	590	600	630
	Min. area reinforcement (cm²)	30.73	29.78	29.78	29.78	29.78	28.91	26.67
	Adopted solution	6φ25 + 1φ16						
	Area (cm²)	31.46						
	w_k (mm) < 0.3 mm	0.210	0.204	0.204	0.204	0.204	0.202	0.193
	$L/250$ (mm) < 28 mm	27.0	27.0	27.5	27.6	27.8	27.9	27.0
P_{50}	Beam height (mm)	580	580	580	590	590	590	600
	Min. area reinforcement (cm²)	30.73	30.73	30.73	29.78	29.78	29.78	28.91
	Adopted solution	6φ25 + 1φ16						
	Area (cm²)	31.46						
	w_k (mm) < 0.3 mm	0.210	0.207	0.207	0.204	0.204	0.204	0.201
	$L/250$ (mm) < 28 mm	27.0	27.5	27.7	27.3	26.8	27.5	27.6

Despite the necessity of a lower area of reinforcement with increasing height, the same area was adopted for all cases as an equivalent criterion. However, environmental impacts and cost could decrease by optimizing this variable.

Almost all cases required an increase to the beams' height with the incorporation of coarse RA. This increase was higher when RA of class B were used due to their greater deformability; using 100 % of it would require 40 mm and 50 mm increases for P_5-designed continuous beams and simply supported beams, respectively (~10 % increase). For P_{50}-designed beams, this increase would be of 20 mm (~5 % increase).

8.4 Concluding Remarks

In spite of the proven technical feasibility of RAC, there have been several obstacles hindering their use in structural applications. One of those barriers is the vague concept handed down by some specifications on the use of RA, which do not contain specific provisions for the expected physical and structural performance of RAC elements. To fill this gap, this chapter contains simple empirical rules from a vast systematic literature review and meta-analysis, previously conducted by the authors, on the effect of using RA on the properties of concrete. The common, albeit faulty, approach of using the composition of RA alone to estimate the performance of concrete was proved inadequate and replaced by a practical and comprehensive performance-based classification developed by the authors. This classification, which was applied in this study, is strongly correlated to the performance of concrete and has allowed the development of formulae that can be used to predict its behaviour. This study, which presents their implementation in parallel with existing codes for structural concrete, shows a practical, conservative, yet sufficiently flexible methodology for a wide range of applications and allowed drawing the following conclusions:

- ULS calculations and bottom steel reinforcement solutions were not affected by RA incorporation, regardless of its quality and content, as all concrete mixes were set out to have the same target strength class. This approach leads to minimum alterations on the cross-section geometry and would generate greater acceptance by the professionals within the construction industry. This reinforces the notion that RAC should be designed with a specific target strength class, by slightly altering its composition, rather than accepting slight performance loss for a fixed mix design. This would also facilitate the job of designers as they would use the existing values for nominal cover in EC2 without changes;
- The use of up to 100 % of coarse high-quality RA (i.e., class A) led to marginal alterations in the structural elements' cross-section geometries in all scenarios. Nevertheless, the production of such RA, without the necessity of energy-intensive treatments, is rare. In cases that use such a beneficiation stage, the outcome is likely one with greater environmental impact than when using NA;
- The use of RA with intermediate quality (i.e., class B) led to an inferior structural behaviour. To guarantee the same load-bearing capacity and service life, the beams' height had to increase with increasing replacement level to compensate for the RA's greater deformability. Nevertheless, these aggregates are much more likely to be encountered as the output of CDW recycling plants and realistic structural design should be carried out having them in mind;
- The use of P_5 or P_{50} design approaches on the modulus of elasticity had an obvious impact on the final element height. Although the use of a P_{50} approach would make more sense from the perspective of avoiding the combination of multiple conservative correction factors, since there is some uncertainty and lack of experience with structural RAC, the use of the P_5 approach should be preferred;
- Despite the aforementioned lower performance of RAC, the practical impact of adding a realistic amount of Class B RA (up to 40 % according to Table 8.2) translates into a maximum increase of 20 mm for beams designed in a "worst-case"

scenario (P_5 approach). This is equivalent to increasing the volumetric amount of the beams by just 3–5 % depending on the support conditions. Though these are reasonably low values, they may have some environmental and cost-related impacts in terms of the mass of concrete required for the construction of a building and should be analysed carefully. Nevertheless, the approach presented in this study is fairly flexible and allows for custom amounts of RA depending on the designers' objective.

8.5 References

[1]   Dhir, R.K., et al., *Sustainable construction materials: Recycled aggregates.* 2019, Duxford, UK: Woodhead Publishing. 652. https://doi.org/10.1016/C2015-0-00466-X

[2]   Ajdukiewicz, A.B. and A.T. Kliszczewicz, *Comparative tests of beams and columns made of recycled aggregate concrete and natural aggregate concrete.* Journal of Advanced Concrete Technology, 2007. 5(2): p. 259–273. https://doi.org/10.3151/jact.5.259

[3]   Pacheco, J., et al., *Dynamic characterization of full-scale structures made with recycled coarse aggregates.* Journal of Cleaner Production, 2017. 142: p. 4195–4205. https://doi.org/10.1016/j.jclepro.2015.08.045

[4]   Pacheco, J., et al., *Flexural load tests of full-scale recycled aggregates concrete structures.* Construction and Building Materials, 2015. 101: p. 65–71. https://doi.org/10.1016/j.conbuildmat.2015.10.023

[5]   Seara-Paz, S., et al., *Long-term flexural performance of reinforced concrete beams with recycled coarse aggregates.* Construction and Building Materials, 2018. 176: p. 593–607. https://doi.org/10.1016/j.conbuildmat.2018.05.069

[6]   Tosic, N., A. de la Fuente, and S. Marinkovic, *Creep of recycled aggregate concrete: Experimental database and creep prediction model according to the fib Model Code 2010.* Construction and Building Materials, 2019. 195: p. 590–599. https://doi.org/10.1016/j.conbuildmat.2018.11.048

[7]   Silva, R.V. and J. de Brito, *Reinforced recycled aggregate concrete slabs: Structural design based on Eurocode 2.* Engineering Structures, 2020. 204: p. 110047. https://doi.org/10.1016/j.engstruct.2019.110047

[8]   Arezoumandi, M., et al., *An experimental study on shear strength of reinforced concrete beams with 100 % recycled concrete aggregate.* Construction and Building Materials, 2014. 53(0): p. 612–620. https://doi.org/10.1016/j.conbuildmat.2013.12.019

[9]   Ma, H., et al., *Cyclic loading tests and shear strength of steel reinforced recycled concrete short columns.* Engineering Structures, 2015. 92: p. 55–68. https://doi.org/10.1016/j.engstruct.2015.03.009

[10] Wang, Y., J. Chen, and Y. Geng, *Testing and analysis of axially loaded normal-strength recycled aggregate concrete filled steel tubular stub columns.* Engineering Structures, 2015. 86: p. 192–212. https://doi.org/10.1016/j.engstruct.2015.01.007

[11] Choi, W.-C., H.-D. Yun, and S.-W. Kim, *Flexural performance of reinforced recycled aggregate concrete beams.* Magazine of Concrete Research, 2012. 64(9): p. 837–848. https://doi.org/10.1680/macr.11.00018

[12] Choi, W.-C. and H.-D. Yun, *Long-term deflection and flexural behavior of reinforced concrete beams with recycled aggregate.* Materials & Design, 2013. 51: p. 742–750. https://doi.org/10.1016/j.matdes.2013.04.044

[13] Dhir, R.K. and K.A. Paine, *Suitability and practicality of using coarse RCA in normal and high strength concrete,* in *1st International Conference on Sustainable Construction: Waste Management.* 2004: Singapore. p. 108–123.

[14] Leite, M.B., *Evaluation of the mechanical properties of concrete produced with recycled aggregates from construction and demolition wastes.* 2001, Federal University of Rio Grande do Sul: Rio Grande do Sul, Brasil. p. 290.

[15] Amorim, P., J. de Brito, and L. Evangelista, *Concrete made with coarse concrete aggregate: Influence of curing on durability.* ACI Materials Journal, 2012. 109(2): p. 195–204. https://doi.org/10.14359/51683706

[16] Evangelista, L. and J. de Brito, *Mechanical behaviour of concrete made with fine recycled concrete aggregates.* Cement and Concrete Composites, 2007. 29(5): p. 397–401. https://doi.org/10.1016/j.cemconcomp.2006.12.004

[17] Evangelista, L. and J. de Brito, *Durability performance of concrete made with fine recycled concrete aggregates.* Cement and Concrete Composites, 2010. 32(1): p. 9–14. https://doi.org/10.1016/j.cemconcomp.2009.09.005

[18] Ferreira, L., J. de Brito, and M. Barra, *Influence of the pre-saturation of recycled coarse concrete aggregates on concrete properties.* Magazine of Concrete Research, 2011. 63(8): p. 617–627. https://doi.org/10.1680/macr.2011.63.8.617

[19] Fonseca, N., J. de Brito, and L. Evangelista, *The influence of curing conditions on the mechanical performance of concrete made with recycled concrete waste.* Cement and Concrete Composites, 2011. 33(6): p. 637–643. https://doi.org/10.1016/j.cemconcomp.2011.04.002

[20] Limbachiya, M., M.S. Meddah, and Y. Ouchagour, *Use of recycled concrete aggregate in fly-ash concrete.* Construction and Building Materials, 2012. 27(1): p. 439–449. https://doi.org/10.1016/j.conbuildmat.2011.07.023

[21] Limbachiya, M., M.S. Meddah, and Y. Ouchagour, *Performance of Portland/silica fume cement concrete produced with recycled concrete aggregate.* ACI Materials Journal, 2012. 109(1): p. 91–100. https://doi.org/10.14359/51683574

[22] Silva, R.V., et al., *Design of reinforced recycled aggregate concrete elements in conformity with Eurocode 2.* Construction and Building Materials, 2016. 105: p. 144–156. https://doi.org/10.1016/j.conbuildmat.2015.12.080

[23] Imbabi, M.S., C. Carrigan, and S. McKenna, *Trends and developments in green cement and concrete technology.* International Journal of Sustainable Built Environment, 2012. 1(2): p. 194–216. https://doi.org/10.1016/j.ijsbe.2013.05.001

[24] EN-933-11, *Tests for geometrical properties of aggregates – Part 11: Classification test for the constituents of coarse recycled aggregate.* 2009, Comité Européen de Normalisation (CEN): Brussels, Belgium. p. 16.

[25] Silva, R.V., J. de Brito, and R.K. Dhir, *Properties and composition of recycled aggregates from construction and demolition waste suitable for concrete production.* Construction and Building Materials, 2014. 65: p. 201–217. https://doi.org/10.1016/j.conbuildmat.2014.04.117

[26] Silva, R.V., J. de Brito, and R.K. Dhir, *The influence of the use of recycled aggregates on the compressive strength of concrete: a review.* European Journal of Environmental and Civil Engineering, 2014. 19(7): p. 825–849. https://doi.org/10.1080/19648 189.2014.974831

[27] Silva, R.V., J. de Brito, and R.K. Dhir, *Establishing a relationship between modulus of elasticity and compressive strength of recycled aggregate concrete.* Journal of Cleaner Production, 2016. 112: p. 2171–2186. https://doi.org/10.1016/j.jclepro.2015.10.064

[28] Silva, R.V., J. de Brito, and R.K. Dhir, *Comparative analysis of existing prediction models on the creep behaviour of recycled aggregate concrete.* Engineering Structures, 2015. 100: p. 31–42. https://doi.org/10.1016/j.engstruct.2015.06.004

[29] Silva, R.V., et al., *Prediction of chloride ion penetration of recycled aggregate concrete.* Materials Research, 2015. 18(2): p. 427–440. https://doi.org/10.1590/1516-1439.000214

[30] Silva, R.V., J. de Brito, and R.K. Dhir, *Tensile strength behaviour of recycled aggregate concrete.* Construction and Building Materials, 2015. 83: p. 108–118. https://doi.org/10.1016/j.conbuildmat.2015.03.034

[31] Silva, R.V., et al., *Carbonation behaviour of recycled aggregate concrete.* Cement and Concrete Composites, 2015. 62: p. 22–32.

[32] Silva, R.V., J. de Brito, and R.K. Dhir, *Availability and processing of recycled aggregates within the construction and demolition supply chain: A review.* Journal of Cleaner Production, 2017. 143: p. 598–614.

[33] EN-1992-1-1:2004+A1:2014, *Eurocode 2 – Design of concrete structures: Part 1-1: General rules and rules for buildings.* 2014, Comité Européen de Normalisation (CEN): Brussels, Belgium. p. 259.

[34] EN-1992-1-1, *Eurocode 2 – Design of concrete structures: Part 1-1: General rules and rules for buildings.* 2008, Comité Européen de Normalisation (CEN): Brussels, Belgium. p. 259.

[35] Silva, R.V. and J. de Brito, *Specification for structural recycled aggregate concrete,* in *IV International Conference Progress of Recycling in the Built Environment, 11–12 October 2018.* 2018: Laboratório Nacional de Engenharia Civil (LNEC), Lisbon, Portugal. p. 8.

[36] EN-206, *Concrete – Specification, performance, production and conformity.* 2013, Comité Européen de Normalisation (CEN): Brussels, Belgium. p. 98.

[37] BS-8500-2:2015+A1:2016, *Concrete. Complementary British Standard to BS EN 206. Specification for constituent materials and concrete.* 2016, British Standards Institution (BSI): United Kingdom. p. 48.

[38] CEN/TR-16349, *Framework for a specification on the avoidance of a damaging Alkali-Silica Reaction (ASR) in concrete.* 2012, Comité Européen de Normalisation (CEN): Brussels, Belgium. p. 14.

[39] EN-933-3, *Tests for geometrical properties of aggregates – Part 3: Determination of particle shape. Flakiness index.* 2012, Comité Européen de Normalisation (CEN): Brussels, Belgium. p. 16.

[40] EN-933-4, *Tests for geometrical properties of aggregates – Part 4: Determination of particle shape. Shape index.* 2008, Comité Européen de Normalisation (CEN): Brussels, Belgium. p. 14.

[41] EN-1097-2, *Tests for mechanical and physical properties of aggregates – Part 2: Methods for the determination of resistance to fragmentation.* 2010, Comité Européen de Normalisation (CEN): Brussels, Belgium. p. 38.

[42] EN-1744-1, *Tests for chemical properties of aggregates – Part 1: Chemical analysis.* 2009, Comité Européen de Normalisation (CEN): Brussels, Belgium. p. 30.

[43] EN-1744-6, *Tests for chemical properties of aggregates – Part 6: Determination of the influence of recycled aggregate extract on the initial setting time of cement.* 2006, Comité Européen de Normalisation (CEN): Brussels, Belgium. p. 12.

[44] Teychenné, D.C., et al., *Design of normal concrete mixes.* 2nd ed. 1997, Watford, UK: Building Research Establishment.

[45] EN-197, *Cement. Composition, specifications and conformity criteria for common cements.* 2011, Comité Européen de Normalisation (CEN): Brussels, Belgium. p. 50.

[46] Gokce, A., et al., *Freezing and thawing resistance of air-entrained concrete incorporating recycled coarse aggregate: The role of air content in demolished concrete.* Cement and Concrete Research, 2004. 34(5): p. 799–806.

[47] De Brito, J. and R.V. Silva, *Design of structural recycled aggregate concrete based on EC2,* in *Seminar "Construction and Demolition Wastes: A value-added resource", Faculdade de Engenharia da Universidade do Porto (FEUP).* 2015: Porto, Portugal. p. 17.

[48] Pacheco, J., et al., *Reliability of the bond strength of recycled coarse aggregate concrete,* in *fib Symposium 2019 – Innovations in Materials, Design and Structures.* 2019: Krakow, Poland. p. 8.

[49] Choi, H.B., et al., *Experimental study on the shear strength of recycled aggregate concrete beams.* Magazine of Concrete Research, 2010. 62(2): p. 103–114.

[50] Pacheco, J.N., et al., *Uncertainty of shear resistance models: Influence of recycled concrete aggregate on beams with and without shear reinforcement.* Engineering Structures, 2019: p. 109905.

Chapter 9

Design Guidelines for Recycled Aggregate Concrete Based on Reliability Analysis

João N. Pacheco, Jorge de Brito

CERIS, IST, CERIS, IST, Universidade de Lisboa, Lisbon, Portugal

9.1 Introduction: Partial Factor Formats and Code Calibration

The structural checks of codes have an intrinsic probabilistic basis, since loads, load models, past load history, material properties, geometry, and resistance modelling are all uncertain to some extent [1]. Due to this uncertainty, codes need to include margins of safety in their formulae.

Before structural design had a scientific basis, the approach of builders (e.g., stonemasons) was a trial-and-error approach based on a similar design to what was already recognised as safe construction. Later, design began to be scientifically based and an allowance for uncertainty in calculations, loads, and material properties was included in the structural checks. At that time, global safety formats were developed. Under such formats, design is deemed safe whenever:

$$S/_R < \gamma_{Global} \tag{9.1}$$

Where γ_{Global} is the global safety factor, S the value of load-effects and R the value of resistance (which was typically defined as a maximum allowed stress). In the initial realizations of this format, γ_{Global} was defined based on backwards calculations from structural design considered as safe [2].

However, global formats are not satisfactory and pose economical and safety issues, since a global margin of error does not take into account the differences in uncertainty of the different parameters of design. For instance, consider two cases of design in which the value of S is the same. For both cases, the resulting structural design will be equivalent (provided γ_{Global} is fixed at exactly the same value). Now consider that the load-effects of the first case of design are due to a single permanent action but those of the second case of design are caused by a combination of permanent and variable actions. Since the uncertainty of variable actions is bigger than that of permanent ones (because the latter are mostly due to self-weight), the second case of design is clearly associated with larger uncertainty. However, the margin of safety provided by γ_{Global} is the same.

Because of such inconsistencies, partial factor formats were developed. In these formats, different partial factors are included for different parameters involved in structural design and the concepts of design value of resistance (R_d) and design value of load-effects (S_d). In partial factor formats, design is safe as long as:

$$R_d - S_d > 0 \qquad (9.2)$$

In the case of reinforced concrete designed with the Eurocodes, the following are the typical partial factors involved in ultimate limit state design:

- $\gamma_p = 1.35$, which accounts for the uncertainty in load-effect modelling and in loading for permanent actions;
- $\gamma_Q = 1.50$, which accounts for the uncertainty in load-effect modelling and in loading for variable actions;
- $\gamma_C = 1.50$, which accounts for the uncertainty in the strength of concrete and in resistance modelling, including geometric uncertainty;
- $\gamma_S = 1.15$, which accounts for the uncertainty in the strength of the reinforcement bars. As in the case of γ_C, this parameter also accounts for uncertainty in resistance modelling, including that due to geometric uncertainty.

With this format, Eurocode design results in reasonably uniform safety for different cases of design. To ensure this, the partial factors of present-day codes are calibrated through reliability analyses.

In reliability contexts, safety is evaluated either through the probability of failure, or through the reliability index (β). They are related as shown in Eq. (9.3), where Φ^{-1} is the inverse of the standard normal distribution:

$$\beta \approx \Phi^{-1} \text{ (probability of failure)} \qquad (9.3)$$

Conceptually, code calibrations are made through iterative processes in which the partial factors under calibration are iteratively changed until the β of representative cases of design is consistently close to a target reliability index (β_{target}) that is considered acceptable by society. Furthermore, the value of each partial factor should be representative of its influence on the reliability of structural design. In practical terms, the calibration of partial factors considers several practical aspects and the process followed is not exactly as just described. This will be understood in Section 9.3 of this chapter.

Due to concerns regarding the heterogeneity of recycled aggregates and their influence on structural safety (therefore, on β), this chapter refers to the reliability-based calibration of design provisions for the design of recycled aggregate concrete (RAC) using prEN1992 [3] concerning both ultimate limit state design and concrete covers for carbonation and chloride ion penetration.

9.2 Recycled Aggregate Concrete

9.2.1 Coarse Recycled Aggregates Produced From Concrete Waste

Today's scientific expertise on recycled aggregate concrete is consensual in stating that coarse recycled aggregates produced from concrete waste (CRCA) have the best properties for structural concrete production. This is also acknowledged in standards and national specifications, which restrict the use of recycled aggregates in concrete to those with a relevant content of concrete and unbound stone waste (typically more than 70 % by weight), in comparison to other types of construction and demolition waste (e.g., ceramics). Since RAC is still scarcely used by the concrete industry, this chapter deals with structural design of RAC made with CRCA. It is understood that before recycled aggregates produced from other types of waste (such as mixed construction and demolition waste or CDW) are implemented, contractors, clients, and designers will opt for CRCA due to the better properties and confidence in this type of recycled aggregate [4]. Furthermore, the majority of research on RAC is made with this type of recycled aggregate, so its influence on most of the properties of concrete is already well understood.

The common production process of CRCA is based on mechanical processes that include:

- Preliminary separation so that the recycled aggregates are produced from concrete waste with minimal contamination of other materials (selective demolition may also be an option to attain this goal, provided its economic feasibility);
- Demolition of the concrete waste into smaller elements, including removal of reinforcement;
- Crushing of the concrete waste, a process that is made with one to three crushers (a single crusher is a common option). Typically, the higher the number of crushers, the better the quality of the CRCA, but the lower the material recovery of the concrete waste as CRCA because a higher volume of fines and fine aggregates will be produced;
- Sieving of the recycled aggregates into the particle fractions intended, rejecting the finer fractions, and recovering only the CRCA.

The typical properties of coarse natural aggregates (CRCA) are inferior to those of CNA. This is due to two reasons:

- CRCA are partly composed of attached mortar, while CNA are totally composed of stone;
- The investment in the equipment used in facilities (or on typical onsite recycling equipment) used to produce CRCA is not as large as the investment made on quarries (e.g., usually the production of CRCA involves a lower number of crushers than those used in quarries to produce CNA). This may not be true in some circumstances, e.g., when CRCA are produced in quarries [5].

Because of the attached mortar, the CRCA are weaker [6, 7], more deformable, more porous, and have higher water absorption and lower density than CNA. The equipment used for production is mainly responsible for CRCA being typically more elongated and flakier than CNA [8].

The overall quality of a CRCA is, on a first approximation, dependent on the strength class of the concrete waste. Concrete of higher strength classes is associated with denser and stronger binder paste and overall better mechanical and durability properties. This directly translates to the effect of the resulting high-quality CRCA on the properties of concrete – research has found that the incorporation of CRCA produced with high-quality concrete rejects has a negligible effect on the propertiesof concrete [9].

The chemical properties of CRCA may not be suitable for the production of concrete (e.g., the CRCA may be contaminated with chlorides). This topic is not treated in this chapter and compliance with relevant standards (e.g. EN12620+A1 [10]) is assumed.

9.2.2 Properties of Concrete Made With Coarse Recycled Concrete Aggregates

Since the properties of CRCA are inferior to those of (CNA), the incorporation of CRCA has detrimental effects on the mechanical and durability properties of concrete.

Compressive Strength

The incorporation of CRCA decreases the compressive strength (f_c) of concrete. This decrease is approximately linearly correlated with the incorporation ratio of CRCA, with the exception of low percentages of medium-quality CRCA (of up to around 20 %) – in this case, the effect on f_c may be negligible [9, 11]. For full (100 %) incorporation of CRCA, the decrease of f_c is usually between 12 % and 25 % [11].

From a designer's perspective, the decrease of f_c is not relevant: concrete is specified in terms of the strength class. Producing a RAC mix that complies with the specified strength class is the responsibility of the concrete producer, not of the designer.

In structural reliability contexts, the most relevant aspect related to f_c is its variability. In [12], five compositions were batched with either natural aggregate concrete (NAC) or RAC with full incorporation of RA. The only change to the mix design was a slight adjustment to the w/b ratio to ensure that the consistency of the NAC and RAC mixes was equivalent. Full incorporation of CRCA did not affect the standard deviation and the coefficient of variation (CoV) of the f_c of the mixes and normal and lognormal distributions suited the f_c data of NAC and RAC equally well.

This suggests that the findings of Rusch [13] are applicable to RAC and the following relationship, used in EN1992 [14] and prEN1992 [3], between specified (f_{ck}) and mean compressive strength (f_{cm}) may be used:

$$f_{cm} = f_{ck} + 8 \text{ MPa} \tag{9.4}$$

Using Eq. (9.4) for RAC is based on two assumptions: i) the incorporation of CRCA has negligible effect on the difference between concrete produced under laboratory conditions and concrete produced in industrial environments; and ii) CRCA also do not affect the difference between concrete cast and cured under laboratory conditions and concrete cast and cured in the construction job. Both assumptions cannot be confirmed at this time, since actual cases of industrial production of RAC are scarce.

Other Mechanical Properties

In typical cases, the incorporation of CRCA leads to inferior performance with regard to all mechanical properties (e.g., tensile strength (f_{ct}), Young's modulus (E), shrinkage, creep). This is mostly due to the attached mortar, which is more deformable and weaker than stone. The extent of the detrimental effect of CRCA on concrete is chiefly dependent on three factors:

- The quality of the CRCA, since CRCA made from concrete rejects of lower strength class and/or with higher content of attached mortar will be weaker and more deformable. The type of CNA used to produce the source concrete is also a relevant factor, as well as the equipment and processes used to convert the concrete waste into recycled aggregates (see Section 9.2.1);
- The mix design and specification of RAC, since, as the strength of concrete increases, the aggregates become a more limiting factor in terms of mechanical behaviour;
- The type of CNA used in an analogue NAC mix design that is used for comparison. Sources of CNA range from sandstone to basalt and have different quality. If the RAC mix is compared with a NAC mix made with lower-quality CNA, the effects of the incorporation of CRCA are smaller than those found when the comparison is made with a NAC mix made with higher-quality CNA.

A meta-analysis [15] that assessed the suitability of the constitutive models between f_c and f_{ct} and between f_{ct} and E of EN1992 [14] found that:

- For the same f_c, the E of RAC is smaller than that of all lithologies of CNA except for sandstone;
- Concerning E, a correction factor α, analogue to that already in use for the constitutive modelling for NAC used in EN1992 [14], was determined for RAC. The most appropriate option was found to be $\alpha = 0.84$. However, due to concerns of construction agents and designers, namely with respect to the possible use of CRCA of a hypothetical low quality, $\alpha = 0.70$ (the value used for sandstone) was recommended in [15];
- For f_{ct}, it was found that full incorporation of CRCA decreases f_{ct} by 15 % in comparison to that of NAC of the same f_c and made with crushed aggregates. Since rounded natural aggregates are also associated with a smaller f_{ct} and the results for RAC are still conservatively estimated by the constitutive model of EN1992, an adaptation of the constitutive model is not necessary;
- The scatter of the constitutive models between f_c and f_{ct} and between f_c and E was not affected by the incorporation of CRCA.

This analysis also found that, as the f_c of concrete increased, the detrimental effects of the use of CRCA on f_{ct} became larger. Therefore, the statements made should only be used for RAC up to the C35/45 strength class. Above this threshold, RAC mix design should include preliminary testing to define suitable f_{ct} for structural design. If other types of recycled aggregate are used (such as fine recycled aggregates and recycled aggregates from other types of CDW), specific testing is recommended as well, because the analysis presented in [15] concerned CRCA only.

Durability

RAC is more prone to the ingress of external agents than NAC [9, 16]. As the incorporation ratio of CRCA increases, the trend is that water absorption by immersion, water absorption by capillarity, carbonation depth, and the migration of chloride ions increase linearly. This effect is justified by two factors [4]. The first is the higher porosity of CRCA in comparison to NAC, which provides additional pathways for water and other substances to penetrate inside RAC. The second is that, in most experimental programmes, the w/b ratio of RAC is larger than that of NAC. This is made to offset the loss of workability caused by the incorporation of CRCA.

The water/binder ratio governs the durability properties of a concrete mix [17], therefore, the increase in water/binder ratio is a relevant factor when a comparison between the durability of NAC and RAC is made. Increases of carbonation depth may be as large as 215 % in comparison to the carbonation of analogue NAC mixes [18], while the incorporation of CRCA has a smaller detrimental effect in the case of chloride-ion penetration. This is because the old cementitious paste present in the attached mortar of the CRCA includes tricalcium aluminates [19], which bind chloride ions and form Friedel's salt [20]. This also occurs in NAC, but to a smaller extent because the cementitious paste is only present in the new mortar. Authors report typical increases in the ingress of chloride ion in the range of 70–150 % [21, 22].

As for the mechanical properties, the increases of carbonation and chloride ion penetration presented in this section concern the full incorporation of CRCA of common quality and RAC with typical concrete mix design. The influence of a specific source of CRCA on a specific concrete mix design depends on several factors and its effect on the properties of concrete may deviate from what was mentioned in the last paragraph. For instance, the effect of CRCA on the durability of high-performance concrete may be smaller than in the case of normal concrete specifications (this is the opposite of what occurs for mechanical properties). This results in low water/binder ratios where the porosity of the new mortar will be low, which means that the CRCA inside of the concrete specimen will not come into contact with external agents; so, the porosity of the CRCA will not contribute relevantly to the ingress of external agents [23].

The practical influence of the increase of carbon dioxide and chloride ion ingress when CRCA are used is not as relevant for designers as initially understood. As presented in the beginning of this section, the incorporation of CRCA decreases f_c. Since both f_c and porosity are positively correlated with the water/binder ratio [17], this means that part of the detrimental effect of the incorporation of CRCA on durability is indirectly accounted for by the decrease of f_c. Therefore, metanalyses have found that the relationships between the 28-day f_c and carbonation depth [18] and between the 28-day f_c and chloride ion migration [19] are unaffected by the incorporation of CRCA.

9.2.3 Uncertainty in Resistance Models and Recycled Aggregate Concrete

Resistance models represent the behaviour of a structural element to the extent possible. Modelling is not exact due to a combination of reasons that, in the case of reinforced concrete design, include:

- Incomplete understanding of the physical or chemical phenomena, at the micro-, meso- and macro-scale;
- Absence of knowledge regarding factors that are specific to the case in question (e.g., past load history, curing conditions of concrete, or differences between the standard f_c measured in cylinders and the actual within member strength of the concrete element [24]);
- The recognition that codes need to be simple and that modelling needs to be simplified at the expense of accuracy. The development of resistance models that depend only on the f_c and neglect other concrete properties or mix design parameters is a common option that is mostly due to this concern.

Code makers recognize that modelling introduces uncertainty. This is the reason why the definition of partial factors γ_P, γ_Q, γ_C, and γ_S specifically includes reference to the uncertainty in modelling. When the partial factors of codes are calibrated, model uncertainties (θ_R) – also termed as bias factors, model errors, or professional factors in other regions, are defined through stochastic modelling and accounted for in reliability analyses. Reference [25] provides comprehensive information on the concept and estimation of θ_R.

Influence of Recycled Aggregate Concrete on the Uncertainties of Mechanical Models

When CRCA are used, the uncertainty in mechanical modelling is affected. This occurs because of several reasons [7, 26, 27]:

- The mesostructure of RAC differs from that of NAC [28, 29], not only because of the attached mortar of CRCA (which is more deformable and weaker than an NA), but also because RAC has three interfacial transition zones between coarse aggregates and mortar. This will change stress-paths, damage propagation, and overall inelastic behaviour [30];
- RAC may have higher statistical and energetic size effects [31] in comparison to NAC because of lower fracture energy [32] and higher material variability [6];
- As presented in Section 9.3.2, the incorporation of CRCA changes the relationships between f_c and other mechanical properties. The resistance models developed for NAC are mostly defined in terms of f_c only, but resistance mechanisms

also depend on other properties (e.g., bond strength is also dependent on f_{ct} and fracture energy [33, 34]). Therefore, a mostly unconservative bias is introduced when models developed for NAC are used for RAC design.

Table 9.1 summarizes the statistics found for the θ_R of resistance models present in prEN1992 [3]. These statistics are discussed in a set of publications of the authors [7, 26, 27]. As shown in the Table 9.1, the incorporation of CRCA decreases the mean θ_R of all resistance models except that for the ultimate moment resistance. This occurs because of the consistent effect of the incorporation of CRCA on the mechanical properties of concrete

Influence of Recycled Aggregate Concrete on the Uncertainty of Durability Modelling

The durability of concrete depends on the ingress of external agents, which is governed by transport mechanisms. These depend on several factors related to concrete mix design that directly affect either porosity (and pore size distribution) or the chemical composition of the hardened binder paste of concrete [20], as well as on climatic factors that govern humidity and temperature [35]. Out of the factors related to concrete mix design that affect transport mechanisms, the water/binder ratio of concrete and the chemical composition of its binder are the most relevant ones [17, 36].

Table 9.1 Model uncertainties of resistance models

Resistance model	Document	CRCA %	Mean	Standard deviation	Coefficient of variation	Reference
Bond strength	*fib* Bulletin 72 / prEN1992	0 %	1.03	0.15	15 %	[7]
		100 %	0.95	0.19	20 %	
Ultimate moment resistance	prEN1992	0 %	1.15	0.12	10 %	[26]
		100 %	1.15	0.12	10 %	
Shear resistance – without shear reinforcement	prEN1992	0 %	0.98	0.09	9 %	[27]
		100 %	0.93	0.12	13 %	
Shear resistance – with shear reinforcement	prEN1992	0 %	1.10	0.25	23 %	[27]
		100 %	1.03	0.23	22 %	

As stated in Section 9.2.2, the durability of concrete is affected when CRCA are used. Part of this is due to the porosity of the attached mortar of CRCA, which affects the transport mechanisms of concrete since: i) it contributes to the network of pores of concrete; and ii) it results in a different interfacial transition zone between aggregates and binder paste when compared to NAC (and is a relevant factor for durability, especially in what concerns chloride ion penetration [17]). Also, the attached mortar changes the chemical composition of concrete.

Nonetheless, these effects are not as relevant for the uncertainty in durability modelling and design as they seem. This is true because NAC durability is already subjected to high uncertainty. This is an acknowledged fact [35, 37] that is based on: i) the need for codes to be practical, and ii) the fact that concrete mix design is seldom known at the time of design. Many aspects that are relevant for durability design are not directly accounted for (e.g., mean porosity, pore distribution, and chemical interactions [17]), and since designers need to have practical guidelines for durability (that is, concrete cover) design, the approach of codes is to seek adequate durability through deemed-to-satisfy provisions. In the case of EN206 [38]/prEN1992 [3] design, this is achieved by defining minimum concrete covers that depend on requirements based on: environmental exposure to external agents, minimum f_c, minimum equivalent cement content, and maximum equivalent w/c ratio. This results in design with heterogeneous β [35, 37] since many factors, such as the type of binder used, are not fully accounted for during design.

Advanced durability models may be used to estimate the actual ingress of carbonation and chloride ions into concrete, as well as to estimate β of durability design. Section 9.3.3 presents the models of the international federation of concrete (*fib*) Bulletin 34 [20] and *fib* Model Code 2010 [39], which convert the carbonation and chloride-ion penetration resistance determined through accelerated laboratory tests to those expected under actual environmental conditions through conversion factors. Then, these models predict the ingress of external agents over time. As in the case of deemed-to-satisfy-provisions used for design, many factors that are relevant for transport mechanisms (such as porosity and pores' network; binder blend) are not directly accounted for. In the probabilistic modelling of [20, 39], the uncertainties due to these simplifications are accounted for in the stochastic modelling used. This means that the uncertainty in NAC design is already governed by very uncertain factors, which are more significantly relevant than those introduced when CRCA is used. Therefore, this chapter assumes that the uncertainty modelling concerning the durability of RAC may be made with the same formulae and factors used for RAC. This hypothesis is confirmed by the direction cosines α (which are defined in Section 9.3.1) of the reliability analyses presented in Section 9.3.3.

prEN1992 [3] also gives designers the option to design concrete covers based on a performance-based method, which requires that durability resistance is defined through specific laboratory testing for the concrete mix design to be used. This is outside the scope of this chapter. The only consideration made here is that, if laboratory testing is performed, the influence of CRCA will be accounted for in the experimental testing and in the resulting environmental resistance class used to define concrete covers. Therefore, NAC and CRCA will be designed following the same criteria and procedures.

9.3 Methodology for Calibration of Design Equations

9.3.1 Code Calibration

Overview of Partial Factor Calibration Procedure

The calibration of a partial factor includes:

- The definition of cases of design addressing the failure mode whose partial factors are under calibration. The cases of design involve the definition of most parameters (which concern load-effects, material properties, and geometry);

- Using the design equation (which compares S_d and R_d), all design parameters are defined so that C_d exactly matches R_d;
- The definition of the limit state equation, which is used to compare actual load-effects and resistance;
- The modelling of each random parameter of the limit state equation with a stochastic model;
- The reliability analysis, which estimates β;
- The comparison of β with a target β and the judgment whether the calibration process has ended or whether the design equation should be changed (by changing the partial factor) and the whole process repeated.

From this description follows that partial factor calibration is an iterative process.

Limit State Equation

The limit state equation – $g(x)$ compares the (random) outcome of load-effects (S) with that of resistance (C). The goal of a reliability analysis is to determine the probability of failure/reliability index and that condition is achieved when:

$$g(x) = R - S \leq 0 \tag{9.5}$$

Therefore, P_f = Probability ($g(x) = R - S \leq 0$).

In the calculation of $g(x)$, all parameters of design that are defined in terms of characteristic values in design equations are replaced with stochastic variables. Furthermore, additional stochastic variables may be introduced – e.g., in the case of geometric parameters as presented in the next section.

The random outcome R is calculated using a resistance model, which is not necessarily the resistance model that served as a basis to define R_d. This is due to the need for code formulae (which include the expression for R_d) to be simple, while the resistance model used to calculate R is intended to be as accurate as possible.

Stochastic Modelling

Stochastic models are used to account for the variability of parameters that are subjected to uncertainty. This is achieved by attributing a mean value, a measure of dispersion (the standard deviation or the CoV) and a probability distribution to the stochastic variables that represent each parameter.

Stochastic models used in reliability analyses are either:

- Conventionally-accepted models that are frequently presented in seminal publications, codes, or their background documents. Such models are usually intended for general modelling;
- Stochastic models that are developed for a specific application (e.g., the θ_R models developed for RAC that are presented in Table 9.1).

The uncertainty of some parameters is most commonly neglected due to their negligible effect on safety and such parameters are assumed deterministic. This is the case with the span of reinforced concrete beams.

Reliability Method

Reliability analyses are performed using software. This chapter resorted to *Comrel v9.00* of the *Strurel* suite [40]. The analyses used the Hasofer/Lind first order reliability method (FORM) with the Rackwitz/Fiessler algorithm [41].

The analysis presented in this chapter is based on two outputs:

- β, which serves to judge safety, namely to check whether β target is complied with or if partial factor calibration is required;
- The squares of the direction cosines (α^2) of the stochastic variables, since these judge the relative relevance of each stochastic variable to the overall uncertainty of a reliability problem.

Object of Calibration

The object of calibration is the set of partial factors that will be calibrated. As few changes as possible to the design procedures of prEN1992 [3] are intended; therefore, the object of calibration of this chapter is a single partial factor for ultimate limit state design and a single deemed-to-satisfy provision for durability design. Section 9.3.2 and Section 9.3.3 discuss this issue.

Target Reliability Index

Usually, code calibration is made by defining a fixed target β for cases of design with similar consequences of failure [42]. However, this was not the approach of this chapter. Since the chapter aims at ensuring that the β of RAC is analogue to that of design for NAC (using the already-calibrated design equations of codes), the criterion was that the β of each case of design defined for RAC was also calculated for an analogue NAC structural element and that the partial factor for RAC design be calibrated resulting in the β of RAC being equal to that of NAC.

The difference between this approach and that commonly used (the *a priori* definition of a target β) is important because β is relatively heterogeneous and the actual β deviates from the target β in some cases of design.

9.3.2 Design Equations for Ultimate Limit State Recycled Aggregate Concrete Design

Presentation of Resistance Models

The resistance models treated in this chapter are the following:

- The bond strength model of *fib* Bulletin 72 [43], which is practically the same as that of prEN1992 [3]:

$$f_{stm} = \left[54\left(f_c/25\right)^{0.25}\left(25/\phi\right)^{0.2}\left(l_b/\phi\right)^{0.55}\right] x \left[\left(c_{min}/\phi\right)^{0.25}\left(c_{max}/c_{min}\right)^{0.1} + k_m k_r\right] \quad (9.6)$$

In which f_{stm} (MPa) is the stress in the reinforcement bar due to bond, f_c between 15 and 110 MPa, l_b the bond length of the reinforcement which must be over 10 times the diameter of the reinforcement (ϕ), c_{min} and c_{max} depend on concrete cover and distance between bars, with c_{min}/ϕ between 0.5 and 3.5 and c_{max}/c_{min} must be smaller than 5.0; ϕ must be below 2.0, k_m and k_{tr} parameters related to transverse reinforcement.

- The ultimate moment resistance is calculated assuming the equivalent rectangular stress block of prEN1992 [3], which differs from that of the current EN1992 [14] only for f_c higher than 50 MPa. Under these conditions:

$$M_R = A_{sl} \cdot f_{sy} \cdot \left[d - 0.5 \cdot \frac{A_{sl} \cdot f_{sy}}{b \cdot \lambda \cdot (\eta \cdot f_c)} \right] \tag{9.7}$$

Where f_{sy} is the tensile yield stress of the reinforcement, d its effective depth and A_{sl} the area of its cross-section, b the width of the beam's cross-section, and λ and η parameters of the equivalent rectangular stress block.

- The shear resistance model of prEN1992 [3] for elements without shear reinforcement:

$$V_R = \left(0.6 \sqrt[3]{100 \rho_{Asl} \cdot f_c \cdot \frac{d_{dg}}{d}} \right) d \cdot b \geq V_{R,c,min} \tag{9.8}$$

In which ρ_{Asl} is the geometric ratio of the longitudinal reinforcement, d_{dg} a parameter related to the maximum aggregate diameter, and $V_{R,c,min}$ the minimum shear resistance.

- The shear resistance model of prEN1992 [3] for elements with shear reinforcement, which is the same as that of EN1992 [14]:

$$V_R = \frac{A_{sw}}{s} \cdot 0.9 \, d \cdot f_{yw} \cdot \cot(\Omega) \leq V_{R,max} \tag{9.9}$$

Where A_{sw} is the area of shear reinforcement, s the longitudinal distance between webs of shear reinforcement; f_{yw} the tensile yield stress of the shear reinforcement; Ω the angle between the longitudinal axis of the beam and the compression strut and between 21.8 and 45°, and V_R the maximum shear resistance, which is limited by the compression strut.

The definition of the limit state equations is straightforward: Eq. (9.5) is used and, in the case of this chapter, S is defined in terms of loads and geometric parameters.

More detailed considerations on the resistance models may be consulted in [7, 26, 27, 43].

Deterministic and Stochastic Models

The deterministic and stochastic models of the reliability analyses made in this subsection are:

- The θ_R of each resistance model, which are stated in Table 9.1. No deterministic models are presented since θ_R are absent from deterministic formulae.
- Those presented in Table 9.2 for load-effects, Table 9.3 for geometric variability, and Table 9.4 for material properties.

Table 9.2 Modelling of the parameters of design: load-effects

| Parameter | Deterministic model | Stochastic model | | | | Reference |
	Symbol	Symbol	Mean	Coefficient of variation	Distribution	
Permanent loading	G_k	G	G_k	10 %	Normal	[42, 44]
Variable loading (50-year extreme)	Q_k	Q	60 % G_k	35 %	Gumbel	[45, 46]
Model uncertainty of load-effects	Absent from deterministic model	θ_E	1.00	5 %	Lognormal	[47, 48]

Table 9.3 Modelling of the parameters of design: geometric variability

| Parameter | Deterministic model | Stochastic model | | | | Reference |
	Symbol	Symbol	Mean	Standard deviation	Distribution	
Deviations in the bottom reinforcement of slabs	Absent from deterministic model	Δ_Y	5 mm	5 mm	Normal	[44, 49]
Deviations in the bottom reinforcement of beams	Absent from deterministic model	Δ_Y	-5 mm	5 mm	Normal	[44, 49]
Deviations in the top reinforcement of slabs and beams	Absent from deterministic model	Δ_Y	5 mm	10 mm	Normal	[44, 49]
Deviation in c_x (bond strength parameter)	Absent from deterministic model	Δc_x	0.45 mm	2.9 mm	Normal	[49]
Deviations in the cover of columns	Absent from deterministic model	Δ_Y	2.5 mm	5 mm	Normal	[44, 49]

* The stochastic model of Δc_x, which concerns the distance between longitudinal bars, is based on that of [49] but mean and standard deviation are assumed as half of those for elements with width below 300 mm.

Since second order effects are not expected to be relevant in the case of common structures, which are the scope of this chapter, the deterministic modelling assumes that geometry is that of design (no deviations are accounted for since geometric deviations are considered in γ_c and γ_s – see Section 9.1). On the other hand, the stochastic modelling of geometry is given by:

$$a = a_{nom} + \Delta_Y \tag{9.10}$$

Where a is the random outcome of a geometric parameter, a_{nom} the nominal/design value of the parameter, and Δ_Y the stochastic model of geometric deviations.

In the case of f_c, the following model is used:

$$f_c = \lambda \times F_2 \times f_{ck} \tag{9.11}$$

Where f_{ck} is deterministic and does not necessarily comply with Eq. (9.1), λ is a stochastic parameter that converts f_{ck} to the delivered strength of ready-mixed concrete measured on standard specimens and is based on Portuguese ready-mixed concrete production [50], and F_2 is the parameter of Bartlett and MacGregor [51] that converts delivered strength measured on standard specimens to the strength of cores taken from actual reinforced concrete elements, which are assumed as representative of the strength within the structural element. Note that prEN1992 [3] affects f_{ck} with a factor ($\eta_{cc,d}$). This factor is omitted in this chapter since its value is equal to 1.0 for the strength classes of the cases of design presented in the chapter (C25/30 and C40/50).

Table 9.4　Modelling of the parameters of design: material properties

Parameter	Deterministic model	Stochastic model				Reference
	Symbol	Symbol	Mean	Standard deviation	Distribution	
Compressive strength of concrete	f_{ck}	See Eq. (9.11)				
Conversion of specified to delivered standard compressive strength	Absent from deterministic model	λ	1.20	0.17	Lognormal	[50]
Conversion of standard delivered compressive strength to compressive strength within structure	Absent from deterministic model	F_2	0.95 (beams and slabs)	0.13 (beams and slabs)	Lognormal	[51]
Tensile yield stress of reinforcement	f_{yk}	f_y	f_{yk} +2 × standard deviation	30 MPa	Lognormal [52]	[49]

The stochastic models for λ and F_2 were based on NAC data. Even though comparisons made for NAC and RAC concerning the within batch variability [24], the standard to onsite strength [53], and the influence of curing conditions on compressive strength [54]

all agree with the assumption that CRCA incorporation does not influence λ and F_2. This hypothesis should be validated whenever sufficient data concerning industrial cases of RAC production are published.

Design Equations to be Calibrated

The design equations to be calibrated are the following:

- That for lap splice length design using prEN1992 [3]:

$$l_{bd} = k_{lb} \times k_{cp} \times \phi \times \left(\frac{f_{yk}}{435}\right)^{n_c} \times \left(\frac{25}{f_{ck}}\right)^{0.5} \times \left(\frac{\phi}{20}\right)^{1/3} \times \left(\frac{1.5\phi}{c_d}\right)^{0.5} \leq 1.5\frac{\sqrt{f_{ck}}}{\gamma_C} \qquad (9.12)$$

This expression is a simplification of (5). Furthermore, Eq. (9.12) is defined in terms of bond length, whereas Eq. (9.6) is defined in terms of reinforcement stress. For good bond conditions and persistent and transient design, k_{lb} = 50 and k_{cp} = 1.0; n_c is 1.0 for reinforcement stress ($\sigma_{sd} = f_{yk}$ in the case presented in this chapter) equal to or below 435 MPa – these conditions are all complied with in the cases of design presented in this chapter. Concerning cover, c_d is a parameter that models the mechanical phenomena covered by c_{max}, c_{min}, k_m, and k_r in Eq. (9.6).

- For flexural design of beams:

$$A_s \times \left(\frac{f_{yk}}{\gamma_S}\right) \times \left[d - \frac{A_s \cdot (f_{yk}/\gamma_S)}{2 \cdot b \cdot (f_{ck}/\gamma_C)}\right] = (\gamma_G \times G_k + \gamma_Q \times Q_k) \times \frac{L^2}{8} \qquad (9.13)$$

- For shear resistance design of elements without shear reinforcement:

$$\left(\frac{0.6}{\gamma_C}\sqrt[3]{100\rho_{Asl+} \times f_{ck} \cdot \frac{d_{dg}}{a_v}}\right) \times d \times b = \gamma_G \times G_k + \gamma_Q \times Q_k \qquad (9.14)$$

- For shear design of elements with shear reinforcement:

$$\left[\frac{A_{sw}}{s} \times \frac{f_{yk}}{\gamma_S} \times 0.9d \times \cot(\Omega)\right] = \gamma_G \times G_k + \gamma_Q \times Q_k \qquad (9.15)$$

Except for bond strength, these expressions are very similar to what would be achieved by equaling Eqs. (9.6–9.9) with load-effects, adding partial factors, and defining most parameters of design through characteristic values. Eqs. (9.14) and (9.15) are not limited by maximum and minimum resistances (which are presented in the respective resistance models – see Eqs. (9.8) and (9.9) – since such boundaries are not limiting for the cases of design presented in this chapter.

As in the case of the resistance models, additional considerations regarding Eqs. (9.12–9.15) may be consulted in [7, 26, 27, 43].

Cases of Design

The cases of design were chosen so that they represent reinforced concrete elements found in common cases of structural design. Two strength classes [3] were analysed: C25/30 and C40/50. Designing reinforced concrete with a strength below C25/30 is not a common option and it is not expected that RAC be designed with a strength above C40/50.

All elements comply with the respective requirements concerning minimum and maximum reinforcement and detailing rules of prEN1992 [3]. Deformability and economic design were not considered when designing the elements, as in other code calibration procedures [47, 52], because the main concern of the calibration was to impose that R_d and S_d are exactly the same to avoid biases that would prevent the proper calibration of the partial factors. All cases of design concern simply supported beams and reinforcement of the S500 class [3].

A detailed description of the cases of design may be consulted in [7, 43, 55]. A brief overview of the cases of design is provided here:

- The reliability of bond length design was analysed for ten slabs and eight beams. The slabs did not have confining reinforcement that was present in the beams (Φ between 6 mm and 12 mm and the bars were spaced between 150 mm and 300 mm). The spacing between spliced bars varied between 100 mm and 260 mm. Concrete cover varied between 20 mm and 40 mm and Φ was either 16 mm or 20 mm in the case of beams and 12 mm and 16 mm in the case of slabs. The width of the beams was equal to 250 mm.
- In the case of elements designed for bending, ten slabs and seven beams were defined. The spans of the elements were either 5 m or 7 m. In the majority of cases, the height of the slabs was 170 mm and that of the beams 45 cm. The width of the beams was fixed at 250 mm. The effective depth was calculated for a cover of 25 mm and longitudinal reinforcement with Φ = 8 mm (slabs) and Φ = 16 mm (beams). Design values for uniformly distributed loads of 12.9 kN/m² (slabs) and 49 kN/m (beams) were defined for most cases of design. Changes in variables concerned: the strength class of concrete, the span of the element, the design value of loads, and the height of the slab. The dimensionless ultimate moment resistance (μ_{MR}) of the slabs varied between 0.08 and 0.18, while the μ_{MR} of beams varied between 0.14 and 0.41;
- Five slabs were designed for the reliability analyses of elements without shear reinforcement. All slabs had a maximum aggregate diameter of 20 mm and a shear span-to-effective depth ratio of 2.5; slab height varied between 140 mm and 215 mm and spans were either 5 m or 6.5 m. The longitudinal reinforcement had Φ of either 12 mm or 16 mm and the longitudinal reinforcement ratio varied between 0.81 % and 0.24 %. The same uniformly distributed loading considered in the design of the slabs for bending was considered for these cases of design;
- A single case of shear design of elements with shear reinforcement was analysed. This option was taken because: i) the resistance side of Eq. (9.15) is a multiplication of deterministic and lognormal stochastic variables (except for geometric uncertainty, which has a negligible effect in this case of design), meaning that R_d will be almost lognormally distributed; ii) load-effects depend on the distributions presented in Table 9.2, whose uncertainty does not depend on the case of design (note that load uncertainty is defined by a fixed CoV, rather than a fixed standard

deviation); iii) from the two previous observations and since a condition for these reliability analyses is that design imposes $R_d = E_d$, it follows that different cases of design will result in virtually the same β. Therefore, the case of design concerns concrete of the C25/30 strength class, $b = 250$ mm and height of 450 mm; concrete cover of 25 mm, longitudinal reinforcement with $\Phi = 16$ mm and transverse reinforcement with $\Phi = 8$ mm (with a 150 m spacing), and $\Omega = 30°$. This corresponds to a geometric ratio of shear reinforcement of 0.27 %.

All parameters stated refer to NAC design. The calibration of partial factors for RAC changes some of the parameters stated – e.g., the reinforcement ratios. This does not have a practical effect on the β and α^2 values that result from the reliability analyses.

Reliability Method

The reliability method used was the one described in Section 9.3.1.

Partial Factors to be Calibrated

As stated in Section 9.3.1, minimal changes to Eurocode design were sought. Therefore, partial factors γ_G, γ_Q, and γ_S are not changed. In all but one case, the partial factor γ_C is not changed either and the calibration presented in this paper concerns an additional partial factor γ_{RAC}. The following format is used:

$$R_d\,(\ldots)_{RAC} = \frac{1}{\gamma_{RAC}} \times R_d\,(\ldots) \tag{9.16}$$

In this format, γ_{RAC} depends on type of failure mode and on the incorporation ratio of CRCA; $R_d\,(\ldots)$ is the design value of resistance calculated for NAC using prEN1992 [3]; and $R_d\,(\ldots)_{RAC}$ is the design value of resistance for RAC.

In the case of shear resistance of elements without shear reinforcement designed with prEN1992 [3], since γ_C is placed outside the brackets of the resistance model, the option was to calibrate γ_C directly and adapting Eq. (9.13) by replacing γ_C with $\gamma_{V,RAC}$:

$$\left(\frac{0.6}{\gamma_{V,RAC}} \sqrt[3]{100\rho_{Asl+} \times f_{ck} \cdot \frac{d_{dg}}{a_v}} \right) \times d \times b = \gamma_G \times G_k + \gamma_Q \times Q_k \tag{9.17}$$

The design formula for bond strength design is defined in terms of bond length, so that the following format is adopted:

$$l_{bd} = \gamma_{RAC} \times \left(\frac{\phi \cdot \sigma_s}{4 \cdot f_{bd}} \right) \tag{9.18}$$

Calibrated Partial Factors for Recycled Aggregate Concrete Design

The procedure just presented resulted in the partial factors presented in Table 9.5. For intermediate incorporation ratios, γ_R may be determined by linear interpolation.

Due to the negligible effect of the incorporation of CRCA on the ultimate moment resistance, the partial factor for the design to bending is equal to 1.0 and may be omitted.

Table 9.5　Partial factors for ultimate limit state design

Type of design	CRCA %	Partial factor	Reference
Bond length	100 %	γ_{RAC}	[7]
Ultimate moment resistance	0 % to 100 %	γ_{RAC}	[26]
Shear resistance – without shear reinforcement	0 %	$\gamma_{V,NAC}$	[43]
	100 %	$\gamma_{V,RAC}$	
Shear resistance – with shear reinforcement	100 %	γ_{RAC}	[43]

Table 9.5 presents a partial factor for NAC design of elements without shear reinforcement (1.40). This was the only partial factor that was calibrated for NAC and this calibration was made because the design equation of prEN1992 [3] was still pending reliability-based calibration at the time this chapter was written.

Fig. 9.1 compares the β of NAC and RAC design with full CRCA incorporation after partial factor calibration for selected cases of design concerning bond length and shear resistance without shear reinforcement. The β calibration for shear resistance with shear reinforcement [43] is similar to that observed for elements without shear reinforcement.

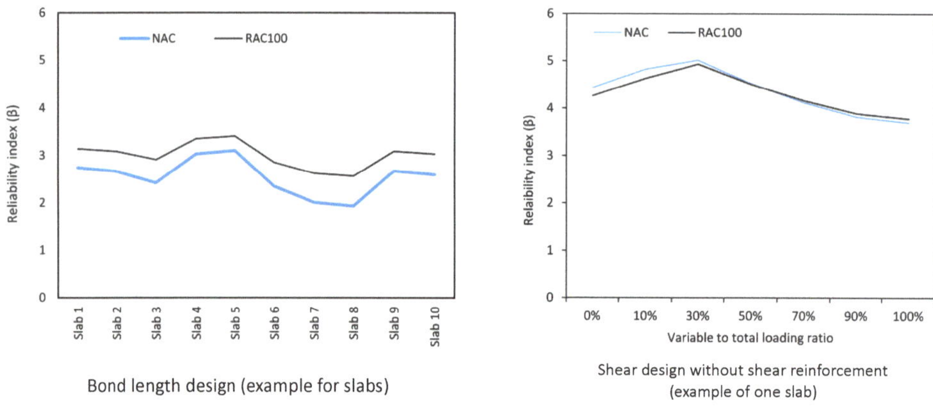

Fig. 9.1　Examples of reliability indices after calibration

The analysis of α^2, which is presented in Fig. 9.2, shows that the uncertainty in θ_R is responsible for most of the uncertainty of the structural design. This was found for all cases of design and incorporation ratios of CRCA. However, larger incorporation ratios of CRCA are associated with larger importance of the uncertainty of θ_R to the overall safety and this fact has a straightforward explanation: the statistics of θ_R worsen (larger standard deviation and smaller mean value) with the incorporation of CRCA (see Table 9.1).

Other noteworthy findings of the analysis of the α^2 are that:

- Geometrical deviations are relatively unimportant for the overall uncertainty;
- The relative importance of Q increases as the ratio of live to total load ratio increases and for larger (more than 70 %) ratios, the partial factor γ_S is smaller than

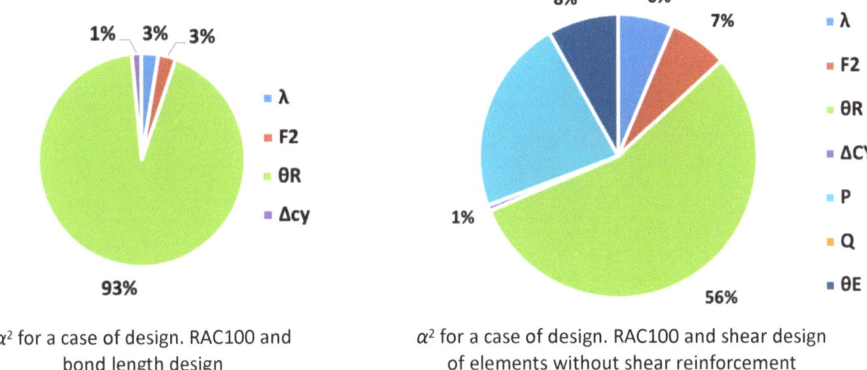

Fig. 9.2 Examples of α^2 of the reliability analysis of ultimate limit state design

intended. This is an acknowledged fact [47, 52] and is the reason for the decrease of β when the variable to total loading ratio presented in Figure 11.1 is high;

- The relative importance of θ_R is higher in the case of bond length design than in all others due to the lower mean value and larger CoV of the respective resistance model in comparison to all others (Table 9.1).

9.3.3 Deemed-to-Satisfy Provisions for Concrete Cover Design of Recycled Aggregate Concrete

Presentation of Resistance Models

The models used to estimate the ingress of carbon dioxide and chloride ions over time are those presented in [20]. For both durability mechanisms, failure is considered as the moment in which depassivation occurs. In the case of carbonation, this happens when the carbonation front (x_c), reaches the concrete cover (a), in the case of chloride ion migration, depassivation happens when a critical chloride content (C_{crit}) is reached at the concrete cover ($C(a, t_{ref}) = C_{crit}$), where $C(a, t_{ref})$ is the chloride content at depth and time t_{ref}.

- The carbonation depth (x_c in mm) is given by:

$$x_c(t_{SL}) = \sqrt{2 \; \{[1 - RH_{real}{}^{fe}]/[1 - RH_{ref}{}^{fe}]\}^{ge} \; \left(\tfrac{t_c}{7}\right)^{bc} \; (k_t \; R_{ACC,0^{-1}} + \varepsilon_t) \; C_s \; \sqrt{t_{SL}} \; (0.0767)^{0.5(p_{SR} \cdot ToW)^{bw}}} \quad (9.19)$$

In which t_{SL} is the exposure time in years, RH_{real} the relative humidity of the layer of carbonated concrete and RH_{ref}= 65 % a reference humidity value; f_e = 2.5, and g_e = 5.0 are deterministic factors; t_c is the curing time in days; b_c and b_w are regression parameters, C_s (ambient concentration of CO_2), p_{SR} = 0.103 is the probability of rain, ToW (number of days in a year in which daily rain is above 2.5 mm), k_t and ε_t (terms for the conversion of experimental accelerated carbonation depths to those under natural carbonation) are stochastic variables presented in Table 9.5. The ToW is deterministic and, for the three carbonation exposure classes treated in this chapter (XC1, XC3 and XC4) is equal to: XC1 - 0.5; XC3 - 0.10; XC4 - 0.18.

$R_{ACC,0}^{-1}$ $(m^2/s/kg \cdot CO^2/m^3)$ is the inverse resistance to carbonation under accelerated conditions. This parameter is calculated from the experimental carbonation depth (d_k) through [20]:

$$R_{ACC,0^{-1}} = 3.154 \cdot 10^{13} \times \left(\frac{d_k}{419,45}\right)^2 \tag{9.20}$$

In this equation, d_k is the experimental carbonation depth and is computed in metres. $R_{ACC,0}^{-1}$ is defined in $mm^2/year/kg\ CO_2/m^3$.

–   The chloride ion content at depth a $(C(x = a,t),$ in (wt.% / cement)) is given by:

$$C(x = a,t) = (C_{S,\Delta x})' \left[1 - \mathrm{erf}\left(\frac{(a)-\Delta_x}{2' \sqrt{k_e \times (D_{RCM,0} \times 3.154 \cdot 10^{13}) \times k_t \times A(t) \times t_{SL}}}\right)\right] \tag{9.21}$$

Where $k_e = \exp\left[b_c\left(\frac{1}{T_{ref}} - \frac{1}{T_{real}}\right)\right]$, $k_t = 1.0$ a regression parameter, $A(t) = \left(\frac{0.0767}{t_{ref}}\right)^{a'}$, and erf the Gauss error function. In the first equation, T_{ref} is a reference temperature (293 K) and T_{real} the real temperature in the concrete element. a' is an ageing exponent and b_c a regression parameter. $C_{S,\Delta x}$ is the superficial chloride ion content of the concrete. Table 9.5 presents the stochastic modelling. $D_{RCM,0}$ is the chloride migration coefficient (m^2/s) and was determined experimentally.

The limit state equations are defined based on Eq. (9.3). Load-effects are calculated using Eqs. (9.20–9.21) and assuming a t_{SL} of 50 years, while resistance is the concrete cover in the case of carbonation and the critical chloride content at the depth of the concrete cover in the case of chloride ion penetration.

Stochastic Models

The stochastic models needed for Eqs. (9.20–9.21) are summarized in Table 9.6, except in the case of d_k and $D_{RCM,0}$, which were both modelled based on experimental results presented and discussed in [4, 56]. The references for each stochastic model can be consulted in [4, 56], along with a detailed description of the resistance models presented in Eqs. (9.20–9.21). Note that the stochastic models for the uncertainty in concrete cover have already been presented in Table 9.3.

Provisions to be Calibrated

No deterministic models or design equations are presented since prEN1992 [3] concrete cover design is based on deemed-to-satisfy provisions, rather than on formulae.

The deemed-to-satisfy provisions of prEN1992 [3] are based on:

–   The determination of environmental exposure classes, which represent the durability "actions" that the concrete element is subjected to;
–   For each environmental exposure class, a minimum water/cement ratio, a minimum strength class, and a minimum cement content are stipulated by EN206 [38];

- The minimum concrete cover is then determined, based on environmental exposure class, type of structural element, design working life, strength class, and other aspects such as the use of corrosion-resistant reinforcement.

Table 9.6 Stochastic modelling for durability

Type of design	Symbol	Distribution	Mean	Standard deviation
Carbonation	b_c	Normal	-0.567	0.024
	k_t	Normal	1.25	0.35
	ε_t	Normal	315.5	48
	C_s	Normal	0.00082	0.0001
	b_w	Normal	0.446	0.163
	RH_{real}	XC1 - deterministic XC3; XC4 - Weibull ($\omega = 1$)	XC1 (dry) - 60 % XC3 - 75 % XC4 - 75 %	XC1 - 0 XC3 - 14 % XC4 - 14 %
Chloride ion penetration	a'	Beta (a=0; b=1)[*1]	0.30[*1]	0.12[*1]
	b_c	Normal	4800	700
	Δ_x	Beta (a=0; b=50)[*2]	10 mm[*2]	5 mm[*2]
	C_{s,Δ_x}	Lognormal	1.5 wt. %/cement	1.13 wt. %/cement
	C_{crit}	Beta (a=0.2; b=2)	0.6 wt. %/cement	0.15 wt. %/cement

[*1] for binders with fly ash: mean = 0.60; standard deviation = 0.15; a = 0; b = 1.0.

[*2] for the XS exposure class: mean = 0.0; standard deviation = 0.0; deterministic.

These provisions are omitted from the chapter since they are well known to designers familiar with Eurocode reinforced concrete design.

The provisions that will be calibrated in this chapter are those related to the minimum concrete cover. All other provisions are kept "as is", to ensure that the changes to the format of prEN1992 [3] are minimal.

Cases of Design

The cases of design analysed covered different environmental exposure classes, types of structural element (slabs and columns, due to differences in both different minimum cover requirements and exposure to rain), and concrete mix design. Five mixes for carbonation and five mixes for chloride ion penetration were produced in laboratory conditions (two of each, NAC and RAC, respectively). The experimental parameters $R_{ACC,0}^{-1}$ for carbonation and $D_{RCM,0}$ for chloride ion penetration were determined from tests [57, 58] on these mixes.

Table 9.7 summarizes the concrete covers of the cases of design. The mixes were: a common C30/37 mix made with a CEM I binder (REF), a C25/30 mix made with a CEM II A\L42,5 (limestone) binder (CEM II), a high-strength concrete mix (HSC), and a cement mix made with 75 % of CEM I blended with 25 % of fly ash (FAC). The mixes had a total binder content of 350 kg/m³, except that made with fly ash, since fly ash (FA) was incorporated by volume (in relation to the REF mix, whose binder content was 350 kg/m³).

Reliability Method

The reliability method used was that described in Section 9.3.1.

Object of Calibration

The object of calibration is the minimum concrete cover, which was increased in order to ensure that the β of RAC concrete cover design was equal to that of NAC.

Table 9.7   Cases of design for durability (NAC and RAC before calibration)

Mix	Carbonation		Chloride ion penetration	
	Exposure class	**a (mm)**	**Exposure class**	**a (mm)**
REF	XC3	Column: 35 Slab: 30	XS1	Column: 45 Slab: 40
CEM II	XC3	Column: 35 Slab: 30	XS1	Column: 45 Slab: 40
HSC	XC4	Column:35 Slab: 30	XS3	Column: 55 Slab: 50
FAC	XC1	Column: 20 Slab: 20	XS1	Column: 45 Slab: 40

Calibration of deemed-to-satisfy provisions

The calibration of the deemed-to-satisfy provisions required that the concrete cover requirements for RAC design are increased by 5 mm in order to ensure that the β of NAC and RAC design are the same. Fig. 9.3 compares the β of NAC and RAC elements after this increase.

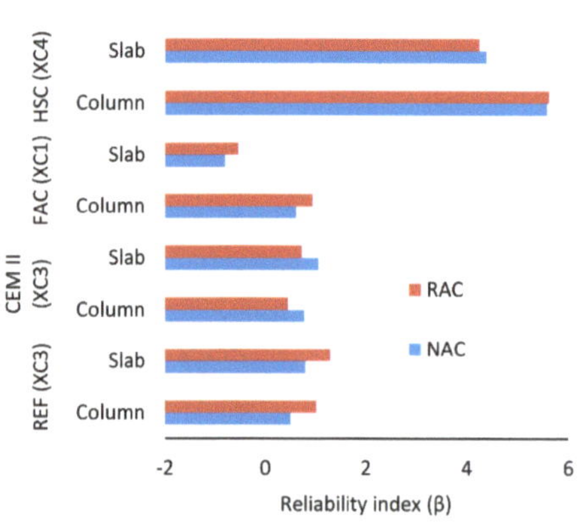

Concrete cover design for carbonation

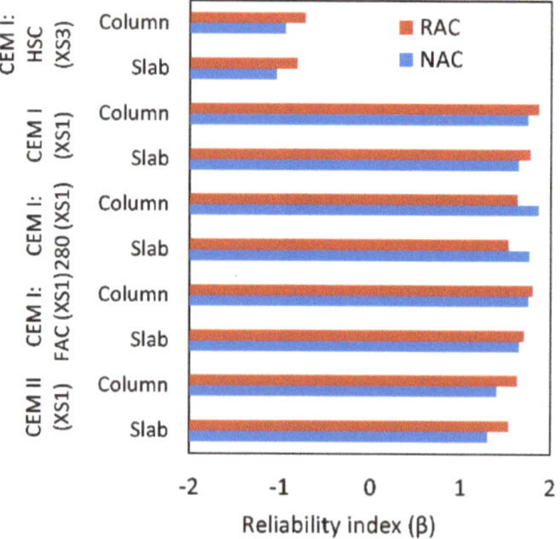

Concrete cover design for chloride ion penetration

Fig. 9.3   Examples of α^2 of the reliability analyses for durability after calibration of the deemed-to-satisfy provisions (increase of RAC cover by 5 mm)

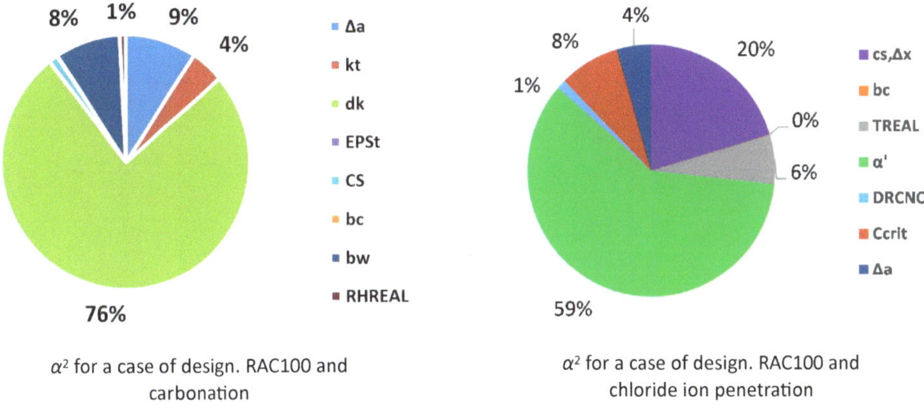

Fig. 9.4 Examples of α^2 of the reliability analysis of concrete cover durability design

Fig. 9.4 shows examples of the distribution of the α^2. As understood from the figure, the experimental carbonation depth (d_k) and ageing (α') are governing for the overall uncertainty of carbonation and chloride ion penetration cover design, respectively.

9.4 Conclusion

This chapter treated the proposal of design guidelines for the structural design of reinforced recycled aggregate concrete with the same safety as the design of reinforced natural aggregate concrete structures. The chapter is about the incorporation of coarse recycled aggregates produced from concrete waste only. This is based on: i) the absence of information regarding the incorporation of other types of recycled aggregates (in what concerns size fraction, type of construction, and demolition waste); ii) the fact that this is the most suitable type of recycled aggregate for structural concrete.

The chapter is based on reliability concepts and its introduction describes how the safety checks of current codes are based on uncertainties and partial factor calibration. After summarizing basic aspects related to the production and properties of recycled aggregates, the chapter presented the influence of recycled aggregates on the mechanical and durability properties of concrete. Then, reasons were put forward for the incorporation of recycled aggregates influencing the accuracy and precision of resistance models (that is, their model uncertainties). This is followed by a description of the concepts needed for code calibration and by the calibration of prEN1992 design, namely:

- Partial factors for ultimate limit state design to bending, shear with shear reinforcement, shear without shear reinforcement and bond failures
- The calibration of the concrete covers of the deemed-to-satisfy provisions for carbonation and chloride ion environments.

A resistance format with an additional partial factor for recycled aggregate concrete ultimate limit state design was proposed and a table with the value of this partial factor was presented. This partial factor is meant for concrete made with full incorporation of recycled aggregates and linear interpolation may be used to find a suitable partial factor for intermediate recycled aggregate incorporation ratios. Concerning durability, a concrete cover increase of 5 mm was found to be suitable for both carbonation and chloride ion concrete cover design.

The following are suggestions of future research:

– The comprehensive testing of the structural behaviour of recycled aggregates produced from other types of construction and demolition waste;
– The development of factors to convert the specified strength to the strength within structural elements made with recycled aggregate concrete (this requires data from recycled aggregate concrete structures produced in real construction environments);
– A comprehensive comparative life-cycle assessment covering different scenarios (e.g., regarding concrete mix, transport distance of raw materials, and structural designs) to assess the conditions under which recycled aggregate concrete is a more environmentally sustainable and/or cost-efficient option in comparison to natural aggregate concrete.

The last suggestion is particularly relevant: the use of partial factors to offset decreases of safety associated with the incorporation of recycled aggregates implies that the volume of concrete and/or steel reinforcement increases. This comes at environmental and economic cost and, under some circumstances, recycled aggregate concrete may be more expensive and less sustainable than natural aggregate concrete.

9.5 Acknowledgements

The authors acknowledge the CERIS research centre, IST, University of Lisbon and the EcoCoRe doctoral programme – grant [PD/BD/113643/2015] of the Portuguese Foundation for Science and Technology (FCT).

9.6 References

[1] Madsen, H.O.K., S.; Lind, N. C., Methods of Structural Safety. 1986, New Jersey, United States of America: Dover Publications.

[2] Ferry-Borges, J. and M. Castanheta, Structural safety. Third ed. 1983, Lisbon, Portugal: Portuguese Laboratory of Civil Engineering.

[3] CEN/TC-250/SC-2, prEN 1992-1-1 D6 Working file (2020-10-05 Rev. 7). 2020.

[4] Albuquerque, A., J. Pacheco and J. de Brito (2022). "Reliability-based recommendations for EN1992 carbonation cover design of concrete with coarse recycled concrete aggregates." Structural Concrete 23: 1873–1889. https://doi.org/10.1002/suco.202100406

[5] de Brito, J., et al., Structural, material, mechanical and durability properties and behaviour of recycled aggregates concrete. Journal of Building Engineering, 2016. 6: p. 1–16. https://doi.org/10.1016/j.jobe.2016.02.003

[6] Pepe, M., et al., Mechanical behaviour of coarse, lightweight, recycled and natural aggregates for concrete. Proceedings of the Institution of Civil Engineers – Construction Materials, 2020. 173(2): p. 70–78. https://doi.org/10.1680/jcoma.17.00081

[7] Pacheco, J., et al., Bond of recycled coarse aggregate concrete: model uncertainty and reliability-based calibration of design equations. Engineering Structures, 2021. 239: 112290. https://doi.org/10.1016/j.engstruct.2021.112290

[8] Pedro, D., J. de Brito, and L. Evangelista, Performance of concrete made with aggregates recycled from precasting industry waste: influence of the crushing process. Materials and Structures, 2014: p. 1–14. https://doi.org/10.1617/s11527-014-0456-7

[9] Soares, D., et al., Use of coarse recycled aggregates from precast concrete rejects: Mechanical and durability performance. Construction and Building Materials, 2014. 71: p. 263–272. https://doi.org/10.1016/j.conbuildmat.2014.08.034

[10] EN-12620+A1, Aggregates for concrete, with 2008 amendment 2008, CEN.

[11] Li, X., Recycling and reuse of waste concrete in China Part I. Material behaviour of recycled aggregate concrete. Resources, Conservation and Recycling, 2008. 53(1–2): p. 36–44. https://doi.org/10.1016/j.resconrec.2008.09.006

[12] Pacheco, J., et al., Experimental investigation on the variability of the main mechanical properties of recycled aggregate concrete. Construction and Building Materials, 2019. 201: p. 110–120. https://doi.org/10.1016/j.conbuildmat.2018.12.200

[13] Rusch, H., R. Sell, and R. Rackwitz, Statistische Analyse der Betonfestigkeit. Berlin, Deutscher Ausschuss für Stahlbeton, 1969.

[14] EN-1992-1-1, Eurocode 2 – Design of concrete structures: Part 1-1: General rules and rules for buildings, 2008, Comité Européen de Normalisation (CEN): Brussels, Belgium.

[15] Pacheco, J., et al., Scatter of constitutive models of the mechanical properties of concrete: comparison of major international codes. Journal of Advanced Concrete Technology, 2019. 17(3): p. 102–125. https://doi.org/10.3151/jact.17.102

[16] Olorunsogo, F.T. and N. Padayachee, Performance of recycled aggregate concrete monitored by durability indexes. Cement and Concrete Research, 2002. 32(2): p. 179–185. https://doi.org/10.1016/s0008-8846(01)00653-6

[17] fib, Bulletin 70: Code-type models for concrete behaviour. 2013: Lausanne, Switzerland.

[18] Silva, R.V., et al., Carbonation behaviour of recycled aggregate concrete. Cement and Concrete Composites, 2015. 62: p. 22–32. https://doi.org/10.1016/j.cemconcomp.2015.04.017

[19] Silva, R.V., et al., Prediction of chloride ion penetration of recycled aggregate concrete. Materials Research, 2015. 18(2): p. 427–440. https://doi.org/10.1590/1516-1439.000214

[20] fib, Bulletin 34: Model Code for Service Life Design. 2006: Lausanne, Switzerland.

[21] Kou, S.C., C.S. Poon, and H.W. Wan, Properties of concrete prepared with low-grade recycled aggregates. Construction and Building Materials, 2012. 36: p. 881–889. https://doi.org/10.1016/j.conbuildmat.2012.06.060

[22] González, A. and M. Etxeberria, Experimental analysis of properties of high performance recycled aggregate concrete. Construction and Building Materials, 2014. 52: p. 227–235. https://doi.org/10.1016/j.conbuildmat.2013.11.054

[23] Thomas, C., et al., Durability of recycled aggregate concrete. Construction and Building Materials, 2013. 40: p. 1054–1065. https://doi.org/10.1016/j.conbuildmat.2012.11.106

[24] Bartlett, F.M. and J.G. MacGregor, Variation of In-Place Concrete Strength in Structures. ACI Materials Journal, 1999. 96(2): p. 261–270. https://doi.org/10.14359/454

[25] Holický, M., J.V. Retief, and M. Sýkora, Assessment of model uncertainties for structural resistance. Probabilistic Engineering Mechanics, 2016. 45: p. 188–197. https://doi.org/10.1016/j.probengmech.2015.09.008

[26] Pacheco, J., et al., Uncertainty models of reinforced concrete beams in bending: code comparison and recycled aggregate incorporation. Journal of Structural Engineering, 2019. 145(4): p. 04019013. https://doi.org/10.1061/(asce)st.1943-541x.0002296

[27] Pacheco, J., et al., Uncertainty of shear resistance models: influence of recycled concrete aggregate on beams with and without shear reinforcement. Engineering Structures, 2020. 204(1): 109905. https://doi.org/10.1016/j.engstruct.2019.109905

[28] Xiao, J., et al., Effects of interfacial transition zones on the stress-strain behavior of modeled recycled aggregate concrete. Cement and Concrete Research, 2013. 52: p. 82–99. https://doi.org/10.1016/j.cemconres.2013.05.004

[29] Xiao, J., et al., Properties of interfacial transition zones in recycled aggregate concrete tested by nanoindentation. Cement and Concrete Composites, 2013. 37: p. 276–292. https://doi.org/10.1016/j.cemconcomp.2013.01.006

[30] Guo, M., et al., Fracture process zone characteristics and identification of the micro-fracture phases in recycled concrete. Engineering Fracture Mechanics, 2017. 181: p. 101–115. https://doi.org/10.1016/j.engfracmech.2017.07.004

[31] Bažant, Z.P. and A. Yavari, Is the cause of size effect on structural strength fractal or energetic-statistical? Engineering Fracture Mechanics, 2005. 72(1): p. 1–31. https://doi.org/10.1016/j.engfracmech.2004.03.004

[32] Casuccio, M., et al., Failure mechanism of recycled aggregate concrete. Construction and Building Materials, 2008. 22(7): p. 1500–1506. https://doi.org/10.1016/j.conbuildmat.2007.03.032

[33] fib, Bulletin 72: Bond and anchorage of embedded reinforcement: Background to the fib Model Code for Concrete Structures 2010 2014: Lausanne, Switzerland. https://doi.org/10.35789/fib.BULL.0072

[34] Zuo, J. and D. Darwin, Splice strength of conventional and high relative rib area bars in normal and high-strength concrete. ACI Structural Journal, 2000. 97(4): p. 630–641. https://doi.org/10.14359/7428

[35] von Greve-Dierfeld, S. and C. Gehlen, Performance based durability design, carbonation part 1 – Benchmarking of European present design rules. Structural Concrete, 2016. 17(3): p. 309–328. https://doi.org/10.1002/suco.201600066

[36] Duracrete, DuraCrete. Final technical report: general guidelines for durability design and redesign, in Document BE95-1347/R17. 2000.

[37] von Greve-Dierfeld, S. and C. Gehlen, Performance-based durability design, carbonation, part 3: PSF approach and a proposal for the revision of deemed-to-satisfy rules. Structural Concrete, 2016. 17(5): p. 718–728. https://doi.org/10.1002/suco.201600085

[38] EN-206, Concrete: Specification, performance, production and conformity. Incorporating corrigendum May 2014. 2013, CEN: Brussels, Belgium.

[39] fib, Bulletin 65: Model Code 2010 final draft. Volume 1. 2010: Lausanne, Switzerland. https://doi.org/10.35789/fib.BULL.0065

[40] Eracons, R.C., Strurel software for reliability analysis -Comrel, Sysrel, Costrel: Users Manual, www.strurel.com. 2016.

[41] Rackwitz, R. and B. Fiessler, Structural reliability under combined random load sequences. Computers & Structures, 1978. 9: p. 489–494. https://doi.org/10.1016/0045-7949(78)90046-9

[42] EN-1990, Eurocode: Basis of structural design. 2002, Comité Européen de Normalisation (CEN): Brussels, Belgium.

[43] Pacheco, J., et al., Eurocode shear design of coarse recycled aggregate concrete: reliability analysis and partial factor calibration. Materials, 2021. 14(15): p. 4081. https://doi.org/10.3390/ma14154081

[44] Gulvanessian, H., J.A. Calgaro, and M. Holicky, Designers' guide to EN1990 Eurocode: Basis of structural design, in Designers' Guide to the Eurocodes. 2002, Thomas Telford: London, United Kingdom. p. 197. https://doi.org/10.1680/dgte.30114

[45] Holický, M.A., Chapter 1 – Self-weight and imposed loads on buildings, in Implementation of Eurocodes. Handbook 3: Action effects for buildings. 2005, Leonardo da Vinci Pilot Project CZ/02/B/F/PP-134007. "Development of skills facilitating implementation of Eurocodes": Aachen, Germany.

[46] Gulvanessian, H. and M. Holicky, Eurocodes: using reliability analysis to combine action effects. Proceedings of the Institution of Civil Engineers – Structures and Buildings, 2005. 158(4): p. 243–252. https://doi.org/10.1680/stbu.2005.158.4.243

[47] BRE, Client report 210297. An independent technical expert review of the SAKO Report, H. Gulvanessian, et al., Editors. 2003, BRE: Watford, United Kingdom.

[48] Gulvanessian, H. and M. Holický, Annex C – Calibration procedure, in Implementation of Eurocodes. Handbook 2: Reliability backgrounds. 2005, Leonardo da Vinci Pilot Project CZ/02/B/F/PP-134007. "Development of skills facilitating implementation of Eurocodes": Prague, Czech Republic.

[49] JCSS, Probabilistic Model Code. Part 3: Material properties. 2001.

[50] Pacheco, J., et al., Statistical analysis of Portuguese ready-mixed concrete production. Construction and Building Materials, 2019. 209: p. 283–294. https://doi.org/10.1016/j.conbuildmat.2019.03.089

[51] Bartlett, F.M. and J.G. MacGregor, Statistical analysis of the compressive strength of concrete in structures. ACI Materials Journal, 1996. 93(2): p. 158–168. https://doi.org/10.14359/1353

[52] NKB/SAKO, NKB1999:01E Basis of design of structures. Proposals for modification of partial safety factors in Eurocodes, I.-B. 1999, Nordic Committee on Building Regulations (NBK): Oslo, Norway.

[53] Soares, D., et al., In situ materials characterization of full-scale recycled aggregates concrete structures. Construction and Building Materials, 2014. 71: p. 237–245. https://doi.org/10.1016/j.conbuildmat.2014.08.025

[54] Fonseca, N., J. de Brito, and L. Evangelista, The influence of curing conditions on the mechanical performance of concrete made with recycled concrete waste. Cement and Concrete Composites, 2011. 33(6): p. 637–643. https://doi.org/10.1016/j.cemconcomp.2011.04.002

[55] Pacheco, J., Reliability analysis of eco-concrete, in Department of Civil Engineering, Architecture and Gerorresources. 2020, Instituto Superior Técnico, University of Lisbon: Lisbon, Portugal.

[56] Albuquerque, A., J.N. Pacheco, and J.d. Brito, Eurocode Design of Recycled Aggregate Concrete for Chloride Environments: Stochastic Modeling of Chloride Migration and Reliability-Based Calibration of Cover. Crystals, 2021. 11(3): 284.

[57] LNEC-E463, Concretes: determination of the chloride diffusion coefficient by the migration test under stationary regime (in Portuguese). 2004, National Laboratory of Civil Engineering (LNEC – Laboratório Nacional de Engenharia Civil) Lisbon, Portugal.

[58] LNEC-E391, Concrete: determination of carbonation resistance (in Portuguese). 1993, National Laboratory of Civil Engineering (LNEC – Laboratório Nacional de Engenharia Civil) Lisbon, Portugal.

Chapter 10

Total Recycling of Concrete Waste Using Accelerated Carbonation

C.S. Poon, P.L. Shen, Y. Jiang, D.X. Xuan

Dept. of Civil and Environmental Engineering and Research Centre for Resources Engineering towards Carbon Neutrality, The Hong Kong Polytechnic University

10.1 Introduction

With accelerated industrialization and urbanization, especially among developing countries and regions, huge amounts of concrete waste are generated. Meanwhile, these regions are also suffering from the depletion of natural mineral resources and shortages of waste disposal sites [1]. Therefore, recycling and reuse of concrete waste is necessary for environmental protection and effective utilization of resources. Recycled concrete aggregate (RCA) is the main product generated from concrete waste after it is processed in recycling facilities.

Fig. 10.1 shows a typical treatment system of a recycling facility. The major process elements are quite similar to those used for the production of natural aggregates (NA) [2]. Nevertheless, the use of RCA to replace natural aggregates in new concrete, unless carefully managed and controlled, is likely to have a negative influence on most concrete properties due to the presence of adhered old cement mortar on the surface of RCA, which could contain a significant amount of hydrated cement paste. To obtain high-quality RCA, improved mechanical crushing processes have been developed by integrating different crushers into the crushing process, such as the use of an impact crusher as indicated in Fig. 10.1. By performing the multiple crushing processes, maximum removal of the adhered mortar can be achieved. Nevertheless, it also significantly increases the generation of fine particles from 40 % to 70 % [2].

After the crushing and screening operations, besides the removal of other contaminants (e.g., timber, plastics, etc.), the concrete waste is converted to three different products, including the coarse RCA (>5 mm), fine RCA (>0.15 mm and <5 mm), and recycled concrete fines (RCF, <0.15 mm). In practice, the fine RCA and RCF with particle sizes smaller than 5 mm have inferior properties with little application values and have to be discarded. This is mainly owing to the increasing amount of attached cement mortar upon the decreasing particle sizes [3].

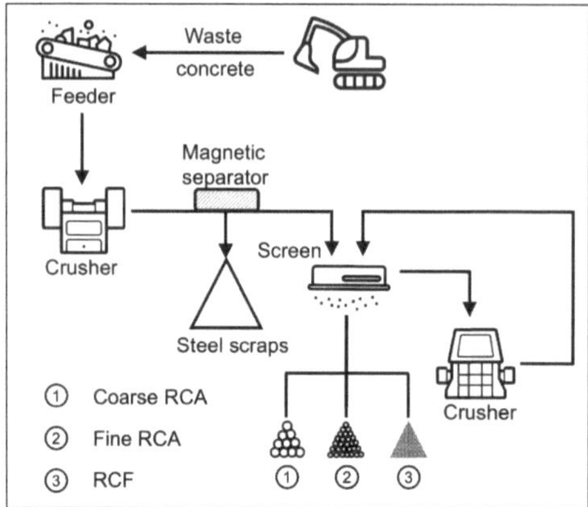

Fig. 10.1 Schematic illustrations of the ordinary treating processes of recycled concrete.

Accelerated carbonation has been reckoned as a promising method to enhance the qualities of RCA and RCF. This is due to the rapid chemical reaction occurring between CO_2 and the hydration products of cement (particularly calcium hydroxide, CH, and calcium silicate hydrates, C-S-H) or even anhydrous cement clinkers contained in the attached mortar, which eventually forms calcium carbonate ($CaCO_3$, Cc) and silica gel after carbonation in accordance with Equations 10.1–10.4. These carbonation products can fill the pore space and densify the microstructure of RCA or activate the reactivity of RCF and impart added values for the products [4, 5].

$$Ca(OH)_2 + CO_2 \rightarrow H_2O + CaCO_3 \tag{10.1}$$

$$xCaO{\cdot}SiO_2 \cdot nH_2O + CO_2 \rightarrow xCaCO_3 + SiO_2 \cdot nH_2O \tag{10.2}$$

$$3CaO \cdot SiO_2 + CO_2 + nH_2O \rightarrow 3CaCO_3 + SiO_2 \cdot nH_2O \tag{10.3}$$

$$2CaO \cdot SiO_2 + CO_2 + n\,H_2O \rightarrow 2CaCO_3 + SiO_2 \cdot nH_2O \tag{10.4}$$

The carbonation reaction rate and the carbonation degree of RCA depend on the diffusion of CO_2 through the attached mortar. The relative humidity (RH), CO_2 pressure and concentration, and temperature are reported to be the critical influential parameters. There are various ways to conduct accelerated carbonation, namely i) the standard gas-solid carbonation method, of which the condition is similar to that for testing the carbonation resistance of concrete [6–8]; ii) the pressurized gas-solid carbonation (PC) method [1, 9], which can accelerate the CO_2 penetration rate into RCA by applying a high CO_2 pressure; iii) flow-through gas-solid carbonation method (FC) [10], for which CO_2 is injected into a carbonation chamber from one side at ambient pressure, allowed to pass through RCA samples and discharged from the opposite side; and (iv) wet carbonation method (WC) [4, 11], in which RCA is placed in water while CO_2 is injected into the water. The schematic illustrations for setups of the above-mentioned carbonation techniques are shown in Fig. 10.2. All these carbonation treatment methods can improve the properties of RCA and the performance of recycled aggregate mortar/concrete (RAC). Therefore, this chapter presents a summary of the carbonation and applications of the three products generated

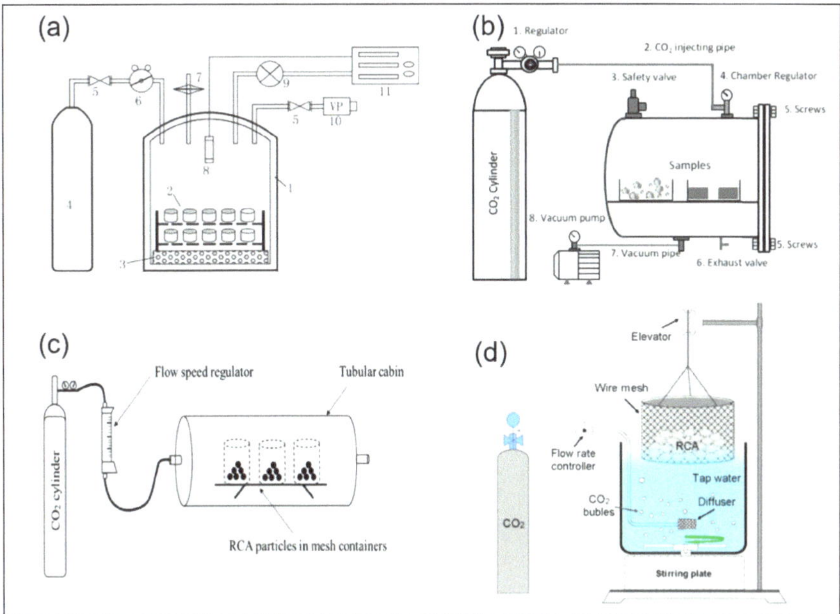

Fig. 10.2 Carbonation process setup illustrations for (a) gas-solid carbonation in an environmental chamber [12], (b) pressurized gas-solid carbonation in an autoclave [1], (c) flow through carbonation [8] and (d) wet carbonation [4].

from the conventional recycling process of concrete waste, i.e., coarse RCA, fine RCA, and RCF, aiming at promoting the total recycling of concrete waste by using accelerated carbonation. The accelerated carbonation processes can not only enhance the properties and values of the recycled concrete waste, but also sequestrate a large amount of CO_2. The research team at the Hong Kong Polytechnic University has conducted research in this area for many years, and this chapter draws heavily on our previously published literature.

10.2 Carbonation and Application of Coarse Recycled Concrete Aggregate

10.2.1 Physical and Microstructural Properties of Carbonated Coarse Recycled Concrete Aggregate

The physical properties of RCA before and after three different carbonation methods (pressurised carbonation, PC; flow-through carbonation, FC; and wet carbonation, WC) are shown in Table 10.1 [13]. The results showed that after using all these carbonation methods for coarse RCA, the water absorption is decreased, and the dry particle density is increased. Moreover, the change in water absorption is more significant than that of dry particle density. Among the three carbonation methods, PC shows higher effectiveness than FC and WC. This is due to the elevated CO_2 pressure facilitating the internal diffusion of CO_2 into the RCA matrix, which enhanced Cc precipitation and led to a denser microstructure.

Table 10.1   Physical properties of RCA before and after carbonation [13]

Treatment method	Particle size (mm)	Water absorption (%)			Dry particle density (g/cm³)		
		Before carbonation	After carbonation	Decreasing rate	Before carbonation	After carbonation	Increasing rate
FC	5–10	7.71	7.02	8.9 %	2.190	2.219	1.3 %
	10–20	6.82	6.39	6.3 %	2.216	2.233	0.8 %
PC	5–10	7.76	6.64	14.4 %	2.207	2.257	2.3 %
	10–20	6.40	5.64	11.9 %	2.252	2.292	1.8 %
WC	5–10	7.83	7.41	5.3 %	2.192	2.203	0.5 %
	10–20	6.99	6.70	4.1 %	2.216	2.228	0.5 %

10.2.2 Microstructural Properties of Carbonated Coarse Recycled Concrete Aggregate

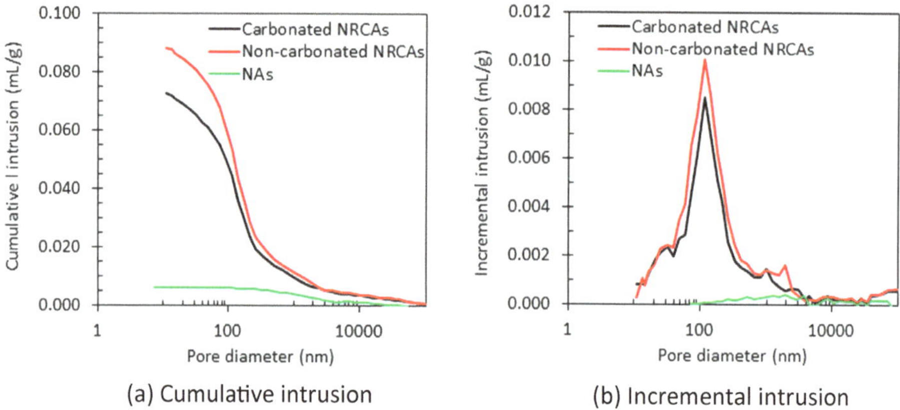

Fig. 10.3   Cumulative pore volume and pore size distribution of coarse RCA [14]. Note: NR-CA=recycled concrete aggregate, NA=natural aggregate.

The pore structures of the raw and carbonated coarse RCA are shown in Fig. 10.3. After carbonation, the amount of pores between 50 nm and 400 nm is reduced. The cumulative porosity of the carbonated RCA is less than that of the raw RCA. The morphologies of RCA before and after carbonation are illustrated in Fig. 10.4, further showing the conversion of the common cement hydration products such as CH, C-S-H, and ettringite to Cc crystals (possibly aragonite and also calcite). Based on the morphology, some fibre-like and cube-like products from Cc (aragonite and calcite) can be found.

Fig. 10.5 also shows the results of the total porosity of raw and carbonated coarse RCA (as embedded in RAC) through the use of image analysis based on Scanning Electron Microscopy-Backscattered Electrons (SEM-BSE) images. The adhered mortar attached in RCA is the area of interest (AOI) highlighted in green boundaries. In the image analysis, the dark pixels detected by the software are regarded as pores or cracks. It is found that

(a) Non-carbonated RCA   (b) Carbonated RCA

Fig. 10.4 Morphologies of RCA before and after carbonation treatment [14].

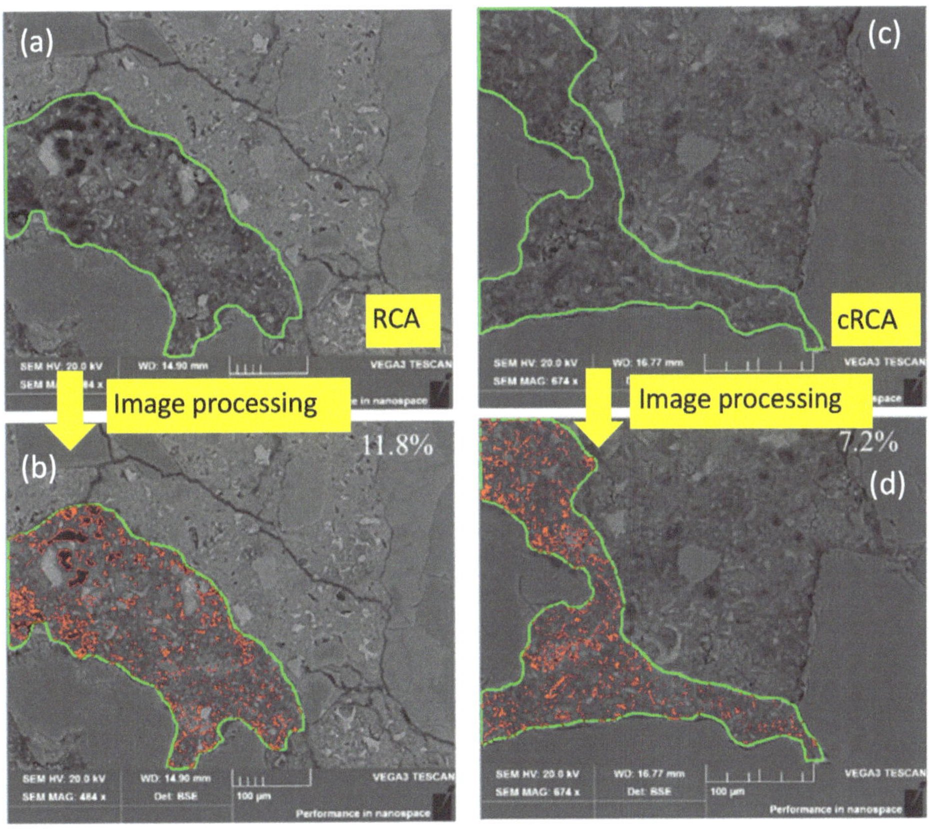

Fig. 10.5 Selection and calculation of the porosity in the adhered mortar of coarse RCA by image analysis [8]: (a) the uncarbonated RCA, (b) the uncarbonated RCA after image processing, (c) the carbonated RCA, (d) the carbonated RCA after image processing, where the red pixel represents the boundaries of the pores and cracks as determined using 'Image Pro Plus'.[8]. Note: cRCA=carbonated recycled concrete aggregate.

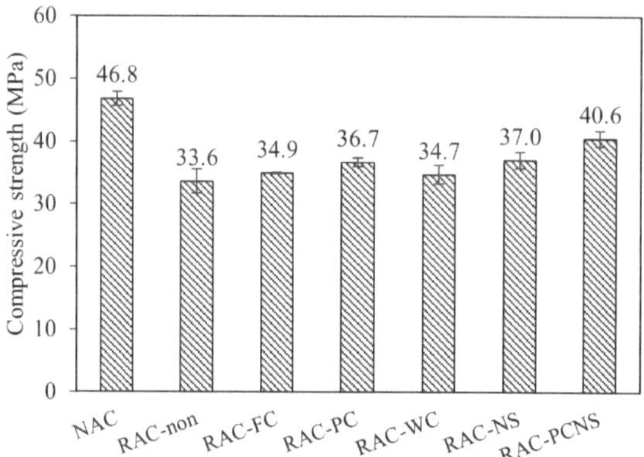

Fig. 10.6 Compressive strength of the RAC [13]. Note: NAC=natural aggregate concrete, RAC-non=concrete made with non-carbonated recycled concrete aggregate, RAC-FC=concrete made with carbonated recycled concrete aggregate via flow-through carbonation, RAC-PC= concrete made with carbonated recycled concrete aggregate via pressurised carbonation, RAC-WC= concrete made with carbonated recycled concrete aggregate via wet carbonation, RAC-NS= concrete made with recycled concrete aggregate treated by nano-silica. RAC-PCNS= concrete made with carbonated recycled concrete aggregate via pressurised carbonation and treated by nano-silica.

the raw RCA has much higher porosity (11.8 %) than the carbonated RCA (7.2 %). This confirms the densification effect of the adhered mortar in RCA after carbonation. This is owing to the formation of Cc upon its conversion from CH. The Cc fills the larger pores, and the whole microstructure becomes denser, demonstrating a pore-filling effect.

10.3 Application of Carbonated Coarse Recycled Concrete Aggregate in Recycled Aggregate Concrete

10.3.1 Compressive Strength of Recycled Aggregate Concrete

The 28 days' compressive strength of the recycled aggregate concrete (RAC) is shown in Fig. 10.6. It shows that the compressive strength of natural aggregate concrete (NAC) is 33.2 % higher than that of RAC with non-treated RCA (RAC-non). After applying carbonation to obtain carbonated RCA, the compressive strength of RAC can be increased. PC shows the highest effectiveness which improves the strength by 9.2 %. By contrast, the improvement is 3.9 % and 3.3 %, respectively, in terms of FC and WC.

10.3.2 Microhardness of Recycled Aggregate Concrete

Microhardness values of concrete at zones near the interface between RCA and the new mortar is tested as shown in Fig. 10.7. The indentation points are located at distances of 80 μm to 200 μm from the interface, which covers the RCA (old mortar), new interfacial

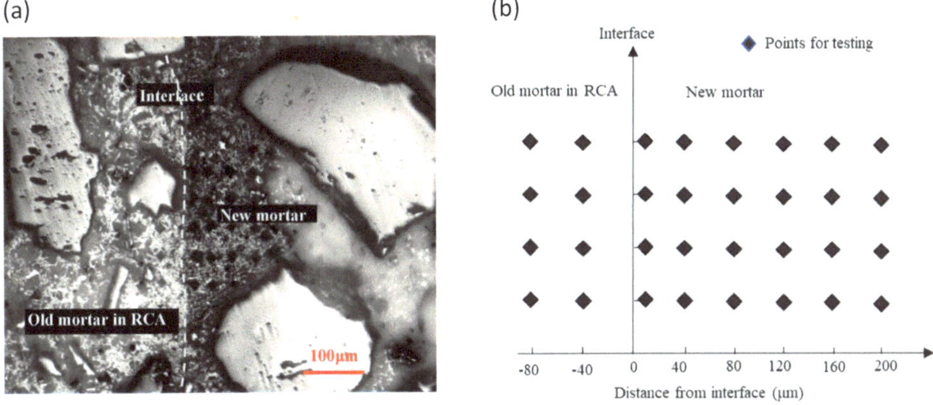

Fig. 10.7 A typical area (a) and distribution of the points (b) for microhardness testing [13].

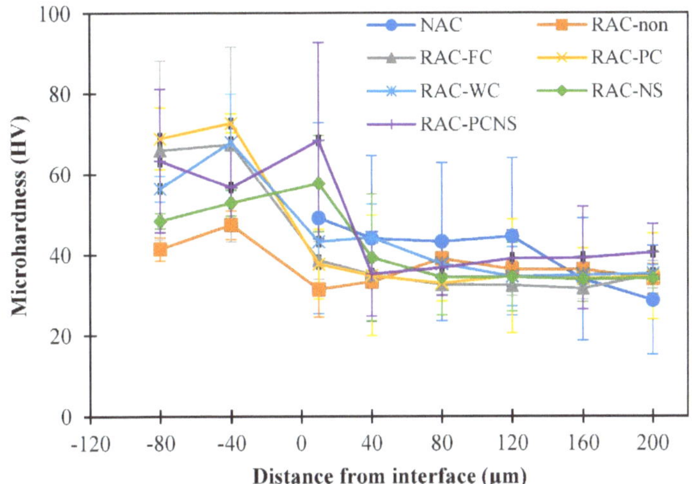

Fig. 10.8 Microhardness at the interfacial transition zone around RCA in RAC [13]. Note: The abbreviations are the same as those in Fig. 10.6.

transition zone (ITZ), and the new mortar. The results are shown in Fig. 10.8. It shows that the microhardness values of the old mortar in RAC are increased regardless of the RCA carbonation methods. This is owing to the densification effect as introduced earlier. The old mortar exhibits higher microhardness values than the new mortar in all RAC. However, the microhardness of the new mortar does not show obvious change after using the different treatment methods. It implies that the influences of these RCA treatment methods on the new mortar are not significant. Besides, the microhardness values of new ITZs at 10 μm from the carbonated RCA is higher than that of the un-carbonated RCA. The above findings are an indication of the beneficial effect of the carbonated RCA on the new ITZ as compared with raw RCA, and this is one of the reasons contributing to the improvement of the compressive strength of RAC prepared with the carbonated RCA.

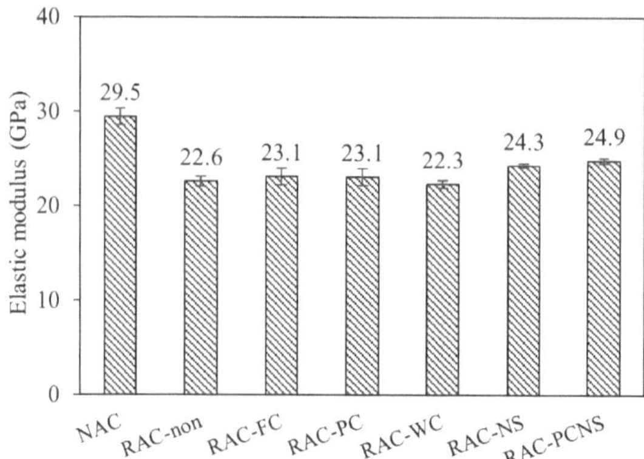

Fig. 10.9 Elastic modulus of the RAC [13]. Note: The abbreviations are the same as those in Fig. 10.6.

10.3.3 Elastic Modulus of Recycled Aggregate Concrete

The elastic moduli of the different concrete samples prepared with the carbonated RCA are shown in Fig. 10.9. It can be seen that the elastic modulus of RAC is significantly reduced when compared with NAC. The reduction rate is even higher than that of compressive strength. Generally, the elastic modulus of RAC increases after using carbonated RCA, but the magnitude of the increase is less than 3 %, lower than that of compressive strength, indicating that the influence of the carbonation on the elastic modulus of RAC is less obvious than that on the compressive strength.

10.3.4 Bulk Electrical Conductivity of Recycled Aggregate Concrete

The bulk electrical conductivities of RAC prepared with the raw and the carbonated RCA are shown in Fig. 10.10. It is affected by the replacement rate of NA by RCA, the type of RCA and the carbonation pressure of RCA. The major results can be summarised as follows: (I) With increasing amounts of RCA in RAC, the bulk electrical conductivity gradually increases from 18 mS/m to 25 mS/m; (II) The bulk electrical conductivity of the RAC decreases with the use of the carbonated RCA; and, (III) Increasing the carbonation pressure during carbonation from 0.1 Bar to 5.0 Bar results in a slightly decreased bulk electrical conductivity.

10.3.5 Water Absorption of Recycled Aggregate Concrete

Fig. 10.11 shows the water absorption values of RAC prepared with different types and dosages of RCA. With the increase in the amount of RCA, the water absorption of RAC almost linearly increases. Meanwhile, the water absorption of RAC decreases when the carbonated RCA are used. It can be noticed that the water absorption of RAC prepared

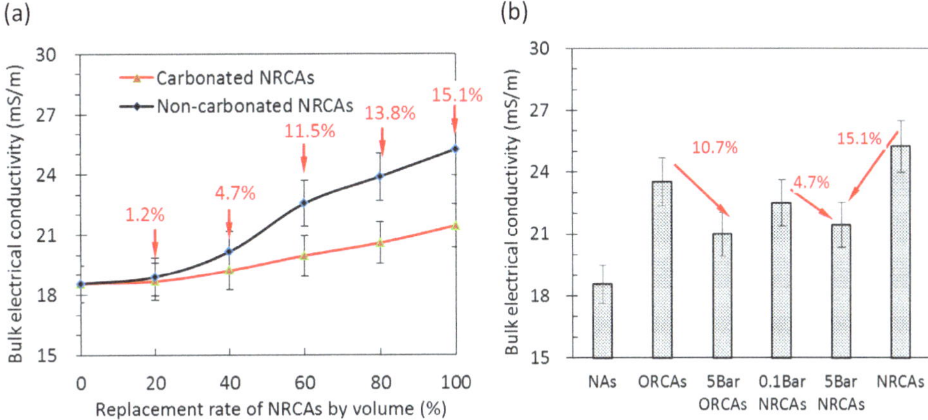

Fig. 10.10 Bulk electrical conductivity of RAC [14].

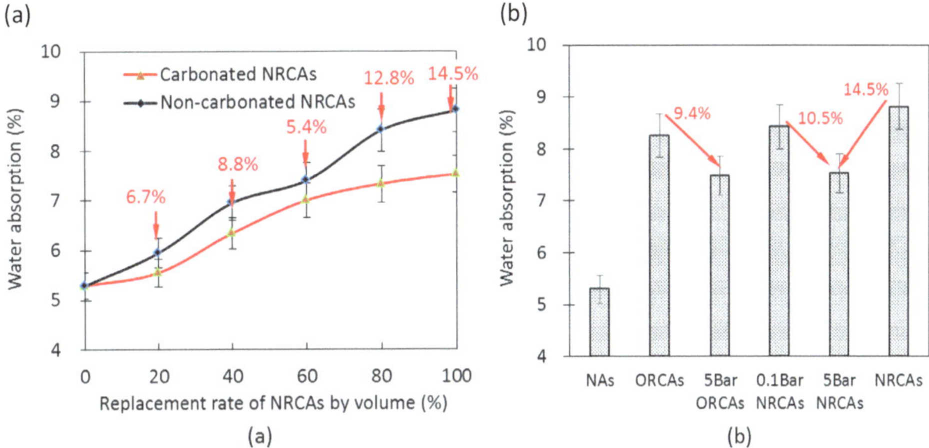

Fig. 10.11 Water absorption of RAC containing RCA at different replacement rates and containing RCA carbonated at different CO2 pressures [14]. Note: The abbreviations are the same as those in Fig. 10.10.

with carbonated RCA is 6.7%-14.5% lower than that of RAC prepared with uncarbonated RCA. Moreover, when increasing the carbonation pressure from 0.1 Bar to 5.0 Bar, the water absorption of RAC decreases by 10.5%, indicating that carbonation at a higher pressure is more effective.

10.3.6 Chloride Penetration Resistance of Recycled Aggregate Concrete

The amount of charge passed through the concrete is a parameter to reflect its chloride penetration resistance, which is another important index of concrete durability. A higher amount of charge passed means a lower chloride penetration resistance. The chloride

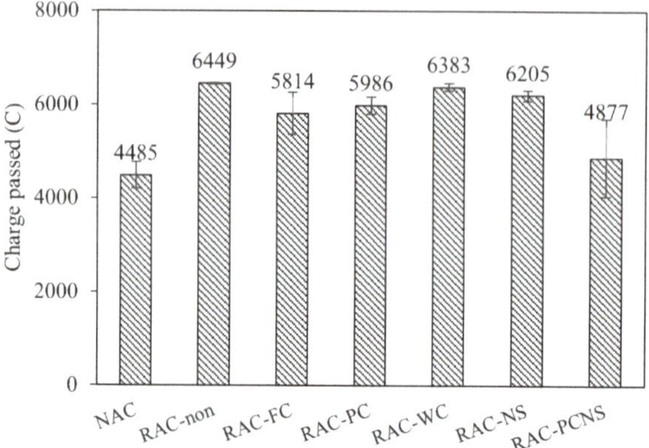

Fig. 10.12 Chloride penetration resistance of hardened concretes [13]. Note: The abbreviations are the same as those in Fig. 10.6.

penetration resistance of the concrete is shown in Fig. 10.12. It can be seen that the charge passed of the RAC with untreated RCA (RAC-non) is much higher than that of NAC. After using the carbonation methods, the charge passed of RAC can be reduced. The densification of the microstructure of RCA as well as the ITZs in concrete, as previously introduced in Section 10.2.2, is responsible for the results. Among the three carbonation methods, the FC and PC are better in improving the chloride penetration resistance of RAC.

10.3.7 Gas Permeability of Recycled Aggregate Concrete

The coefficients of intrinsic gas permeability of RAC are shown in Fig. 10.13. Similar to other test results, the following findings are highlighted: (I) The gas permeability of RAC decreases significantly with the carbonated RCA; (II) Increasing the carbonation pressure from 0.1 Bar to 5.0 Bar helps to reduce the gas permeability of RAC prepared with RCA.

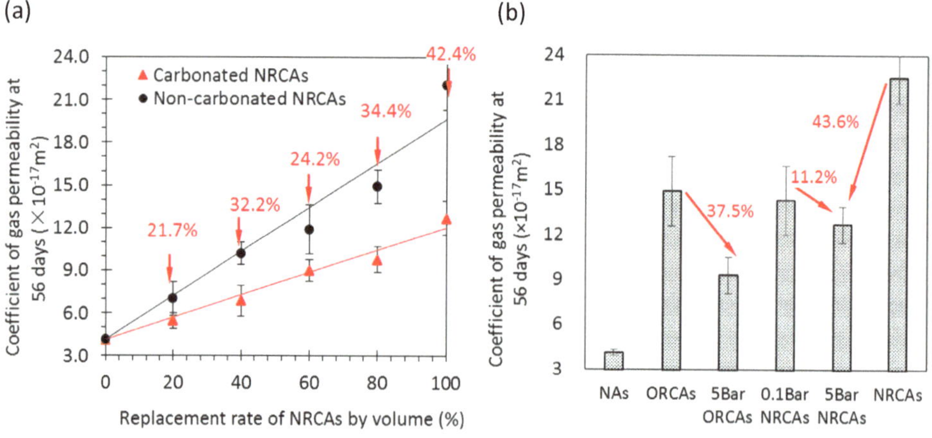

Fig. 10.13 Intrinsic gas permeability of RAC [14]. Note: The abbreviations are the same as those in Fig. 10.10.

10.3.8 Carbonation Resistance of Recycled Aggregate Concrete

The carbonation depth of concrete is an indicator of its carbonation resistance. Lower carbonation depth means better carbonation resistance. The photos of the concrete subjected to 28 days' carbonation after spraying it with phenolphthalein solution are shown in Fig. 10.14. According to the colour, it can be found that in the inner part of RAC that have not been carbonated, many carbonated RCA particles are observed when using FC and PC. However, no obvious carbonated RCA particle is observed after using WC.

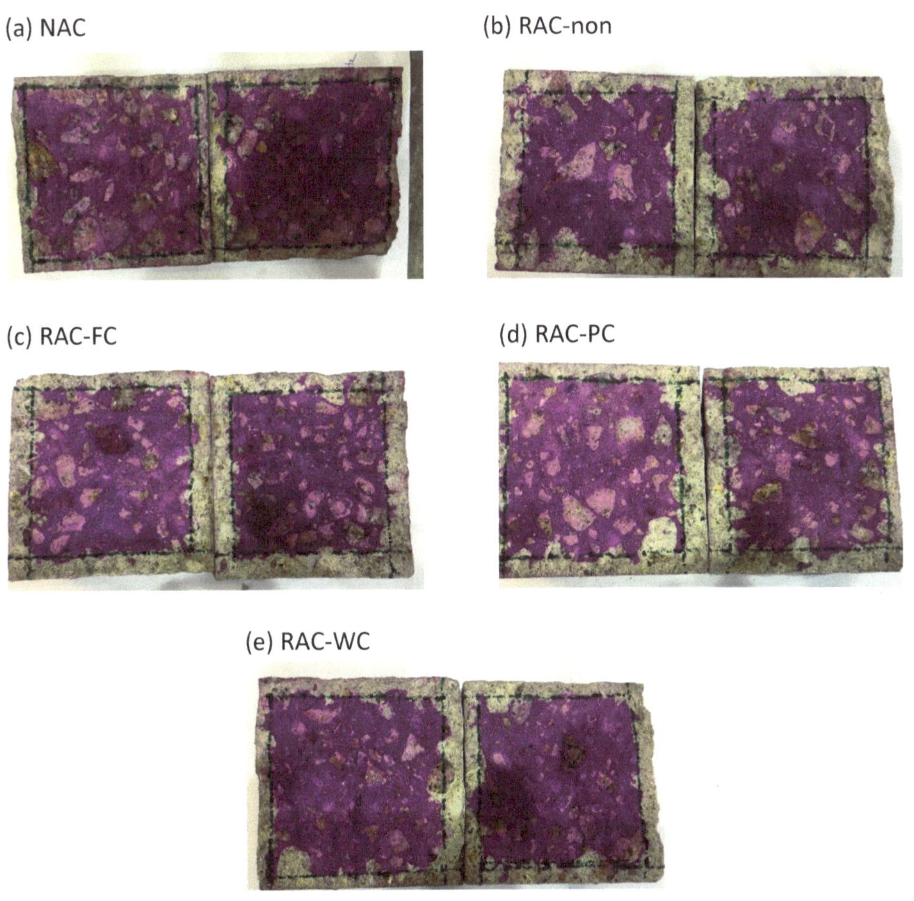

Fig. 10.14 Photos of concretes subjected to 28 days of carbonation after spraying phenolphthalein solution [13]. Note: The abbreviations are the same as those in Fig. 10.6.

The 7 days and 28 days carbonation depths of concretes are shown in Fig. 10.15. The results show that the 7 days and 28 days carbonation depths of RAC-non are much higher than that of NAC. After using FC and PC, the 7 days carbonation depths of RAC are fur-

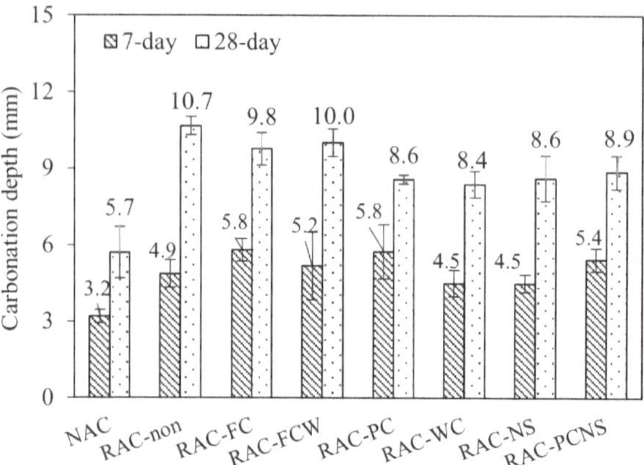

Fig. 10.15 Carbonation resistance of the RAC [13]. Note: The abbreviations are the same as those in Fig. 10.6.

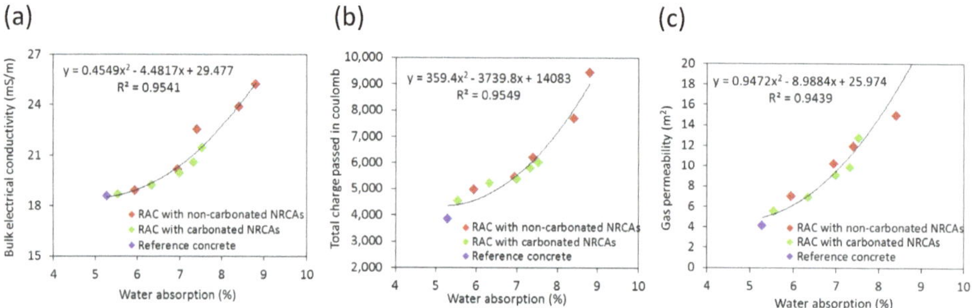

Fig. 10.16 Relationship between water absorption of RAC and its bulk electrical conductivity, chloride-ion penetration and gas permeability [14].

ther increased. That is because the RCA have been carbonated already, which increases the average carbonation depth when the carbonation depth is small. However, the 28 days carbonation depths of these RAC are lower than that of RAC-non. That is because the RCA are densified after carbonation, which reduces the penetration rate of CO_2. In contrast, the 7 days and 28 days carbonation depths of RAC-WC are both reduced compared to RAC-non. That is because only the surface layer of the RCA is carbonated after using WC. At the same time, many nano-$CaCO_3$ particles are formed on the surface of RCA, which can densify the new ITZ. As a result, the penetration rate of CO_2 is reduced.

10.3.9 Relation Between Water Absorption and Permeability of Recycled Aggregate Concrete

After carbonation treatment, two of the important changes in the physical properties of the carbonated RCA are the water absorption and permeability. Fig. 10.16 illustrates the relationship between water absorption and permeability indicators (bulk electrical conductivity, chloride-ion penetration, and gas permeability) of RAC. The general trend is that all the permeability indicators have good correlation with the water absorption values of RAC. It can be concluded that the water absorption value of RAC may be considered as a simple and good criterion to indicate the permeability of RAC.

10.4 Carbonation and Recycling of Fine Recycled Concrete Aggregate

10.4.1 Physical and Microstructural Compositions of Carbonated Fine Recycled Concrete Aggregate

The density and water absorption values of the fine RCA before and after subjecting them to different pretreatment methods and pressurized carbonation (PC) are shown in Table 10.2. Similar to coarse RCA, the typical PC could result in 12.6 % decrease in water absorption and 1.4 % increase in density as compared with uncarbonated fine RCA, thus confirming the beneficial effect of PC to the enhancement of fine RCA. Nevertheless, the effect is not significant enough, possibly due to the natural carbonation already taking place under ambient conditions during storage, particularly for the fine RCA with a higher amount of residual cement paste. Therefore, there are also different pre-treatment methods available to amplify the beneficial effect by soaking the RCA in or spraying with additional CH and CH-containing wastewater (e.g., wastewater from concrete batching plants). The combined effects of using the pretreatment methods and carbonation are also shown in Table 10.2. It can be seen that the CH soaking method results in similar density and water absorption values to those after the CH spray treatment. The wastewater spray method is equally effective in reducing the water absorption compared to the CH soaking and spray methods. However, the use of the pretreatment methods significantly densifies the fine RCA and decreases the water absorption value when compared with those carbonated without any treatment.

Table 10.2 Density and water absorption values of fine RCA with different treatment methods before pressurized carbonation (PC) [15]

Treatment method	Particle size (mm)	Water absorption (%)			Density (kg/m³)		
		Before treatment	After treatment	Decreasing rate	Before treatment	After treatment	Increasing rate
PC	2.36–5		19.4 ± 0.2	12.6 %		1631.6 ± 53	1.4 %
CH spray+ PC	2.36–5	22.2 ± 0.3	17.6 ± 0.8	20.7 %	1609.6 ± 51	1736.8 ± 45	7.9 %
CH soaking + PC	2.36–5		17.2 ± 0.7	22.5 %		1743.3 ± 48	8.3 %
Wastewater spray + PC	2.36–5		17.1 ± 0.9	23.0 %		1824.7 ± 60	13.4 %

As reflected by the physical properties, the microstructure of the fine RCA is densified. Experimental evidence in terms of the pore volume and pore size distribution before and after carbonation confirm this result. Exemplary pore structure evaluations taken from [4] aiming at comparing the effect of PC and wet carbonation (Fig. 10.2d) are shown in Fig. 10.17. It is observed that the total pore volume is reduced after both PC and WC, indicating their refinement ability of pore structure. The total pore volume decreased from 0.12 ml/g to 0.08 ml/g after the 24-h PC and to ~0.09 ml/g after 1-h WC. For the pore size distribution curves, the coarser pores larger than 10 nm decreased significantly after the carbonation treatment, which is due to the precipitation of carbonation products in the internal pore space of the fine RCA. Moreover, it is noticed that PC generally shows a higher ability of carbonating the RCA as imparted by the elevated CO_2 pressure, while WC has a higher efficiency in carbonation.

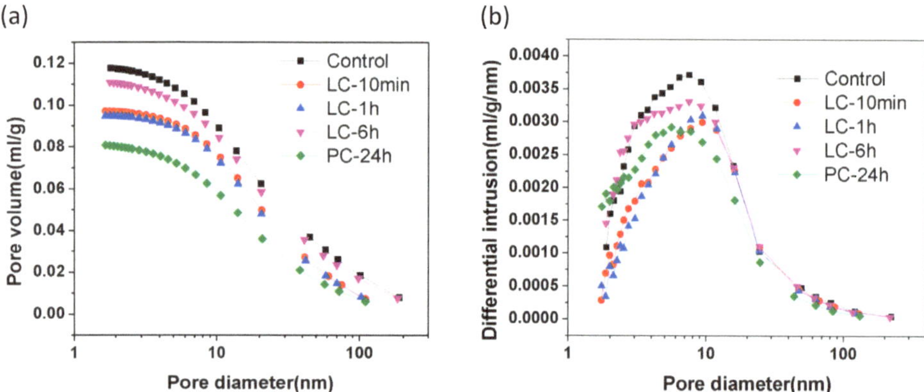

Fig. 10.17   (a) The total pore volume and (b) pore size distribution of the fine RCA before and after PC and WC [4]. Note: LC annotated in the figure means wet carbonation (WC).

10.4.2 Mineralogical Compositions and Chemical Reactivity of the Carbonated Fine Recycled Concrete Aggregate

After carbonation, the mineralogical compositions of the fine RCA are significantly altered. Clear evidence can be seen from the SEM images of the RCA surfaces as shown in Fig. 10.18. It is found that the RCA surfaces are covered by an extensive layer of calcite that is possibly either in rhombohedral morphology (for WC) or scalenohedral morphology (for PC).

Fig. 10.19 shows the X-ray powder diffraction (XRD) patterns and Table 10.3 lists the amount of CH and Cc present before and after carbonation in four different particle fractions of the fine RCA. They show similar XRD patterns, and the diffraction peaks of CH, externally added corundum as an internal standard, as well as Cc are detected from the raw RCA. Compared with the raw RCA, an increase in the peak intensity of Cc is noticed from the carbonated RCA. After carbonation, the mass fractions of CH, ranging from 14.1 wt% to 14.9 wt%, are still significant in all RCA samples. The reduction of

(a)   (b)

Fig. 10.18 Surface morphologies of carbonated fine RCA (a) after PC and (b) after WC [16].

CH content is from 0.5 wt% to 1.8 wt% and the increase of the Cc is from 2.6 wt% to 3.5 wt%, indicating that the hydration products other than CH are also carbonated. The wet carbonation increases the content of Cc as the carbonation product that is believed to be able to fill the pores of RCA and therefore improve its quality. Because of the pore-filling effect of the carbonation products, the water absorption of the fine RCA is reduced by about 2 % after the WC. Due to the short carbonation duration, the decrease in CH content is not significant. Nevertheless, the wet carbonation still effectively increases the amount of carbonation products formed in the fine RCA.

Table 10.3 Amounts of calcium hydroxide (CH) and calcium carbonate (Cc) in the fine RCA [17]

Particle size	CH (wt%)		Cc (wt%)	
	raw	carbonated	raw	carbonated
0.15–0.6 mm	15.4	14.9	11.2	14.7
0.6–1.18 mm	15.4	14.1	10.3	13.6
1.18–2.36 mm	16.2	14.4	10.8	13.6
2.36–5 mm	15.0	14.1	13.6	16.2

The thermogravimetric analysis results of the raw and carbonated fine RCA are shown in Fig. 10.20. The mass loss below 500 °C is mainly due to the dehydration of calcium aluminate hydrates (AFt, AFm, etc.), C-S-H and CH. The mass loss between 500 and 800 °C is mainly associated with the decarbonization of Cc formed during the carbonation. After carbonation treatment, it is clearly noticed that Cc is formed, and CH is consumed in all the carbonated samples. It can be found that with the increase of WC duration, more Cc is formed, but the content is still less than that in the samples subjected to PC. Nevertheless, the results indicated that the WC reaction is fast, and 83.4 % of Cc is formed within 10 min.

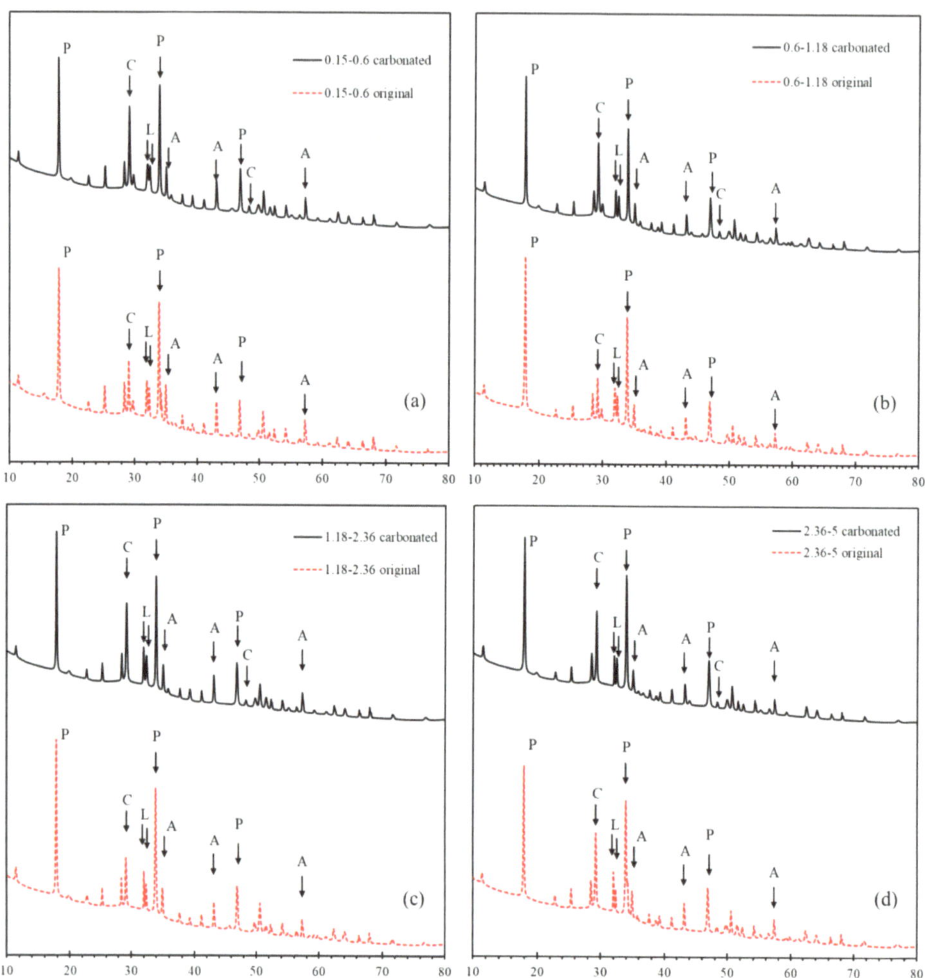

Fig. 10.19 XRD patterns of fine RCA: (a) 0.15–0.6 mm; (b) 0.6–1.18 mm; (c) 1.18–2.36 mm; (d) 2.36–5 mm [17]. (P = portlandite C = calcite, L = dicalcium silicate, A = corundum) X-axis: CuKα, 2θ (degree). Note: the fine RCA for XRD is prepared from well-hydrated pure cement paste for quantitative and comparative studies.

The Fourier-transform infrared spectroscopy (FTIR) is used to detect structural changes of the amorphous silica gel formed after the carbonation treatment. As shown in Fig. 10.21, the broad absorption between 900 cm^{-1} and 1200 cm^{-1} corresponds to the asymmetrical stretching vibration (v3) of the Si-O bond. After the carbonation treatment, the wavenumber of Si-O shifted to a higher value compared to the control sample (from 956 cm^{-1} to 1033 cm^{-1}), indicating that the polymerization degree of Si-O increases and highly polymerized silica gel forms after both PC and WC. Moreover, new bands are found at 1420 and 870 cm^{-1}, corresponding to the asymmetrical stretching vibration (v3) and out-of-plane bending vibration (v4) of the C-O bond, respectively. With the increase of WC duration, the band intensity of C-O increases significantly, suggesting more calcium carbonate is formed. Those results are consistent with the TGA results. Overall, after the

WC, the performance of RCA could be enhanced in a short time due to fast chemical reactions in the solution compared to the PC.

The hydration heat evolution curves of the mortar prepared by mixing ordinary Portland cement (OPC) with the carbonated and non-carbonated fine RCA are shown in Fig. 10.22. Compared with the control sample, both WC and PC carbonated RCA accelerate the cement hydration at an early age. More specifically, in the samples of WC-6h and PC-24h, the induction period is decreased from 3.4 h (control) to 2.4 h, and the time to reach the maximum heat flow is shortened from 13.7 h to 10.7 h, while the maximum heat flow is similar. The accelerated hydration of the new cement is mainly attributed to the Cc formed during the carbonation treatment, which acts as nucleation sites for the nucleation and growth of hydration products [18]. It was reported that the nucleation effect of Cc on the hydration of cement is mainly affected by its particle size, addition dosage, and microstructure. With the increase in the amount of fine Cc, the rate of hydration is increased, and the total hydration heat of the mixtures is also higher. It should be noticed that the acceleration effect of the new mortar incorporated with the WC carbonated fine RCA is similar to that of the PC RCA although the Cc formed after PC (39.05 %) is much higher than that of WC (24.09 %), suggesting that Cc formed by the WC would be more reactive to accelerate the cement hydration in the new mortar. Moreover, a more pronounced shoulder on the primary heat flow peak could be ob-

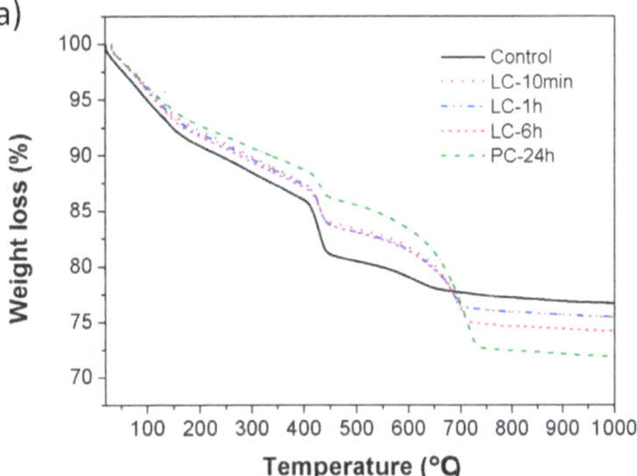

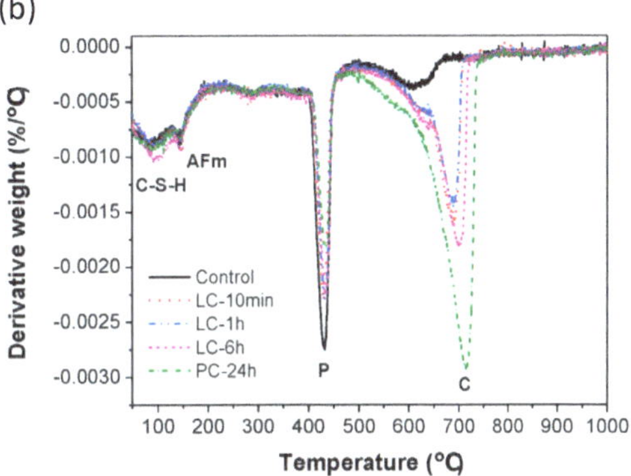

Fig. 10.20 Thermogravimetric analysis (TGA) of the raw and carbonated fine RCA: (a) TGA curves and (b) the differential curves (DTG) [4]. Note: LC annotated in the figure means wet carbonation (WC).

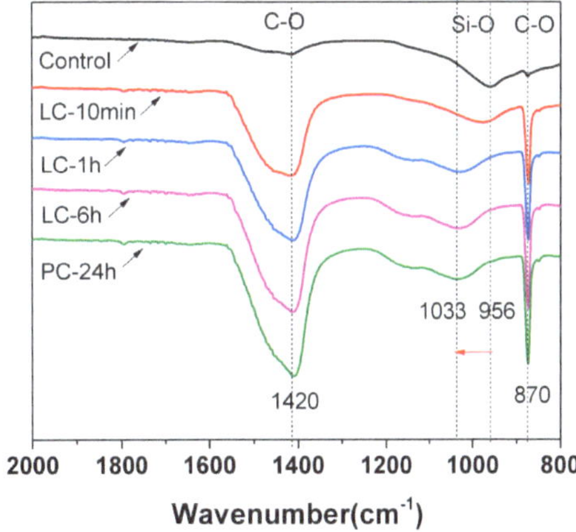

Fig. 10.21 FTIR spectra of the fine RCA before and after PC and WC [4]. Note: LC annotated in the figure means wet carbonation (WC).

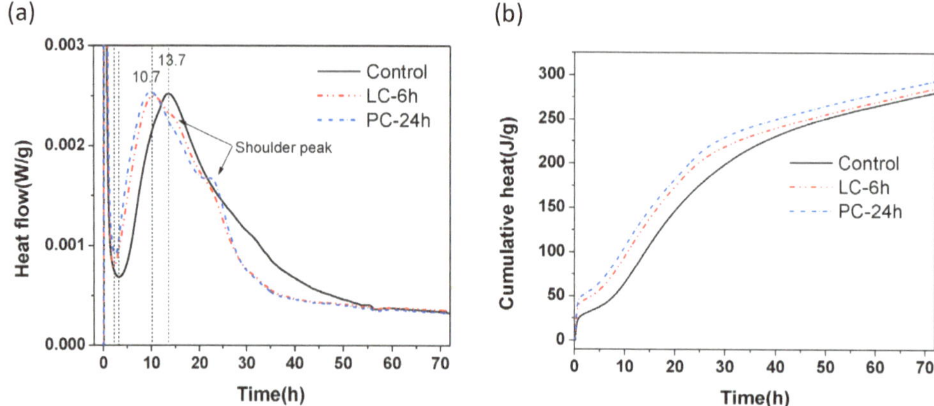

Fig. 10.22   (a) hydration heat flow rate and (b) cumulative heat released of OPC-fine RCA mixes [4]. Note: LC annotated in the figure means wet carbonation (WC).

served for both WC-6h and PC-24h samples, indicating the chemical reaction between Cc and new cement pastes may also occur to form additional hydration products.

The XRD patterns of the fine RCA samples and the OPC-RCA mortar samples are shown in Fig. 10.23. The crystalline phases identified in the noncarbonated RCA sample include some typical cement hydration products, such as CH, AFm, and a small amount of Cc due to natural carbonation. After the carbonation treatment, the diffraction peak intensities of CH and AFm decrease considerably, and the diffraction peak intensity of calcite increase significantly. The diffraction peak intensity of calcite in the samples of PC-24h is higher than that of LC-6h. These results are consistent with the TGA results. There are no other polymorphs of Cc (vaterite and aragonite) detected under both WC and PC treatments. For the OPC-RCA sample cured for 28 days, there is very little change in the crystalline phase for the control sample compared to the original RCA, indicating that no or few other types of crystalline hydration products formed. However, in the OPC-RCA samples of WC-6h and PC-24h, a noticeable difference in XRD patterns is observed, in particular, the appearance of the diffraction peaks of monocarbonate (11.7∘ 2θ) and ettringite (9.1∘ 2θ) as shown in Fig. 10.14b. The presence of ettringite in the OPC-RCA sample might be associated with the stabilizing effect of Cc due to the reaction between Cc and the new cement paste to form mono-carbonate, which inhibits the transformation of ettringite (AFt) to mono-sulfate (AFm). Also, the newly formed mono-carbonate (Mc) and the stabilized AFt phase proved that the Cc could react with the new cement paste to form other types of hydration products. Furthermore, the higher diffraction peak intensity of Mc in the sample of WC-6h also indicated that the reactivity of Cc formed after WC could be higher than that of the PC. In addition to Cc, the highly polymerized silica gel can also be formed after carbonation treatment. The silica gel could rapidly react with CH to form additional C-S-H [19]. The newly formed Cc and the silica gel are thus responsible for the reactivity of carbonated fine RCA, thus contributing to performance improvement when applying these carbonated RCA in new mortars and concrete [20].

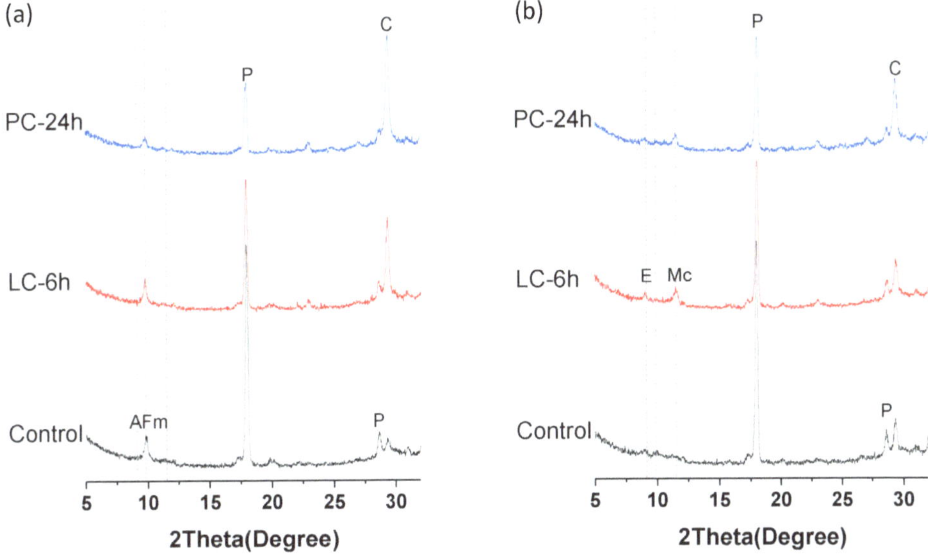

Fig. 10.23 XRD patterns of (a) carbonated and non-carbonated RCA samples; (b) OPC-RCA mortars. The main peaks of Cc (C), ettringite (E), AFm phase (AFm), CH (P), and monocarbonate (Mc) are indicated [4]. Note: LC annotated in the figure means wet carbonation (WC).

10.4.3 Influence of Carbonated Fine Recycled Concrete Aggregate on the Properties of Recycled Aggregate Mortar

The comparison of 28-day compressive strength and drying shrinkage of mortars prepared with different types of fine RCA are shown in Fig. 10.24 and Fig. 10.25 respectively. As can be seen, the reference group with 100 % natural quartz sand achieves the highest strength while the mortar prepared with 100 % raw fine RCA attains the lowest strength. The specimens prepared with 50 % sand replaced by carbonated fine RCA shows similar strength to the 100 % sand mortar mixture. However, replacing 50 % sand with raw fine RCA results in a noticeable strength reduction. On the other hand, the specimens prepared with 100 % carbonated RCA show a significant strength increase compared with those prepared with 100 % raw RCA, suggesting that WC is able to enhance the quality of fine RCA and thus improve the mechanical performance of the mortar specimens. The carbonation products formed in the carbonated RCA improve the properties of RCA by increasing its density and reducing its water absorption. This eventually leads to the improvement of the compressive strength of the produced new mortars. Besides, as Fig. 10.25 shows, the reference mortar specimens show the lowest shrinkage values while the 100 % raw RCA specimens have the highest shrinkage value. The mortar specimens prepared with 100 % carbonated RCA show a lower shrinkage value, indicating the beneficial property improvement after WC.

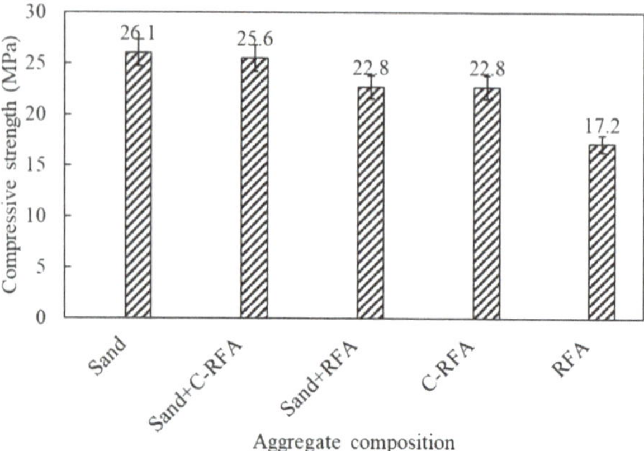

Fig. 10.24   Compressive strength values of mortar specimens [17]. Note: the RFA means fine recycled concrete aggregate, and C-RFA means carbonated fine recycled concrete aggregate.

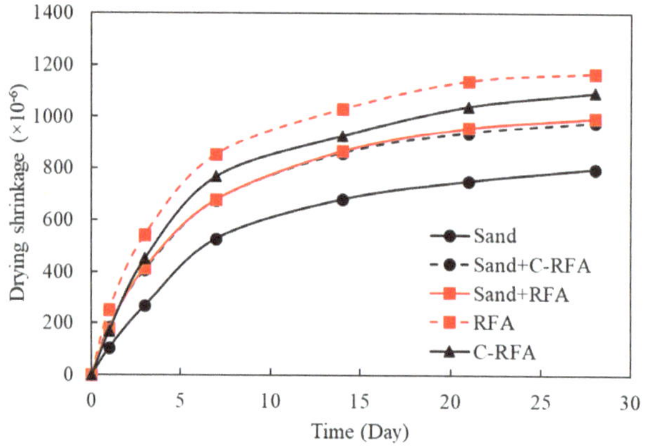

Fig. 10.25   Drying shrinkage of mortar specimens [17]. Note: The abbreviations are the same as those in Fig. 10.24.

10.4.4 Bond Strength Between Recycled Concrete Aggregates and New Cement Mortar

The effect of carbonation on the interfacial bonding between the carbonated RCA and the new mortar is of interest. It is experimentally evaluated by bonding tests as illustrated in Fig. 10.26, and the results are shown in Fig. 10.27. It is found that the frictional bonding strength and the tensile bonding strength between the (modelled) carbonated RCA and new cement mortar are significantly improved as compared with that of uncarbonated RCA. The beneficial effect is attributed to i) the calcite-aluminate reaction which forms monocarbonate (Mc), ii) the preferential C-S-H formation on the surface of carbonated RCA due to the calcite acting as nuclei, and iii) the pozzolanic reaction having occurred between CH and silica gel produced from the carbonation reaction.

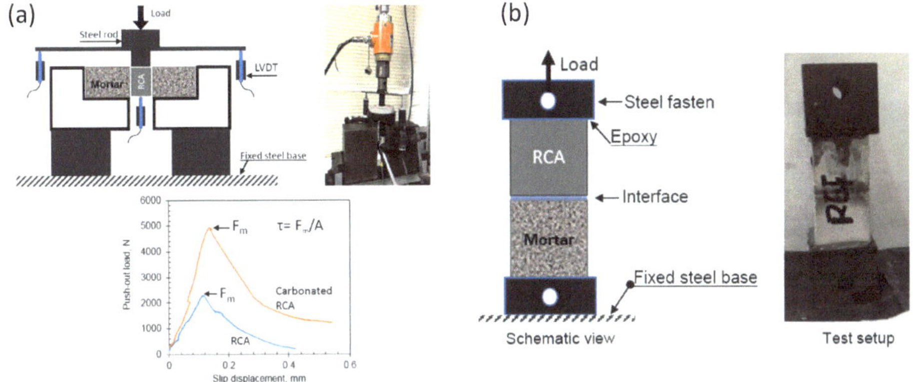

Fig. 10.26   Testing illustrations and experimental setups for (a) frictional bonding strength and (b) tensile bonding strength.

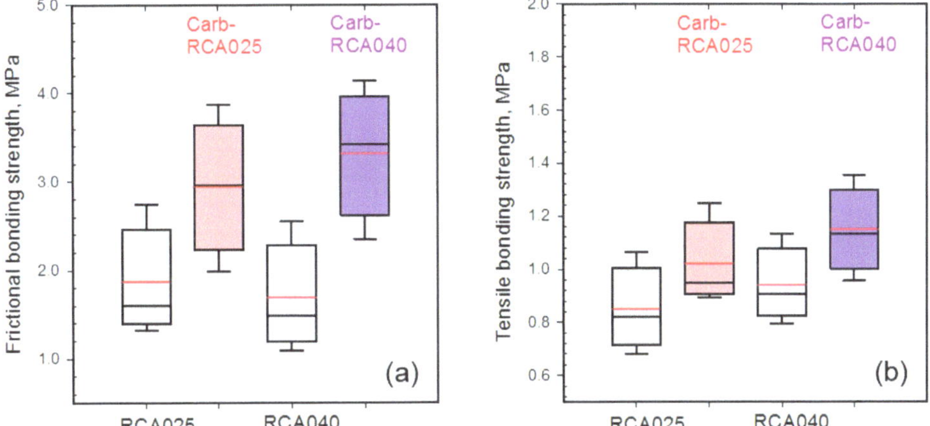

Fig. 10.27   Change of bond strength between new mortar and RCA subjected to carbonation treatment: (a) frictional bond strength; (b) tensile bond strength [20]. (Red line in the box represents the mean value.)

10.5 Carbonation and Recycling of Recycled Concrete Fines Powder (RCF, <150 um)

10.5.1 Phase Assemblance and Microstructure Evolutions of Carbonated Recycled Concrete Fines

Calcium Phase Evolution in Carbonation Products

RCF powder is inevitably produced by the crushing operation of the concrete recycling process. Fig. 10.28 shows the development of crystalline phases of RCF after subjecting them to different durations of wet carbonation. The main crystalline phases in the non-carbonated RCF are CH, AFm, and a small amount of calcite. With the ongoing

WC, the CH peaks disappear after 5 min, and the AFm disappears after 30 min. Meanwhile, Mc could be observed between 5 min and 20 min, which is regarded as a kind of carbonate-AFm [21], being related to the carbonation of AFm. This new carbonation product disappears after further carbonation until 30 min. Furthermore, C_2S is carbonated continually with the increase in carbonation time, but it is still present even when the carbonation time has reached 360 min. The primary carbonation product is found to be calcite, unlike the polymorphs of Cc formed under the gas-solid carbonation that may contain aragonite as well [22].

Table 10.4 shows the quantification of CH, calcite, AFm, and unhydrated phases (C_2S and C_4AF) in the carbonated RCF. The calcite content increases from 6.60 % to 60.06 % with the progress of carbonation from 0 min to 360 min. The CH content decreases to 0 % after 5 min, while the AFm decreases from 2.37 % to 0 % within 30 min. However, there remain about 0.38 % of C_2S and 1.26 % of C_4AF in the RCF even after 360 min of carbonation.

Table 10.4 QXRD data of the content of CH, C2S, AFm, C4AF, and calcite in carbonated RCF [5]

NO.	CH wt%	Calcite wt%	C_2S	AFm	C_4AF
Reference	15.79	9.08	7.01	2.37	2.56
C5min	/	28.65	5.92	0.51	0.94
C10min	/	34.39	5.63	0.48	1.12
C20min	/	49.61	4.04	0.07	1.33
C30min	/	56.49	1.06	/	1.38
C60min	/	58.25	0.79	/	1.12
C360min	/	60.06	0.38	/	1.26

Silicon Phase Evolution in Carbonated RCF

The ^{29}Si Nuclear Magnetic Resonance (NMR) spectra of the carbonated RCF are shown in Table 10.5 and Fig. 10.29. The uncarbonated RCF sample shows that all the silicon is present in Q^0, Q^1 and Q^2 units. A rapid decrease of Q^1 intensity is observed after carbonation, and it completely disappears after 30 min, indicating that the Q^1 is transformed to Q^2 and Q^3 units. The intensity of Q^2 increases within the first 20 min and then continuously decreases for the remaining carbonation period. A new Q^{3b} unit located at -92.4 ppm is recorded in the carbonated RCF at 5 min, and its amount increases from 5.59 % to 33.21 % after 30 min and then decreases gradually. As the Q^{3b} could be detected in C-S-H with a Ca/Si ratio of 0.66 to 1 [23], the appearance of Q^{3b} demonstrates that the decalcification of C-S-H is triggered immediately after injecting CO_2, leading to an increase in the mean chain length (MCL) of C-S-H. However, there is still 29.53 % Q^{3b} remaining after carbonation for 360 min. The carbonation of C-S-H gel (Q^{3b}) progressed slowly in the aqueous environment, so the carbonation degree only increases slowly after 30 min and much C-S-H gel with a low Ca/Si ratio is still present even after 360 min. The quick formation of Q^3 and Q^4 especially between 20 and 30 min demonstrates decalcification of C-S-H gel and the clinker phases, and also the quick polymerization of silicates. For the samples carbonated for 20 min and 30 min, the proportion of Q^{3b} site is higher than that of

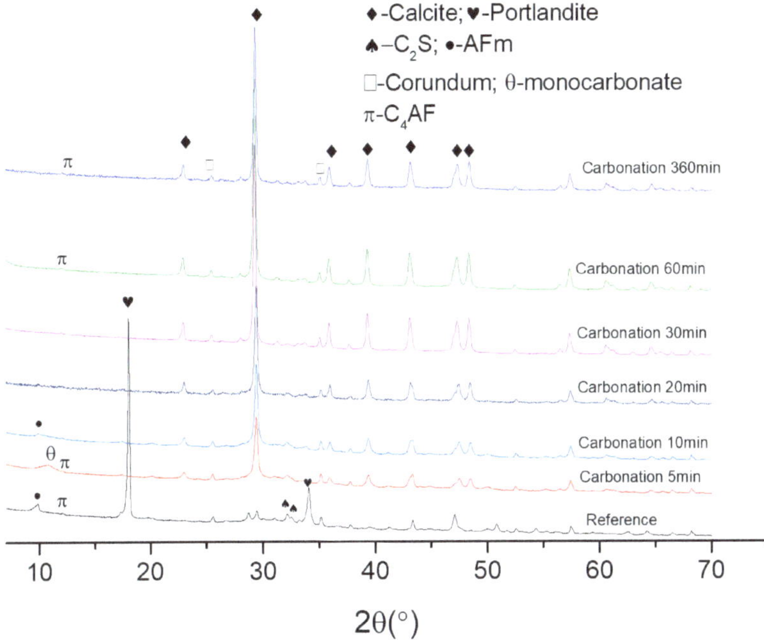

Fig. 10.28 XRD spectra of carbonated RCF [5].

Q^3 resonance, whereas the proportion of Q^3 and Q^4 became dominant after 60 min. This shift indicates that the cross-linked chains of SiO_4 tetrahedra in C-S-H are transformed to Q^{3b} ((SiO-)$_3$Si*-1/2 Ca), Q^3((SiO)$_3$-Si-OH) and Q^4 ((SiO-)$_4$Si) units [24].

Table 10.5 Deconvolution data of 29Si NMR spectra of carbonated RCF[5]

	Q^0	Q^1	Q^2	Q^4(Al)	Q^{3b}	Q^3	Q^4	MCL
Reference	-71.7 ppm 20.57 %	-80.2 ppm 37.52 %	-85.5 ppm 41.91 %	/ /	-93.4 ppm 0 %	-101 ppm		4.23
Carbonation 5 min	-71.7 ppm 18.19 %	-80.2 ppm 34.80 %	-85.5 ppm 41.42 %	/ /	-93.4 ppm 5.59 %			4.70
Carbonation 10 min	-71.7 ppm 17.68 %	-80.2 ppm 31.61 %	-85.5 ppm 43.57 %	/ /	-93.4 ppm 7.1 %			6.38
Carbonation 20 min	-71.7 ppm 11.81 %	-80.2 ppm 20.05 %	-85.5 ppm 50.09 %	/ /	-93.4 ppm 12.89 %	-101 ppm 5.16 %		8.28
Carbonation 30 min	-71.7 ppm 3.52 %	-80.2 ppm 0.00 %	-85.5 ppm Q^2+Q^4(Al) 17.86 %		-93.4 ppm 33.21 %	-101 ppm 29.19 %	-110 ppm 16.22 %	/
Carbonation 60 min	-71.7 ppm 0.00 %	-80.2 ppm 0.00 %	-85.5 ppm Q^2+Q^4(Al) 10.63 %		-93.4 ppm 32.22 %	-101 ppm 37.10 %	-110 ppm 20.02 %	/
Carbonation 360 min	-71.7 ppm 0.00 %	-80.2 ppm 0.00 %	-85 ppm Q^2+Q^4(Al) 10.10 %		-93.4 ppm 24.53 %	-101 ppm 40.09 %	-110 ppm 25.28 %	/

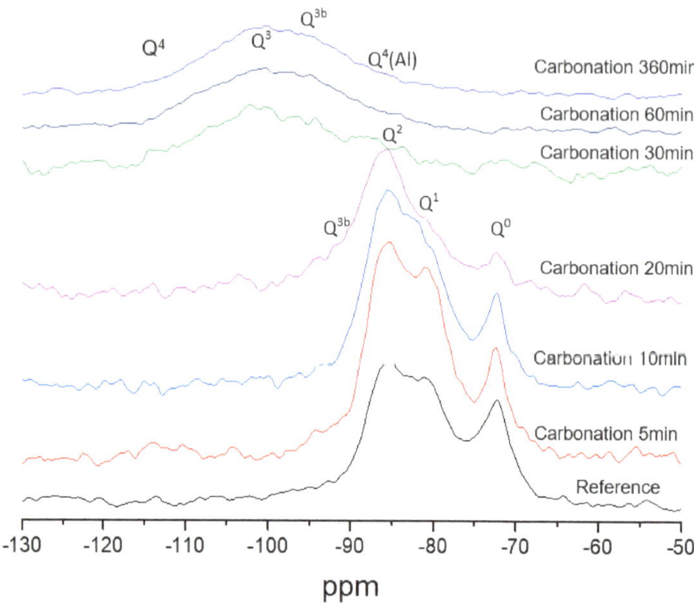

Fig. 10.29 ^{29}Si NMR spectra of RCF after subjecting to different carbonation times [5].

Overall, the silicon phases in the RCF which has been carbonated for 360 min mainly consist of Q^2, $Q^4(Al)$, Q^{3b} and Q^3, and Q^4 sites. The fraction of $Q^4(Al)$ and Q^2, Q^{3b}, Q^3, as well as Q^4 sites are 10.10 %, 24.53 %, 40.09 % and 25.28 %, respectively. Due to the limited amounts of Na_2O and K_2O in RCF, little alumina-silica gel could be formed according to its stoichiometric chemistry [25].

Aluminium Phase Evolution in Carbonated RCF

Fig. 10.30 shows the ^{27}Al NMR spectra of RCF subjected to different carbonation durations. The broad resonances centred at 72 ppm and 11 ppm are attributed to the aluminium substituting silicon in the C-S-H, the presence of AlO_6 in AFm, and other calcium aluminate hydrates [24, 26], respectively. The intensity of AlO_4 resonance incorporated in the bridging sites of the silicate chains decreases immediately after carbonation starts. Meanwhile, the resonance at 11 ppm remains almost unaffected during the first 10 min of carbonation. The resonance at 72 ppm disappears after carbonating for 30 min, and a new resonance centred at about 58 ppm is observed. This resonance shifts to 56 ppm gradually with increasing carbonation time. From the ^{27}Al NMR analysis, almost all the aluminium from the carbonation of C-A-S-H and calcium aluminate hydrated phase are present in the tetrahedral resonance at 56 ppm. This resonance might be attributed to aluminium incorporated in the alumina-silica phase [27] and $Al(OH)_3$ gel [28] and $-OAl(OSi)_4$ site in zeolite [29]. ^{29}Si NMR has confirmed the presence of alumina-silica gel. Besides, the resonances centred at 56 ppm show that the AlO_4 is connected with Si and protons, which includes two species involving in the resonance at 58 ppm and resonance at 65 ppm. The AlO_4 with the resonance at about 58 ppm can be referred to the aluminium sites in

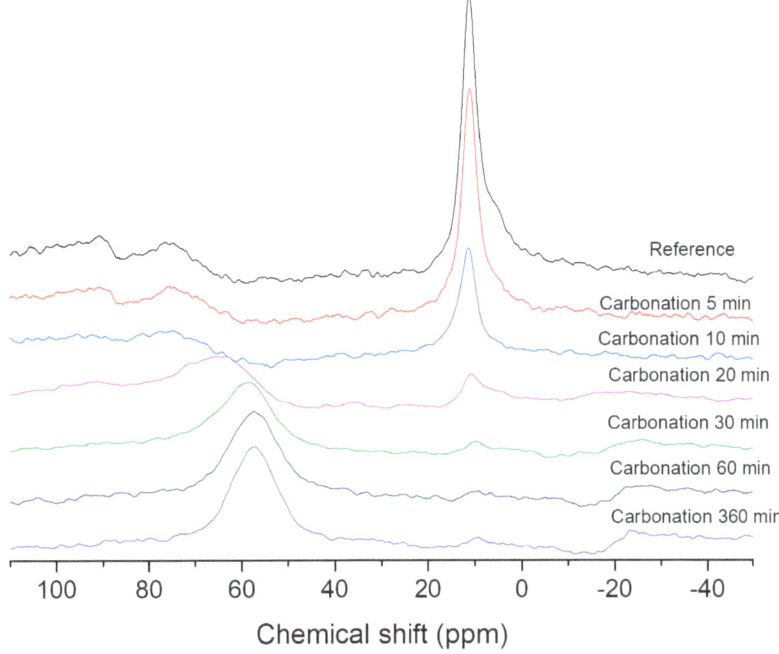

Fig. 10.30 ^{27}Al NMR spectra of carbonated RCF [5].

the alumina-silica gel [30–32]. The AlO_4 with the resonance at about 65 ppm might be caused by AlO_4 in the alumina gel [28, 33]. Therefore, the aluminium phases in RCF are transformed to the alumina-silica gel and alumina gel incorporated AlO_4 sites.

Microstructure Development

The morphological evolution of carbonated RCF is shown in Fig. 10.31. It can be observed in Fig. 10.31a that after 5 min of carbonation, the RCF are coated by a layer of calcite. This calcium carbonate seems to be amorphous, with fibroid and lumpy shapes. When the carbonation time reaches 10 min, the calcite grains on the surface of the particles are largely aggregated. After 20 min, the calcite grains with a layer structure covers the cement particle surface completely. However, a remarkable difference in morphology could be observed when the RCF are carbonated for more than 60 min. The RCF particles are disintegrated, and aggregated calcite grains with an average size of about 0.5 μm are observed. The individual calcite aggregation is the dominant carbonation product.

Fig. 10.32 shows the development of the pore structure of the carbonated RCF. The cumulative porosity increased from 0.076 m^3/g to 0.078 m^3/g as the carbonation time increased from 0 to 60 min, but it increased rapidly to 0.125 m^3/g at 360 min. This increase in total porosity is attributed to the decalcification of the calcium-bearing hydrates and formation of porous silica- or alumina-bearing gel. Meanwhile, the amount of gel pores smaller than 10 nm increased continuously between 60 min and 360 min. Also, based on the relationship between the average particle diameter and Brunner-Emmett-Teller (BET) surface area of RCF [34], the 360 min carbonated RCF have a very fine particle size.

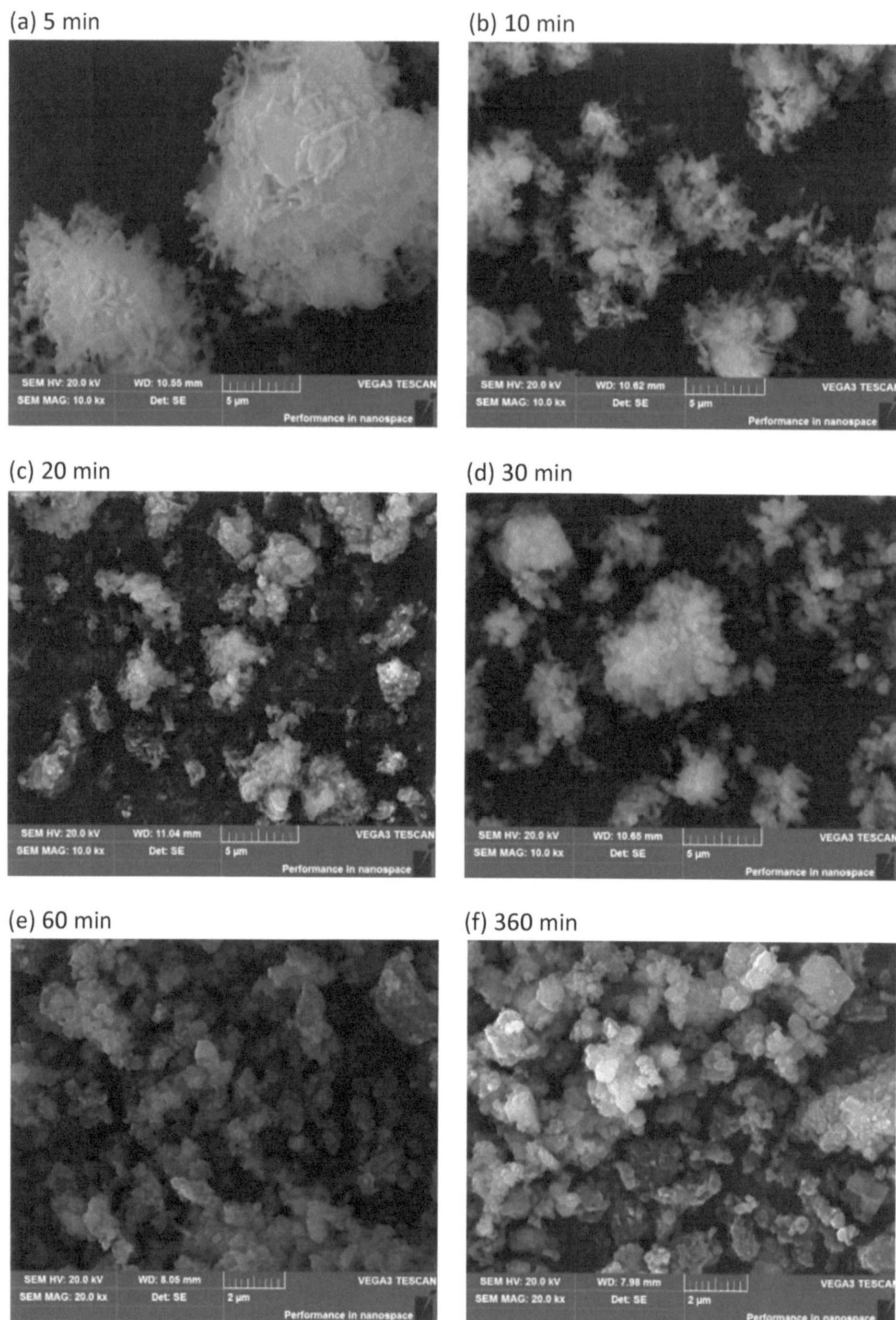

Fig. 10.31   Morphology of carbonated RCF [5].

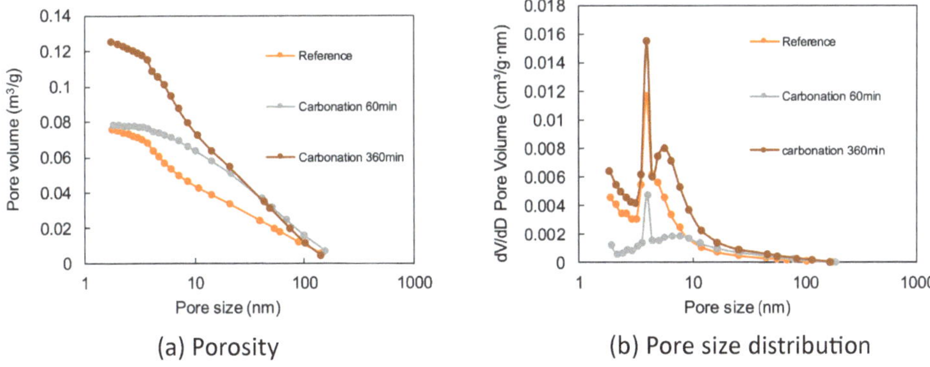

(a) Porosity   (b) Pore size distribution

Fig. 10.32   Development of pore structure of carbonated RCF [5].

Potential Application of Carbonated RCF as High-Reactive SCMs

The heat evolution of the carbonated RCF and CH blends (water to binder ratio of 0.4) is used to indicate the pozzolanic reactivity of carbonated RCF (shown in Fig. 10.33). For the blends containing different amounts of CH, most of the hydration heat is released during the first 6 h. This indicates that the amorphous gel possesses a significantly high pozzolanic reactivity. Thus, the produced carbonated RCF possess a high pozzolanic reactivity and fine particle size that can not only be used as a filler but also as a supplementary cementitious material (SCM) in cement-based materials directly.

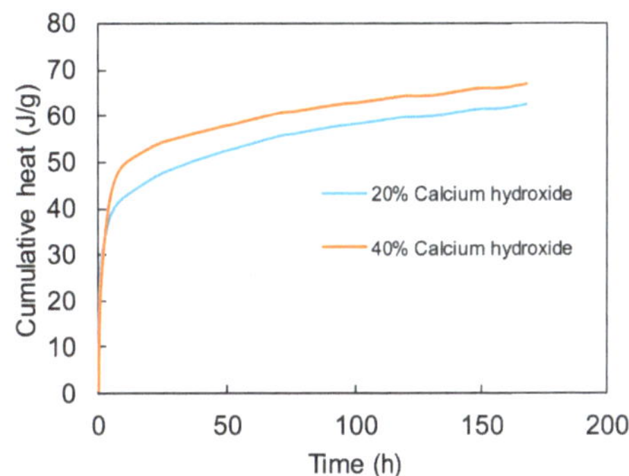

Fig. 10.33   Hydration heat of blends containing carbonated RCF and CH[5]

10.5.2 Synthesis of Amorphous Nano-Silica from Recycled Concrete Fines

Design of Two-Step Carbonation Process for Synthesizing Amorphous Nano-Silica

A "two-step carbonation" process is designed to convert the RCF into amorphous silica gel (Fig. 10.34). The two-step carbonation process is shown as follows:

Step 1: The wet carbonation of RCF in water

The first step carbonation refers to the wet carbonation of RCF in water. Firstly, 15 g of RCF powders are added into 300 ml of deionized water and the mixture is mechanically stirred with a laboratory stirrer at a speed of 200 rpm for 10 min at room temperature.

Then, a commercially sourced CO_2 with a concentration of 99.99 vol% is injected into the aqueous system at a flow rate of 0.2 l/min to induce carbonation. After the carbonation reaction, the suspended solid particles in the solution are collected by filtration (2.5 μm filter paper), which mainly consists of calcium carbonate and silica gel [11]. The main carbonation reaction of the silicate phase of cement hydrates in RCF can be described by Equation 10.5. The suspended solid is then dried in a freeze dryer under vacuum.

$$C - S - H \text{ (calcium silicate hydrates)} + CO_2 \rightarrow CaCO_3 + SiO_2 \text{ (gel)} + H_2O \quad (10.5)$$

Step 2: Extraction of silica phase by wet carbonation

The carbonated RCF residue obtained at the end of Step 1 is placed into a prepared NaOH solution with different concentrations for 2 hours at 40 °C with a stirring speed of 200 rpm. This is to dissolve the silica gel (shown in Equation 10.6). The resulting solution is collected again after filtration (2.5 μm filter paper). The collected solutions are used for further carbonation. The same commercially sourced CO_2 is injected into the solution at a flow rate of 0.05 l/min/100 ml to precipitate silica [24]. During the carbonation, the solution is stirred at a speed of 200 rpm at room temperature. The amorphous silica gel could be extracted by lowering its alkalinity by injecting CO_2 (shown in Equation 10.7). The precipitation in solution is collected by a centrifugation process at a speed of 5000 rpm/ min for 5 min, followed by deionized water washing. Finally, to further improve the purity of the product, the precipitate is further washed by a diluted HCl solution and DI water.

$$SiO_2 \text{ (gel)} + 2NaOH \rightarrow Na_2SiO_3 + H_2O \quad (10.6)$$

$$Na_2SiO_3 + 2CO_2 + H_2O \rightarrow 2NaHCO_3 + SiO_2 \text{ (gel)} \quad (10.7)$$

Compositions of Carbonation Products

The oxide compositions of silica-bearing gel and residue after NaOH treatment are determined by XRF and are shown in Table 10.6. The S-1M refer to the precipitation products after carbonating for 60 min, which are collected from the solution prepared with 1 mol/l NaOH concentrations. The S-washed is the precipitation product of S-1M after washing by a diluted HCl solution and DI water, and the residue is filtrated from the suspension after reacting in 1 mol/l NaOH solution for 2 hours at 40 °C.

The S-1M is a kind of alumina-silica gel incorporating with alkalis. The presences of Al_2O_3 and SiO_2 indicates that the silicon and aluminium phases in carbonated RCF are dissolved. As the obtained gel is adequately washed by DI water, the sodium is likely to be chemically adhered in the gel. The gel obtained from the second step carbonation is confirmed as alumina-silica gel incorporating with alkalis. However, the purity of the gel is obviously enhanced after washing by HCl solution. The aluminium and sodium are washed off with SiO_2 remaining. Therefore, a silica gel with a high purity of 98.8 % is extracted from RCF by the two-step carbonation process. Meanwhile, the residue after NaOH treatment is collected, the oxide composition shows it is a kind of Ca-rich residue, which mainly consists of calcite and therefore could be used to replace cement clinker as an additive for OPC.

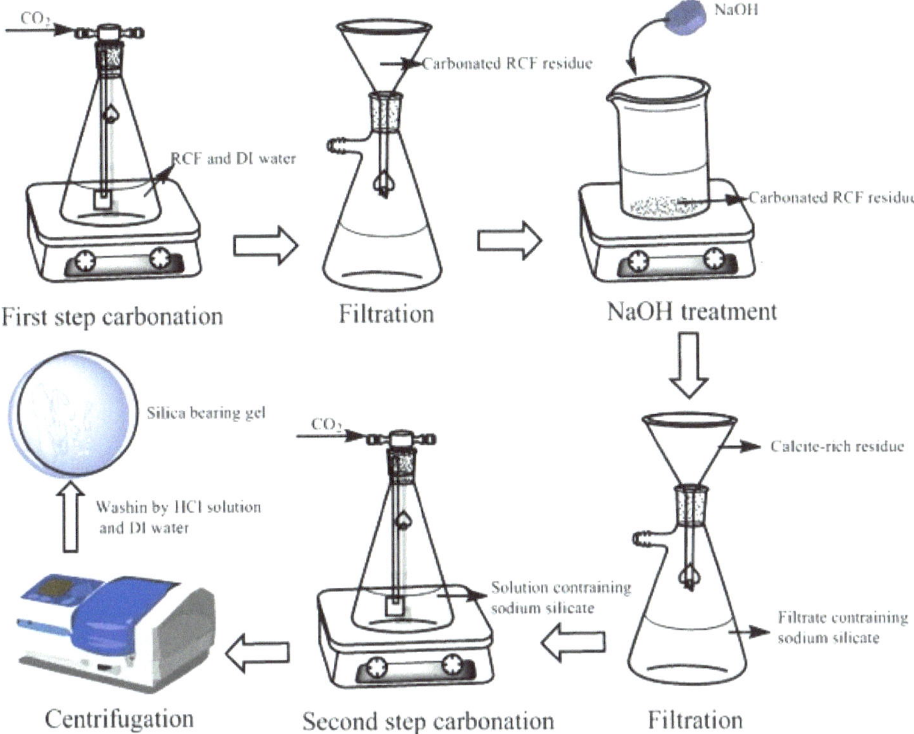

Fig. 10.34   Illustration of two-step wet carbonation process [35].

Table 10.6   The chemical composition of silica bearing gel and residue [35]

No.	Na$_2$O	MgO	Al$_2$O$_3$	SiO$_2$	K$_2$O	CaO	Fe$_2$O$_3$	P$_2$O$_5$
S-1M	12.8	0.202	24.9	60.6	1.01	0.264	0.121	0.055
S-washed	/	/	0.56	98.8	0.16	0.29	0.072	/
Residue	5.5	1.05	3.32	14.2	0.173	70.3	3.43	0.165

The ^{29}Si NMR spectra of the silica-bearing gels (S-1M and S-washed) are shown in Fig. 10.35. The main peak is observed at about -85 ppm, which is quite different from amorphous silica [36] and silica gel in the carbonated RCF. This peak is usually attributed to Q^4 (Al) units in the silica-bearing gel due to the presence of aluminium. This was related to amorphous sodium aluminosilicate gel, and it possesses a molar SiO$_2$/Al$_2$O$_3$ ratio of 2 [25, 31]. The resonances at about -110 ppm and -102 ppm correspond to Q^4 units (SiO$_4$ tetrahedra) and Q^3 sites (hydroxylated surface sites, (SiO)$_3$-Si-OH) in amorphous silica phase [24, 37]. To understand the compositions of carbonated silicate phases, the NMR spectra are deconvoluted by fitting the intensity of silicate using Peakfit software. The fraction of Q^4(Al), Q^3 and Q^4 units in S-1M are 52.2 %, 19.7 % and 28.1 %, respectively. When the silica-bearing gel is washed by HCl, the Q^4(Al) units are disappearing as the sodium and aluminium in the silica structure are washed off. As the aluminium does not exist, only the Q^3 resonance at -102 ppm and Q^4 resonance at -110 ppm could

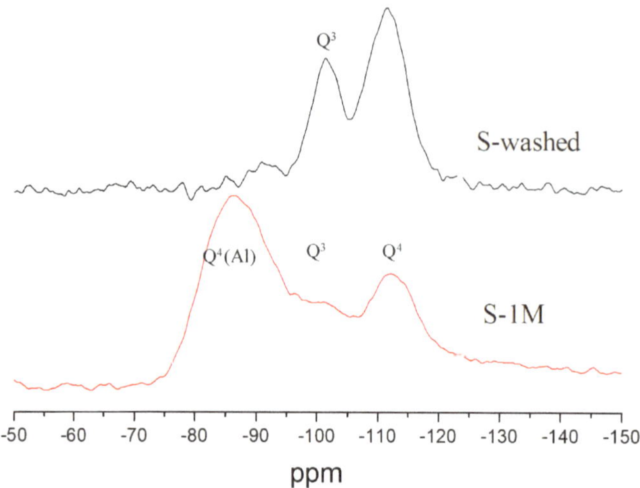

Fig. 10.35   The ^{29}Si NMR spectra of silica-bearing gel [35].

be observed, which possess fractions of 39.2 % and 60.8 %, respectively. The Q^4 units are dominated in S-washed, which indicates that it is a kind of highly polymerized amorphous silica. Thus, it can be concluded that a silica-bearing gel (S-1M) consisting of sodium aluminosilicate gel and amorphous silica gel is prepared after the two-step wet carbonation. An amorphous silica with high purity is prepared after washing by HCl, where the co-existence of aluminium and sodium disappear.

Microstructure of Carbonation Products

Fig. 10.36 shows the SEM and Transmission Electron Microscopy (TEM) images of silica-bearing gels with and without washing by the diluted HCl solution. It can be observed from SEM images that aggregated nanoparticles are present in S-1M and S-washed. The agglomerated nanoparticles with a diameter of about 200 nm could be observed in S-1M and S-washed, which is attributed to the attraction between nanoparticles. The agglomeration influences the observed particle size of these two gels, but the single gel particle seems to be nano-size. Furthermore, the TEM micrographs are captured to indicate the morphology of the nanostructure of silica-bearing gel (shown in Fig. 10.36c and d). The particle sizes of these two gels are nano-scale. The particle size of S-1M ranges between 10 nm and 50 nm and is further decreased to smaller than 10 nm after washing by HCl solution. The small particle size is similar to the nano-silica produced from alkali-extracted rice husk ash [38], and it possesses a high BET surface area of 634 m^2/g. Therefore, it confirms that a significant decrease in particle size after washing by HCl solution and a nano-silica with a narrow size distribution is synthesized by using the two-step wet carbonation.

The particle characteristics of RCF carbonated for 60 min and silica-bearing gels (S-1M and S-washed) are determined by using the BET method. The nitrogen adsorption isotherms and pore area of the carbonated RCF and silica-bearing gel could be

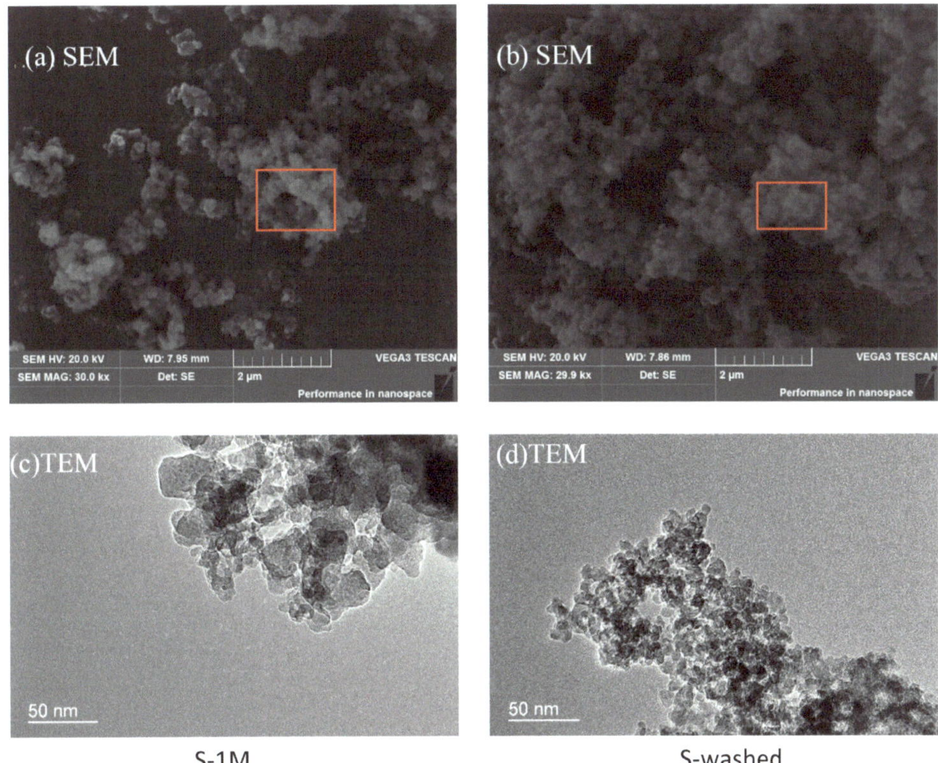

Fig. 10.36   SEM and TEM images of silica-bearing gel [35].

obtained and is shown in Fig. 10.37. It can be observed that the nitrogen adsorption isotherms show rapid increases of absorbed nitrogen content at the relative pressure of 0.5–1, 0.7–1 and 0.9–1, which corresponds to the carbonated RCF, S-1M and S-washed, respectively. From the results, the range of relative pressure for the rapid increase of absorbed nitrogen content becomes narrow, demonstrating that the S-washed showed the narrowest pore size distribution [39]. The BET surface areas of carbonated RCF, S-1M, and S-washed are 18.98 m²/g, 85.11 m²/g, and 662.47 m²/g, respectively. The average particle diameter can be calculated assuming nonporous spherical particles and density of silica-bearing gel. So the average particle diameter of carbonated RCF and gel can be calculated by Equation 10.8 [34]. The calculated average particle diameters are 129.03 nm, 37.10 nm, and 4.12 nm corresponding to carbonated RCF, S-1M and S-washed, respectively. Therefore, it can be concluded that the silica-bearing gel has a nano-scale particle size.

$$d_{BET} = \frac{6000}{S_{BET} \times \rho} \tag{10.8}$$

where d_{BET} is the particle diameter (nm); S_{BET} is the BET surface area (m²/g); ρ is the density of tested materials (g/cm³).

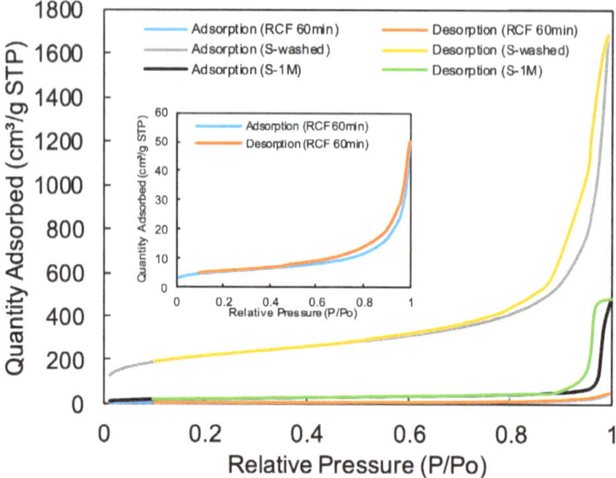

(a) BET adsorption and desorption isotherm

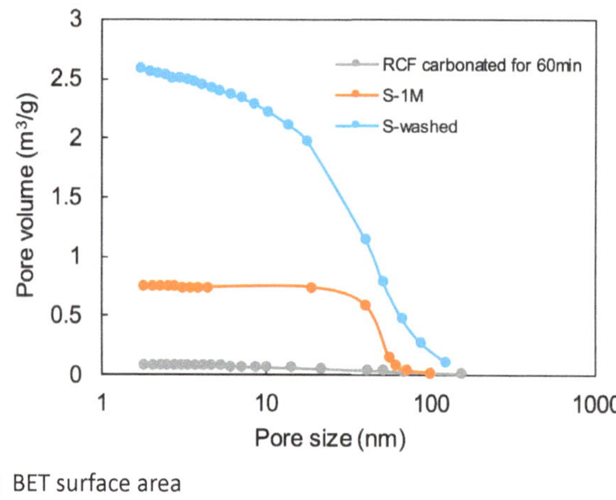

(b) BET surface area

Fig. 10.37   Particle characterization of carbonated RCF and silica-bearing gels [35].

CO$_2$ Consumption and Yields of Carbonation Process

According to the above test results, a large amount of CO$_2$ is consumed by the two-step carbonation process. The consumption process of CO$_2$ can be shown in Fig. 10.38. The results show about 0.23 g CO$_2$ is consumed by 1 g RCF during the first step carbonation, and the second step carbonation consumes a further 0.04 g CO$_2$. Therefore, to extract the silica phase from 1 g of RCF by the two-step wet carbonation process, in total about 0.27 g of CO$_2$ is consumed. Furthermore, it can be calculated that using this process, 1 g of RCF can produce 1.13 g of calcium carbonate-rich materials and 0.050 g of silica gel. As there is a huge amount of RCF being generated consistently in the world, this two-step carbonation process would produce large quantities of nano-silica and calcium carbonate-rich materials. Besides, the calcium carbonate-rich residues could be

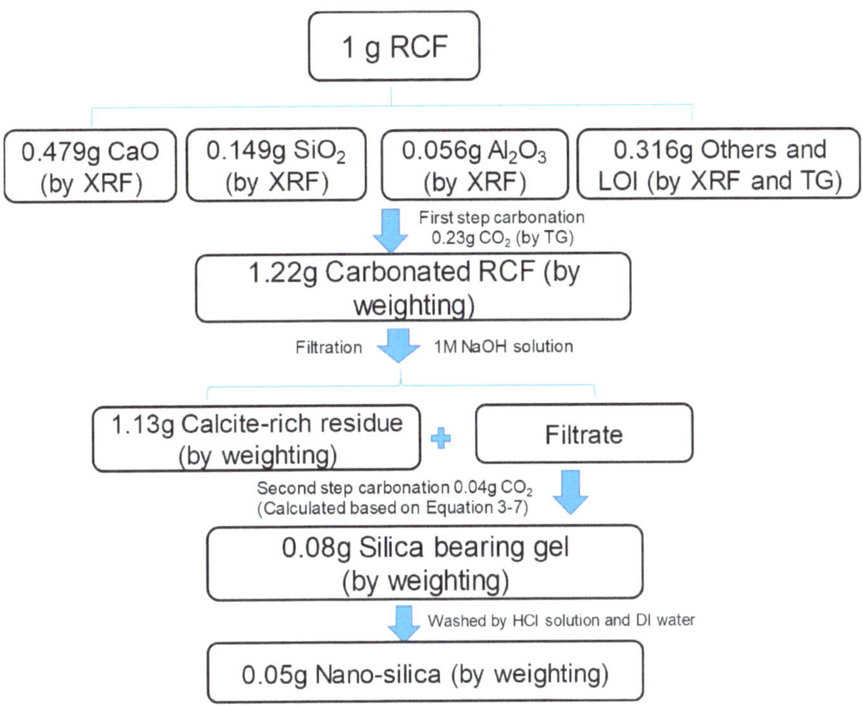

Fig. 10.38 CO_2 consumption and yields of carbonation products during carbonation [35].

used to replace clinker as supplementary cementitious materials or as a filler in cement and concrete production, which could further reduce the CO_2 emission from cement production [34].

Besides, the waste solution generated from this process (i.e., sodium carbonate/ sodium bicarbonate) could also be used to carbonate RCF to produce supplementary cementitious materials [11]. Moreover, modified carbonation processes involving the enforced decalcification and then carbonation were also available where the additives (e.g., NH_4Cl) could be recycled, making the processes more sustainable [41].

10.6 Summary

Recycled products from a typical concrete recycling process including the coarse RCA, fine RCA and RCF (as indicated in Fig. 10.1) can be valorized further by applying suitable accelerated carbonation processes. The properties of the coarse RCA can be enhanced for use in structural concrete and the fine RCA can be used in mortars. In particular, the developed two-step wet carbonation process can turn the conventional waste RCF into two value-added products (i.e., micro-sized calcium carbonate and nano-sized silica gel) that potentially have high values for industrial or construction applications.

10.7 Acknowledgements

The authors wish to thank the Strategic Public Policy Research Funding Scheme from the Hong Kong Government and the Hong Kong Construction Industry Council for funding support.

10.8 References

[1]   D. Xuan, B. Zhan, C.S. Poon, Assessment of mechanical properties of concrete incorporating carbonated recycled concrete aggregates, Cement and Concrete Composites, 65 (2016) 67–74. https://doi.org/10.1016/j.cemconcomp.2015.10.018

[2]   M. Quattrone, S.C. Angulo, V.M. John, Energy and CO2 from high performance recycled aggregate production, Resources, Conservation and Recycling, 90 (2014) 21–33. https://doi.org/10.1016/j.resconrec.2014.06.003

[3]   P. Belin, G. Habert, M. Thiery, N. Roussel, Cement paste content and water absorption of recycled concrete coarse aggregates, Materials and Structures, 47 (2014) 1451–1465. https://doi.org/10.1617/s11527-013-0128-z

[4]   S. Liu, P. Shen, D. Xuan, L. Li, A. Sojobi, B. Zhan, C.S. Poon, A comparison of liquid-solid and gas-solid accelerated carbonation for enhancement of recycled concrete aggregate, Cement and Concrete Composites, 118 (2021) 103988. https://doi.org/10.1016/j.cemconcomp.2021.103988

[5]   P. Shen, Y. Zhang, Y. Jiang, B. Zhan, J. Lu, S. Zhang, D. Xuan, C.S. Poon, Phase assemblance evolution during wet carbonation of recycled concrete fines, Cement and Concrete Research, 154 (2022) 106733. https://doi.org/10.1016/j.cemconres.2022.106733

[6]   C. Shi, Z. Wu, Z. Cao, T.C. Ling, J. Zheng, Performance of mortar prepared with recycled concrete aggregate enhanced by CO2 and pozzolan slurry, Cement and Concrete Composites, 86 (2018) 130–138. https://doi.org/10.1016/j.cemconcomp.2017.10.013

[7]   J. Zhang, C. Shi, Y. Li, X. Pan, C.-S. Poon, Z. Xie, Influence of carbonated recycled concrete aggregate on properties of cement mortar, Construction and Building Materials, 98 (2015) 1–7. https://doi.org/10.1016/j.conbuildmat.2015.08.087

[8]   X. Fang, B. Zhan, C.S. Poon, Enhancement of recycled aggregates and concrete by combined treatment of spraying Ca2+ rich wastewater and flow-through carbonation, Construction and Building Materials, 277 (2021) 122202. https://doi.org/10.1016/j.conbuildmat.2020.122202

[9]   B.J. Zhan, D.X. Xuan, W. Zeng, C.S. Poon, Carbonation treatment of recycled concrete aggregate: Effect on transport properties and steel corrosion of recycled aggregate concrete, Cement and Concrete Composites, 104 (2019) 103360. https://doi.org/10.1016/j.cemconcomp.2019.103360

[10] S. Kashef-Haghighi, S. Ghoshal, CO2 Sequestration in Concrete through Accelerated Carbonation Curing in a Flow-through Reactor, Industrial & Engineering Chemistry Research, 49 (2010) 1143–1149. https://doi.org/10.1021/ie900703d

[11] M. Zajac, J. Skibsted, J. Skocek, P. Durdzinski, F. Bullerjahn, M. Ben Haha, Phase assemblage and microstructure of cement paste subjected to enforced, wet carbonation, Cement and Concrete Research, 130 (2020) 105990. https://doi.org/10.1016/j.cemconres.2020.105990

[12] C. Wang, J. Xiao, G. Zhang, L. Li, Interfacial properties of modeled recycled aggregate concrete modified by carbonation, Construction and Building Materials, 105 (2016) 307–320. https://doi.org/10.1016/j.conbuildmat.2015.12.077

[13] L. Li, D. Xuan, A.O. Sojobi, S. Liu, C.S. Poon, Efficiencies of carbonation and nano silica treatment methods in enhancing the performance of recycled aggregate concrete, Construction and Building Materials, 308 (2021) 125080. https://doi.org/10.1016/j.conbuildmat.2021.125080

[14] D. Xuan, B. Zhan, C.S. Poon, Durability of recycled aggregate concrete prepared with carbonated recycled concrete aggregates, Cement and Concrete Composites, 84 (2017) 214–221. https://doi.org/10.1016/j.cemconcomp.2017.09.015

[15] X. Fang, B. Zhan, C.S. Poon, Enhancing the accelerated carbonation of recycled concrete aggregates by using reclaimed wastewater from concrete batching plants, Construction and Building Materials, 239 (2020) 117810. https://doi.org/10.1016/j.conbuildmat.2019.117810

[16] Y. Jiang, P. Shen, C.S. Poon, Improving the bonding capacity of recycled concrete aggregate by creating a reactive shell with aqueous carbonation, Construction and Building Materials, 315 (2022) 125733. https://doi.org/10.1016/j.conbuildmat.2021.125733

[17] X. Fang, D. Xuan, P. Shen, C.S. Poon, Fast enhancement of recycled fine aggregates properties by wet carbonation, Journal of Cleaner Production, 313 (2021) 127867. https://doi.org/10.1016/j.jclepro.2021.127867

[18] Y. Briki, M. Zajac, M.B. Haha, K. Scrivener, Impact of limestone fineness on cement hydration at early age, Cement and Concrete Research, 147 (2021) 106515. https://doi.org/10.1016/j.cemconres.2021.106515

[19] R. Zarzuela, M. Luna, L.M. Carrascosa, M.P. Yeste, I. Garcia-Lodeiro, M.T. Blanco-Varela, M.A. Cauqui, J.M. Rodríguez-Izquierdo, M.J. Mosquera, Producing C-S-H gel by reaction between silica oligomers and portlandite: A promising approach to repair cementitious materials, Cement and Concrete Research, 130 (2020) 106008. https://doi.org/10.1016/j.cemconres.2020.106008

[20] B.J. Zhan, D.X. Xuan, C.S. Poon, K.L. Scrivener, Characterization of interfacial transition zone in concrete prepared with carbonated modeled recycled concrete aggregates, Cement and Concrete Research, 136 (2020) 106175. https://doi.org/10.1016/j.cemconres.2020.106175

[21] J. Goergens, T. Manninger, F. Goetz-Neunhoeffer, In-situ XRD study of the temperature-dependent early hydration of calcium aluminate cement in a mix with calcite, Cement and Concrete Research, 136 (2020) 106160. https://doi.org/10.1016/j.cemconres.2020.106160

[22] V. Shah, K. Scrivener, B. Bhattacharjee, S. Bishnoi, Changes in microstructure characteristics of cement paste on carbonation, Cement and Concrete Research, 109 (2018) 184–197. https://doi.org/10.1016/j.cemconres.2018.04.016

[23] E. Tajuelo Rodriguez, I. Richardson, L. Black, E. Boehm-Courjault, A. Nonat, J. Skibsted, Composition, silicate anion structure and morphology of calcium silicate hydrates (CSH) synthesised by silica-lime reaction and by controlled hydration of tricalcium silicate (C_3S), Advances in Applied Ceramics, 114 (2015) 362–371. https://doi.org/10.1179/1743676115y.0000000038

[24] T.F. Sevelsted, J. Skibsted, Carbonation of C-S-H and C-A-S-H samples studied by ^{13}C, ^{27}Al and ^{29}Si MAS NMR spectroscopy, Cement and concrete Research, 71 (2015) 56–65. https://doi.org/10.1016/j.cemconres.2015.01.019

[25] S. Greiser, G.J. Gluth, P. Sturm, C. Jäger, $^{29}Si\{^{27}Al\}$, $^{27}Al\{^{29}Si\}$ and $^{27}Al\{^{1}H\}$ double-resonance NMR spectroscopy study of cementitious sodium aluminosilicate gels (geopolymers) and gel-zeolite composites, RSC Advances, 8 (2018) 40164–40171. https://doi.org/10.1039/d3ra90078a

[26] P. Shen, L. Lu, W. Chen, F. Wang, S. Hu, Efficiency of metakaolin in steam cured high strength concrete, Construction and Building Materials, 152 (2017) 357–366. https://doi.org/10.1016/j.conbuildmat.2017.07.006

[27] J.F. Stebbins, Nuclear magnetic resonance spectroscopy of silicates and oxides in geochemistry and geophysics, in Thomas J. Ahrens (ed.), Mineral Physics & Crystallography: A Handbook of Physical Constants, vol. 2 (1995) 303–332. https://doi.org/10.1029/rf002p0303

[28] M.I. Macedo, C.C. Osawa, C.A. Bertran, Sol-gel synthesis of transparent alumina gel and pure gamma alumina by urea hydrolysis of aluminum nitrate, Journal of Sol-gel Science and Technology, 30 (2004) 135–140. https://doi.org/10.1023/b:jsst.0000039497.46154.8f

[29] K. De Weerdt, G. Plusquellec, A.B. Revert, M. Geiker, B. Lothenbach, Effect of carbonation on the pore solution of mortar, Cement and Concrete Research, 118 (2019) 38–56. https://doi.org/10.1016/j.cemconres.2019.02.004

[30] R.A. Fletcher, K.J. MacKenzie, C.L. Nicholson, S. Shimada, The composition range of aluminosilicate geopolymers, Journal of the European Ceramic Society, 25 (2005) 1471–1477. https://doi.org/10.1016/j.jeurceramsoc.2004.06.001

[31] P. Sturm, S. Greiser, G. Gluth, C. Jäger, H. Brouwers, Degree of reaction and phase content of silica-based one-part geopolymers investigated using chemical and NMR spectroscopic methods, Journal of Materials Science, 50 (2015) 6768–6778. https://doi.org/10.1007/s10853-015-9232-5

[32] B. Walkley, R. San Nicolas, M.-A. Sani, J.D. Gehman, J.S. van Deventer, J.L. Provis, Phase evolution of Na_2O–Al_2O_3–SiO_2–H_2O gels in synthetic aluminosilicate binders, Dalton Transactions, 45 (2016) 5521–5535. https://doi.org/10.1039/c5dt04878h

[33] T. Isobe, T. Watanabe, J.D.E. De La Caillerie, A. Legrand, D. Massiot, Solid-state ^{1}H and ^{27}Al NMR studies of amorphous aluminum hydroxides, Journal of Colloid and Interface Science, 261 (2003) 320–324. https://doi.org/10.1016/s0021-9797(03)00144-9

[34] L. Jiqiao, H. Baiyun, Particle size characterization of ultrafine tungsten powder, International Journal of Refractory Metals and Hard Materials, 19 (2001) 89–99. https://doi.org/10.1016/S0263-4368(00)00051-2

[35] P. Shen, Y. Sun, S. Liu, Y. Jiang, H. Zheng, D. Xuan, J. Lu, C.S. Poon, Synthesis of amorphous nano-silica from recycled concrete fines by two-step wet carbonation, Cement and Concrete Research, 147 (2021) 106526. https://doi.org/10.1016/j.cemconres.2021.106526

[36] X. Fang, D. Xuan, B. Zhan, W. Li, C.S. Poon, A novel upcycling technique of recycled cement paste powder by a two-step carbonation process, Journal of Cleaner Production, 290 (2021) 125192. https://doi.org/10.1016/j.jclepro.2020.125192

[37] P. Colombet, A. R. Grimmer, H, Zanni, P. Sozzani, Nuclear Magnetic Resonance Spectroscopy of Cement-Based Materials, 2012, Springer Science & Business Media. https://doi.org/10.1007/978-3-642-80432-8

[38] T.-H. Liou, C.-C. Yang, Synthesis and surface characteristics of nanosilica produced from alkali-extracted rice husk ash, Materials Science and Engineering: B, 176 (2011) 521–529. https://doi.org/10.1016/j.mseb.2011.01.007

[39] N. Venkatathri, M. Nookaraju, A. Rajini, A.A. Kumar, I. Reddy, Influence of Organic Amines on Size and Properties of Amorphous Porous Nano Silica, Advanced Porous Materials, 1 (2013) 224–228. https://doi.org/10.1166/apm.2013.1017

[40] H. Klee, Briefing: The Cement Sustainability Initiative, Proceedings of the Institution of Civil Engineers – Engineering Sustainability, Thomas Telford Ltd, 157 (2004), pp. 9–11. https://doi.org/10.1680/ensu.2004.157.1.9

[41] H. Mehdizadeh, K.H. Mo, T.-C. Ling, CO2-fixing and recovery of high-purity vaterite CaCO3 from recycled concrete fines, Resources, Conservation and Recycling, 188 (2023) 106695. https://doi.org/10.1016/j.resconrec.2022.106695

Chapter 11

Large-Scale Structural Applications of Recycled Aggregates Concrete – Shandong Shanghe Industrial Base

Jianzhuang Xiao[1], Zhangyi Cheng[1], Yubo Pan[2], Haibo Fang[2]

1 Tongji University, No. 1239, Siping Road, Shanghai, 200082, China
2 17F, Unit 1, Phase II office building, Zhongjian cultural city, No. 16, Wenhua East Road, Lixia District, Jinan City, Shandong Province, 250014, China

11.1 Introduction

With the development of the global construction industry, there is a huge demand for building materials in the construction of infrastructure and large-scale projects. According to a study of Monteiro et al. [1], concrete is the most consumed material in the construction industry. However, the rising price of aggregate and low recycling rate of waste concrete has become the main problems in the concrete industry. Velay-Lizancos et al. [2] believed that economic benefits could be achieved by using recycled aggregate concrete (RAC) made of construction waste. The use of recycled aggregate not only alleviates the shortage of natural aggregate, but also provides an appropriate treatment method for construction solid waste. Then, a stable and economic bond between "demolition" and "construction" can be established.

So far, many scholars have tested the mechanical properties, durability, and construction performance of recycled concrete. Most of the test results show that the performance of RAC is inferior to that of ordinary concrete, and it is affected by some comprehensive factors such as the replacement ratio, the type of recycled aggregates, the mix proportion, and the mixing process. However, by selecting excellent recycled aggregates and using correct modification methods, the performance of RAC can reach or exceed that of ordinary concrete and satisfy the demands of projects. On this basis, the usage scope of recycled concrete has gradually changed from road landfill material to structural components such as beams, slabs, and columns, and finally to large-scale structures. Xiao [3] introduced the research status and development of a recycled concrete structure from the aspects of constitutive relationship, bonding performance, members, and structural performance of recycled concrete. Many countries and regions also have successively issued specifications and standards related to recycled concrete materials and constructions, which will help the recycled concrete industry to develop in the direction of industrialization and high quality. This paper introduces the Shandong Shanghe industrial base. It is an example of large-scale structures in which RAC are used as construction materials.

11.2 Design Summary

In the Shandong Shanghe Industrial Base, the comprehensive office building is an example of the large-scale application of recycled concrete. The office building has 5 floors above ground and a structural height of 19.3 m. The building is supported on a raft foundation, and the structural system is a concrete-filled steel tubular frame support structure.

Fig. 11.1 Appearance of office building

This project used all ready-mixed recycled concrete. It is mixed with class II recycled coarse aggregate. The "Recycled coarse aggregate for concrete" [4] specified the characteristics of class II recycled aggregate with a maximum water absorption rate of 7 %, a maximum porosity of 50 %, and a minimum apparent density of 2350 kg/m³. The replacement rate of recycled coarse aggregate was as follows: 50 % for the fifth floor and the machine room floor, and 30 % for the rest. The concrete strength grade is shown in Table 11.1.

Table 11.1 Strength grade of recycled concrete

Component parts	Strength grade	Remark
Foundation cushion	C20	Requirements for impermeable concrete: P6 impermeable concrete shall be used for concrete components in contact with water or soil such as balcony, kitchen, toilet board, roof slab and basement roof. P6 impermeable concrete shall be used for the exterior wall, foundation, and waterproof ground slab of the basement.
Foundation	C35	
Column, wall	C35	
Beams, slabs, stairs	C35	
Ring beam, structural column, cast-in-place lintel	C30	
Post pouring belt	C40 Micro-expansion	

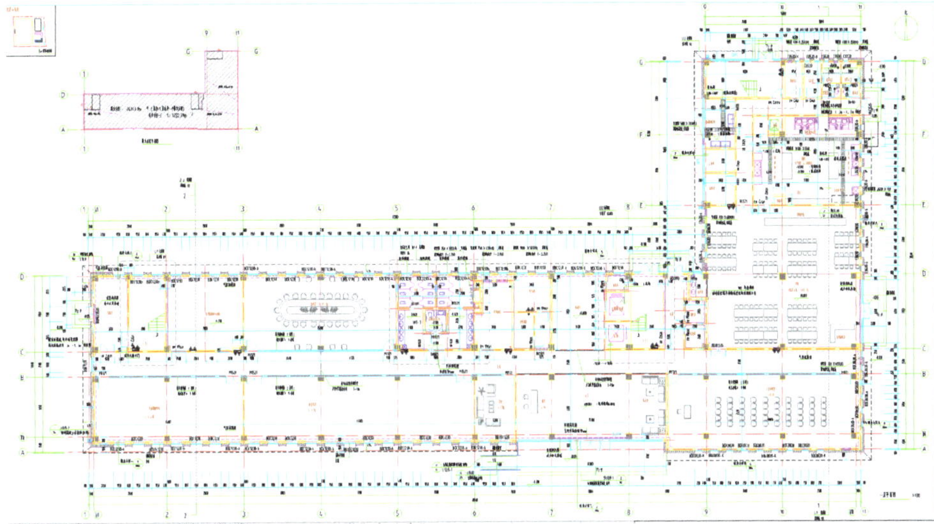

Fig. 11.2 Floor plan of office building

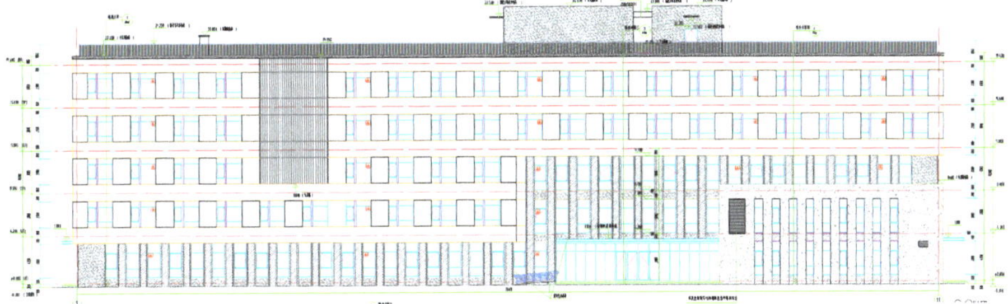

Fig. 11.3 Elevation of office building

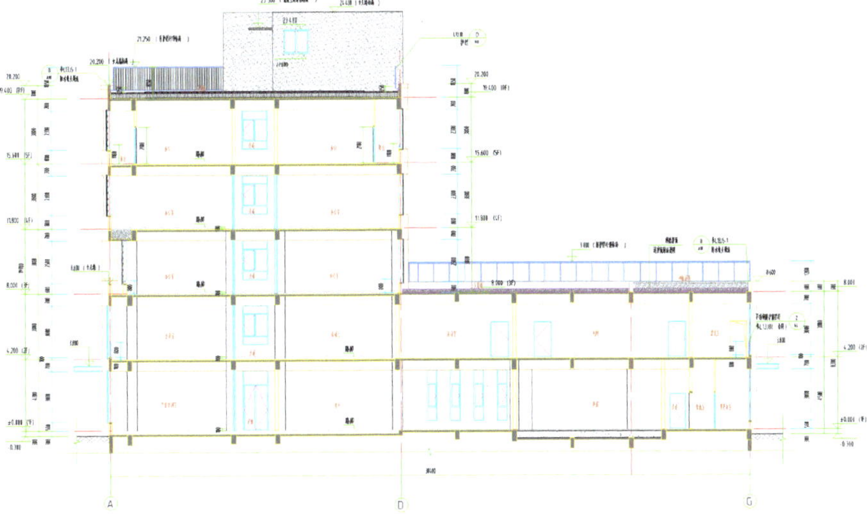

Fig. 11.4 Sectional view of office building

11.3 Application of Recycled Concrete

11.3.1 Mix Proportion Design

The mix proportion of recycled concrete was designed according to different strength grades and different replacement rates of recycled coarse aggregate, and the 7-day strength and 28-day strength were verified, as shown in the tables below.

11.3.2 Construction Process

Construction Process Record

During the construction with recycled concrete, the pouring time, pouring position, pouring method, pouring equipment, pumping height, humidity, wind speed, departure time, slump (including delivery, arrival, and after pumping), expansion, pumping pressure, moulding temperature, initial, and final setting time after pumping of each train were to be recorded to form construction records.

Table 11.2   Mix proportion of recycled concrete with 30 % replacement rate of recycled aggregate

Grade	Cement	Slag	Fly ash	Sand	RA	Gravel	Water	AD	volume-weight
C20	220	65	65	820	285	657	195	6.2	2313
C25	240	85	40	814	287	669	185	7.2	2327
C30	255	95	40	790	290	675	175	7.7	2328
C35	280	90	60	758	290	672	175	9.1	2334
C40	310	110	50	704	301	680	175	9.7	2340

Table 11.3   Mix proportion of recycled concrete with 50 % replacement rate of recycled aggregate

Grade	Cement	Slag	Fly ash	Sand	RA	Gravel	Water	AD	volume-weight
C20	230	65	65	800	455	455	200	6.0	2276
C25	250	85	40	813	450	450	190	7.0	2285
C30	260	95	50	779	460	460	180	7.5	2292
C35	290	90	60	735	467	468	180	9.0	2299
C40	330	100	50	695	470	470	180	9.3	2304

Setting Time of Recycled Concrete

The initial setting and the final setting time of recycled concrete after pump were monitored by in-situ marking. From the data, the initial setting time of recycled concrete after pump was about 35 min, which was shorter than that of ordinary concrete.

Fig. 11.5 Construction process record

Sample Wall

A sampling wall of recycled concrete was made on site, and the changes of concrete strength and carbonation depth under different curing states (natural curing, film curing, and watering curing) were continuously monitored through on-site measurement and coring. According to the data, the strength of the three maintenance methods in 8 days exceeded 100 %, of which the film coating maintenance strength was the highest, the watering maintenance was the second, and the natural maintenance strength was the lowest.

Table 11.4 Setting time record

Wagon number	departure time	Arrival time	Initial setting time after pump	Final setting time after pump
AB2179	15:51	17:10	30 min	210 min
AE8977	15:48	17:15	35 min	220 min
AB1895	16:21	17:40	35 min	210 min
AE0065	16:03	17:30	32 min	200 min
AX3132	16:49	18:20	31 min	230 min
AE1285	17:18	19:05	34 min	210 min
AE1875	18:04	19:30	32 min	200 min
AE6728	18:21	19:38	30 min	220 min
AC0165	18:33	19:40	35 min	200 min
AX7230	19:19	21:40	36 min	220 min
AE1368	19:52	21:50	34 min	220 min
AX2576	20:06	21:55	32 min	210 min
AV7129	20:32	22:00	30 min	230 min
AE6079	21:10	22:20	32 min	210 min

Table 11.5   Sample wall record

Position	Strength grade	Production date	Test date	Age / d	Curing method	Specimen specification / mm	Strength value / MPa	Reach the design strength grade / %
Sample 1	C35	2020.9.21	2020.9.29	8	Sprinkling water	Diameter 75, height 75	37.9	108.3 %
Sample 2	C35	2020.9.21	2020.9.29	8	Tectorial membrane	Diameter 75, height 75	42.2	120.6 %
Sample 3	C35	2020.9.21	2020.9.29	8	Natural curing	Diameter 75, height 75	37.3	106.6 %

Fig. 11.6   Sample wall

11.4 Application Monitoring of Recycled Concrete

A set of automated monitoring schemes have been developed for this office building, in which a sensor neural network was systematically constructed to comprehensively analyse the safety status of the building and provide early warning for areas exceeding the safety status. The monitoring data covered a period of 17 months from July 2021 to November 2022.

11.4.1 Dynamic Monitoring

The structural dynamic characteristic parameters (frequency, vibration mode, etc.) and vibration level (vibration intensity and amplitude) were the markers of the overall safety of the structure. The strength degradation of structural materials would change the vibration characteristics of the structure. For example, the reduction of structural stiffness would reduce the natural vibration frequency of the structure, and the change of a local vibration mode of the structure may indicate the local damage of the structure. Therefore, the monitoring of structural dynamic characteristics and vibration levels could fulfil the purpose of monitoring the structural health status as a whole.

The layout of structural dynamic characteristic monitoring points is shown in Fig. 11.7. A total of 4 vibration monitoring points were arranged at the corner of the first floor and the top floor, respectively. Each vibration monitoring point was equipped with a two-way measuring point (transverse + longitudinal), with a total of 8 integrated vibration monitoring systems. The specific point layout could be adjusted according to the actual situation on site.

During the monitoring period, the overall vibration acceleration amplitude of the office building was relatively small, with a maximum amplitude of approximately 0.10

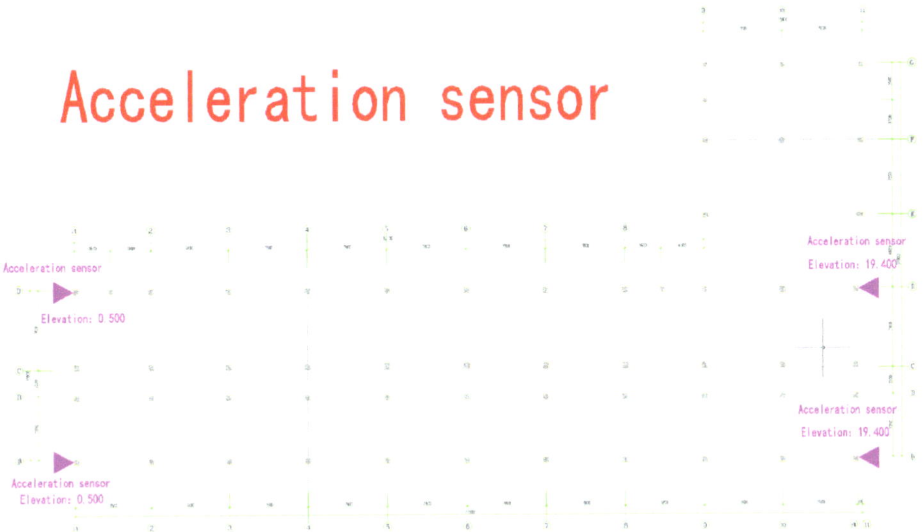

Fig. 11.7   Location map of dynamic characteristic monitoring points

gal in the X and Y directions, which was less than the standard limit of 0.25 gal and met the comfort requirements.

11.4.2 Settlement Monitoring and Creep Monitoring

The layout of structural settlement monitoring points is shown in Fig. 11.8. There were 10 monitoring points in total, and a base point was arranged at the location where the surrounding was stable and did not exceed the range.

The layout of structural creep monitoring points is also shown in Fig. 11.8. There were 3 creep monitoring points in total. Each creep monitoring point was equipped with 4 embedded strain gauges (including standby strain gauges), with a total of 12 embedded strain gauges.

During the monitoring period, the settlement difference between the settlement monitoring point and the base point did not exceed 15 mm, which was less than the allowable limit of 15.6 mm in the standards. As time went on, the creep of the concrete gradually increased, and the current maximum creep value was about 150 με.

11.4.3 Deflection Monitoring for Beams

Four beams were selected at the top of the second floor (elevation 7.940 m) for deflection monitoring. The deflection monitoring points were arranged at both ends and middle of each beam, respectively. The locations of the 12 deflection monitoring points are shown in Fig. 11.9.

During the monitoring period, the measured values of the monitored beam deflection were all less than the L/200 limit requirements in the standards.

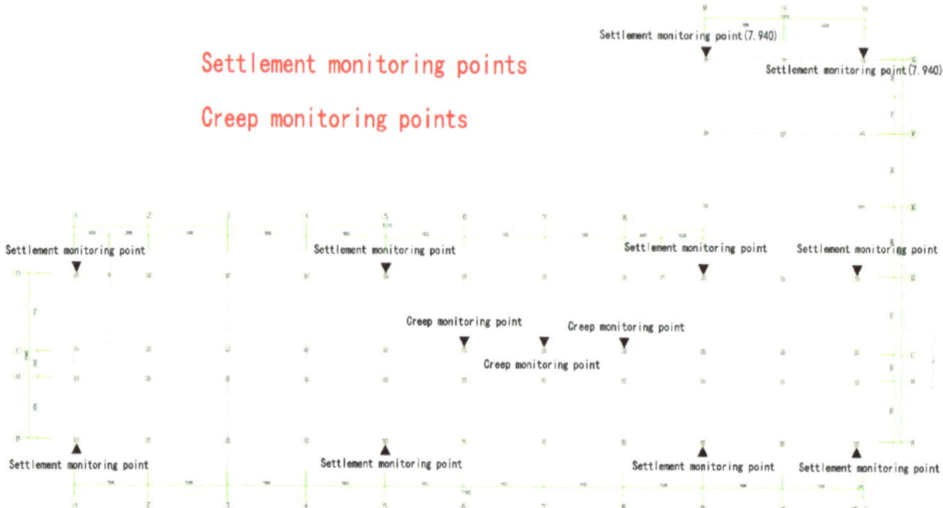

Fig. 11.8   Location map of settlement and creep monitoring points

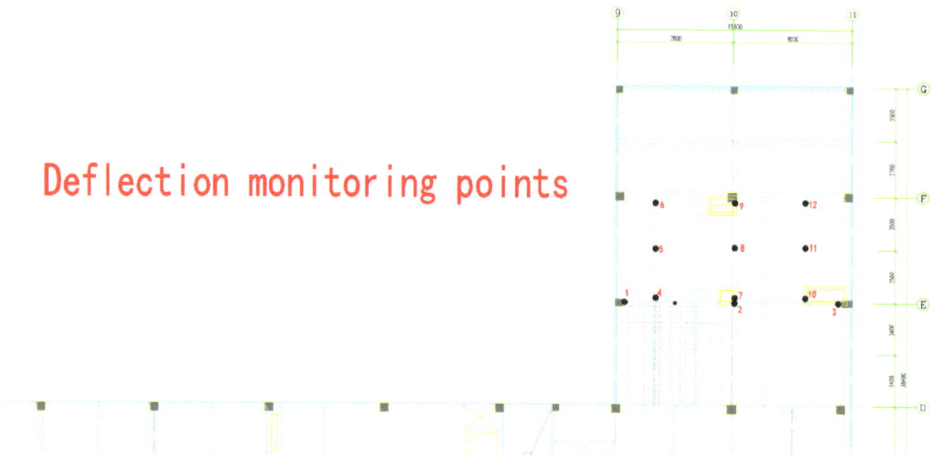

Fig. 11.9   Location map of deflection monitoring points

11.4.4 Shrinkage Monitoring of Recycled Concrete

The layout of shrinkage monitoring points of recycled concrete at the top of the third floor (structural elevation 11.740 m) is shown in Fig. 11.11, with a total of 2 shrinkage monitoring points, 4 embedded strain gauges for each shrinkage monitoring point, and thus, a total of 8 embedded strain gauges. During the monitoring period, the shrinkage of the recycled concrete gradually stabilized in the later stage.

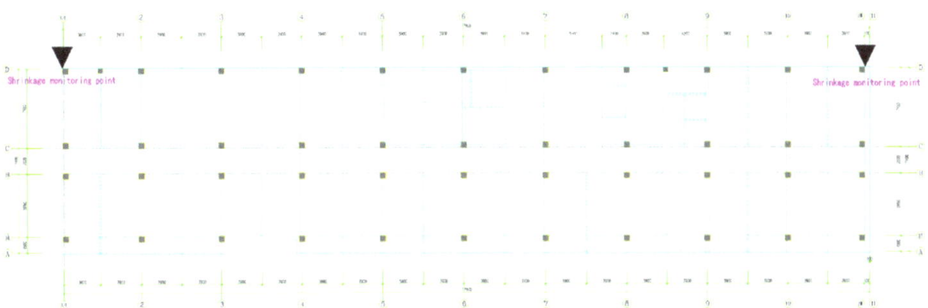

Fig. 11.10 Location map of shrinkage monitoring points of recycled concrete on the fourth floor

11.4.5 Crack and Deflection Monitoring for Slab

Four recycled concrete slabs were selected at the top of the fourth floor and the roof floor for crack monitoring. Two crack monitoring points were selected for each detection board where the cracks are obvious, a total of 8 crack monitoring points and a total of 8 crack metres were arranged. Three deflection monitoring points were selected for each detection board, with a total of 12 deflection monitoring points. The layout of monitoring points is shown in Fig. 11.12 and Fig. 11.13. During the monitoring period, the measured deflection value of the recycled concrete slab was less than the standard limit of 19.5 mm. The variation amplitude of cracks was within 0.06 mm, which was less than the allowable limit in the standards.

11.5 Summary and Prospect

This chapter introduced the comprehensive office building project case in Shandong Shanghe Industrial Base. This was a large-scale structural application of recycled aggregates concrete. Since the mid to late 20th century, there had been application cases of recycled concrete as a building material both domestically and internationally. However, in actual engineering building construction, recycled concrete was often used as auxiliary structural members, and its structural form was relatively simple. This chapter took a project case of a large-scale structure of recycled concrete, which provided a reference for the application of recycled concrete as the main component of the structure. The safety of large-scale structure of recycled concrete was verified through a year and a half of automatic monitoring. In the future, recycled concrete can be more widely used in large-scale structures, and its application prospects are very broad.

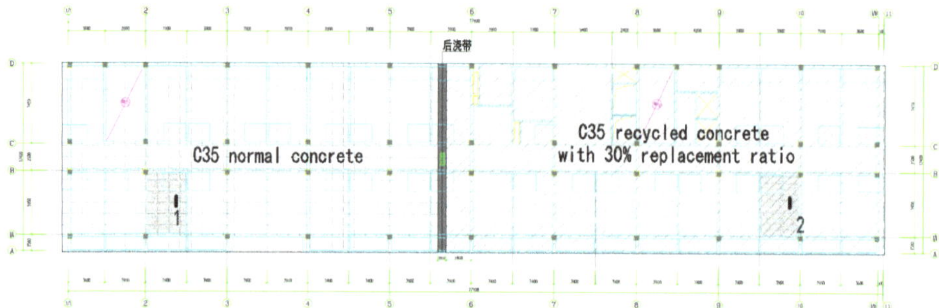

Fig. 11.11 Location map of crack monitoring points of recycled concrete slabs

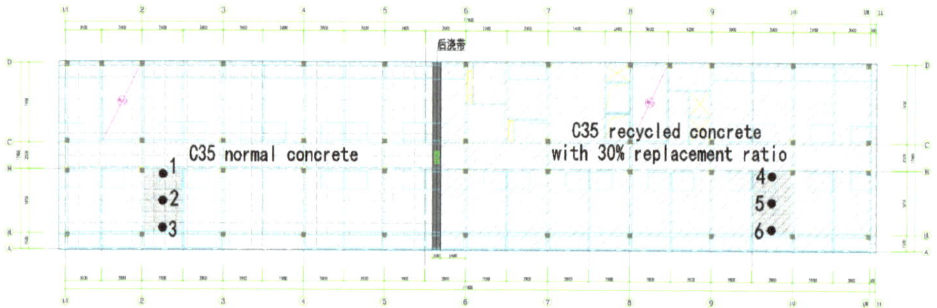

Fig. 11.12 Location map of deflection monitoring points of recycled concrete slabs

11.6 Acknowledgements

The authors acknowledge the part played by colleagues, project co-collaborators and engineers in stimulating their thinking on this recycled concrete project.

11.7 References

[1]   Monteiro P, Miller S A, Horvath A. Towards sustainable concrete. Nature Materials, 2017, 16(7):698. https://doi.org/10.1038/nmat4930

[2]   Velay-Lizancos M, Martinez-Lage I, Azenha M, et al. Concrete with fine and coarse recycled aggregates: E-modulus evolution, compressive strength and non-destructive testing at early ages. Construction and Building Materials, 2018, 193(DEC.30):323–331. https://doi.org/10.1016/j.conbuildmat.2018.10.209

[3]   Jianzhuang Xiao, Innovative research and development of recycled aggregate concrete[M],2020:553

[4]   GBT 25177-2010, Recycled coarse aggregate for concrete[S], 2010.

Chapter 12

Electric Arc Furnace Slag Aggregates in Concrete

Flora Faleschini[1], Mariano Angelo Zanini[1], Vanesa Ortega-Lopez[2]

1 University of Padova, via Marzolo 9, 35121 Padova, Italy
2 University of Burgos, c/Villadiego s/n, 09001, Burgos, Spain

12.1 Introduction

The need for a sustainable use of natural resources has clearly emerged in all of the most important fields of the market, including the construction sector. This is reflected by the implementation of a set of priorities, known as the Sustainable Development Goals (SDGs), which are positively accepted by communities from countries worldwide [1]. Further, in Europe, governments adopted the New Green Deal, a strategy that allows the European Union to transform into a competitive, resource efficient, and independent economy. Thus, carbon neutrality has become for many countries a golden standard to be achieved in a relatively short time window, *i.e.,* in some decades. Different strategies can be used to target such a goal, but generally almost all of them follow the so-called waste hierarchy principle, first included in the Waste Framework Directive 2008/98/EC [2] which defines the priority order of operations to be followed in the management of waste streams. From the most to the least preferable, the following operations can be carried out: prevention, preparing for reuse, recycling, other recovery (including energy recovery), and disposal. Applying these operations seems to be the solution that could allow the built environment to contribute to the sustainable development, making the construction processes compliant with circular economy principles.

However, large amounts of industrial waste are still produced worldwide, which are often simply landfilled. Such poor waste management strategy leads to high environmental risks and financial impacts. Among the different types of industrial waste, the residues produced during both iron and steelmaking, for example, are often dumped. Depending on each country's regulations, and in some cases depending on authorization procedures released by local authorities, slags produced during iron production and steelmaking can be classified as by-products or as waste. In this last case, slags should reach the so-called End-of-Waste status to be introduced into the market and to be used in cement-based materials.

Despite of well-established literature on the reuse of electric arc furnace slag (EAFS) in concrete, the complex normative state, the absence of specific policies aimed at increasing the recyclability of these materials, together with low confidence in waste reuse in concrete have limited, until now, a wide exploitation of slags in the construction market. This chapter seeks to identify which are the main features of EAFS and EAFS concretes, the latter in terms of main mechanical and elastic properties, durability, and even considering associated environmental impacts. The variability of EAFS concrete mechanical properties is analysed too: here, the coefficients to modify the probabilistic distributions of compressive strength and density of reference concrete are proposed. Some high-value applications, such as in self-compacting mixes, fibre-reinforced concrete, and radiation shielding are described. Lastly, a case study on the casting of reinforced concrete (RC) structural elements is presented: for this application, the realised mix designs and the obtained mechanical properties are shown.

12.2 Steelmaking Industry and Slags Formation

The iron and steel industry is among the most fundamental sectors for the global economy. China is the greatest steel producer worldwide, with more than 1000 Mt/y, representing more than half of the global produced steel, followed by the European Union (EU27), with more than 160 Mt/y, representing about 9 % [3]. Two main processes can be applied to produce the steel: in one case, the Linz-Donowitz process (known also as Basic Oxygen) is used, in the other the Electric Arc Furnace process. They differ from the raw materials and the furnace type: in the Basic Oxygen Furnace, pig iron is converted into steel starting from iron-ore, melted in the Blast Furnace; on the other hand, ferrous scrap is the main input material in the Electric Arc Furnace. The latter process is generally recognised as being more sustainable than the first one.

Currently, almost all the European steel producers are faced with the development of decarbonisation strategies and running pilot plants to evaluate the feasibility of alternative production processes. Some of the most interesting strategies are: the implementation of blast furnace/blast oxygen furnace efficiency programmes to improve efficiency and/ or reduce the production losses; the use of biomass as an alternative reductant or fuel; carbon capture and usage to create new products for chemical industry; the increase of share of scrap-based EAFs to maximise secondary flows and recycling by melting more scrap or direct reduced iron (DRI) in EAFs; the optimization of usage of DRI and the use of green hydrogen-based DRI and scrap in combination with an EAF to produce high-quality products in an EAF process [4].

Different types of slags are generated depending on the process. When iron-ore is used to generate pig iron, adding into the Blast Furnace coking coal and fluxes (generally limestone), the combustion process leads to the transformation of iron oxide into metallic iron. From this process, other than pig iron with a Fe content of approximately 90 % and a carbon content of 3.5–4 %, Blast Furnace slag (BFS) is generated. For each ton of pig iron, about 290 kg of BFS is formed [5]. Its composition depends on the iron-ore and fluxes, but in average, it is mainly made by CaO (30–50 %), SiO_2 (28–38 %), Al_2O_3 (8–18 %), and

MgO (1–18%). Based on the cooling method, BFS may display good hydraulic activity, showing mostly an amorphous structure that makes it a suitable supplementary cementing material when properly graded [6].

Once pig iron is obtained, it has to be converted into the steel inside a Basic Oxygen Furnace, through a process that removes carbon, phosphorus, and sulphur to reach its final composition. During this stage, another slag type is generated, the Basic Oxygen Furnace slag (BOFS), which is formed thanks to chemical reactions between liquid oxides of carbon, silicon, manganese, phosphorus, and iron with lime or dolomitic lime (fluxes). For each ton of steel, about 110 kg of BOFS are generated [5]. The composition and structure of BOFS completely differ from that of BFS: CaO ranges between 45 and 54%, Fe_xO_y between 14 and 22%, SiO_2 between 11 and 18%, Al_2O_3 between 1 and 5%, MgO between 1 and 6%, and other constituents between 1 and 5% [7]. Regardless of the cooling method, BOFS hardly vitrifies, and conversely it displays a complex crystalline structure.

When steel is produced in the Electric Arc Furnace, on the other hand, the primary metallurgical process relates to the melting of recycled scraps with fluxes (principally lime and fluorspar) and to the addition of oxygen for steel purification. In this way, a first by-product is generated, the so-called Electric Arc Furnace slag (EAFS), in a ratio of 120–180 kg/t of steel produced. EAFS has again a different composition from the above slags, which differs due to the flux type, the composition of the scraps and the impurities inside the furnace (e.g., coming from the refractories). In average, the composition of EAFS can be summarised as: CaO (25–35%), Fe_xO_y (17–50%), SiO_2 (10–20%), Al_2O_3 3–10%, MgO 2–9%, and MnO less than 6% [8]. Its morphology and the main physical properties depend on the quenching method, which influences particularly its porosity and hardness.

Both in cases of steelmaking in Basic Oxygen and in Electric Arc Furnaces, a secondary metallurgical process is required that aims to refine the molten steel in a ladle, which further undergoes a continuous casting process to form semi-finished products such as bloom, billet, and slabs. Such secondary metallurgical slags include the Ladle Furnace Slag (LFS), which is a basic slag that contains high amounts of lime and magnesia, followed by silica and alumina. It is produced at a ratio of about 70 kg/t of produced steel [9]. Among its constituents, it is worth to mention that free CaO and free MgO can be present in a range of 0–20% and 0–10% [10], respectively, which make this material poorly volumetrically stable. The mineralogy of LFS is quite complex, and different varieties of dicalcium silicate, calcium aluminates, silico-aluminates, and calcium-magnesium compounds [10] can be recognised through the X-Ray diffraction technique.

Following the principles of circular economy, each of these slags should find an alternative to being landfilled. LFS can be used for the stabilization of clayey soils, road supporting bases, and, finally, fillers in bituminous mixes, whereas some other applications are still under research [11]. Contrary to BFS, that typically exhibits pozzolanic activity when finely ground and thus can be used as a supplementary cementing material or to realize blended cements, BOFS and EAFS have been mostly used as aggregates, owing to their relatively good mechanical performances but virtual absence of amorphous phases. Concrete is made by about 60–70% in volume of aggregates, and concrete is the most used construction material to create structures and infrastructures. In fact, the construction sector is one of the largest consumers of natural resources, and about 40 billion tons of natural aggregates were produced in 2016, with an estimated annual growth rate of 5 % [12]. Some countries, such as China, are characterised by an even larger annual growth rate, which is only counterbalanced by some market contractions in Europe and North America [13]. Even if construction and demolition waste are seen as the primary source to produce Recycled Aggregates [14], EAFS can be a valid alternative, particularly when high performances in terms of strength and durability are required of the green concrete.

12.3 Electric Arc Furnace Slag Concrete

When designing a structure with a non-conventional material, it is essential to understand which are the key properties that govern its mechanical and elastic behaviour. To this aim, physical properties, morphology, chemical composition, and mineralogy of EAFS are presented. Further, the risk against environmental contamination should be prevented, thus it is fundamental to identify if the leachate contains ions or heavy metals at high concentration. Then, the main properties of EAFS concrete are reviewed: workability, density, compressive and indirect tensile strength, elastic modulus, durability against carbonation, chloride penetration, and aggressive environmental conditions, e.g., thermal cycles and freezing/thawing. The associated environmental impacts are also shown. Lastly, the properties of some innovative applications are reviewed, e.g., self-compacting mixtures, fibre-reinforced concretes, or heavy-weight mixes used for radiations shielding.

12.3.1 Characteristics of Electric Arc Furnace Slag: Physical Properties, Chemical Composition, and Mineralogy

EAFS have a mineral nature, and they appear as a set of hard grains, similar to dark-grey and black crushed stones, having a rough texture and some cavities with some inclusions of metallic iron particles (Fig. 12.1). There are several physical properties to be considered for their characterisation, the most relevant of them summarised in Table 12.1:

Shape: depending on the process used to transform the slag into an aggregate (sieving, crushing, etc.), EAFS display typically high angularity, a rough texture, and a low flakiness index;

Density: EAFS have high density, which ranges between 3200 and 4300 kg/m^3. This value is about 25–30 % higher than the density of natural aggregates, similar to that of barite;

Porosity: EAFS porosity depends significantly on the quenching and cooling methods during slag treatment. Accordingly, this parameter may display a wide variability of results, e.g., from less than 1 % to more than 8 %, and it depends also on the carbonation process occurring at slag surface that modifies the distribution of pore sizes due to the presence of calcium carbonation deposits [15]. It should be recalled that the porosity may significantly affect the volumetric stability and water absorption of the slag, as well as the mechanical strength and the durability of the concrete where it is used;

Fig. 12.1 EAFS aggregate

Water absorption: it is generally higher than in natural aggregates, but significantly lower than in other recycled products, and it ranges between 0.5–4 % [16];

Crushing resistance: EAFS are typically very cohesive, non-friable, with a reduced Los Angeles (LA) abrasion coefficient than natural aggregates. However, also in this case the cooling method influences the LA value: water-cooled EAFS display LA of about 14, air-cooled about 16, while slow cooling e.g. in water pools between 18 and 25 [8];

Electrical conductivity: EAFS have higher electrical conductivity than natural aggregates, because of the high iron content [17]. Such difference should be considered when using rapid chloride ion permeability test methods for EAFS concrete that are based on the evaluation of electric current flow, because they may lead to an incorrect estimation of durability against chlorides;

Thermal conductivity, expansion, diffusivity, and specific heat: for all these properties, EAFS display a reduction compared to natural aggregates, due to some factors linked to density, vacuolar porosity, moisture content, and mineralogic composition [18–19];

Volumetric stability: this is a key property that needs to be addressed for EAFS. Indeed, the potential expansiveness of EAFS is one of the problems which has hindered a large-scale application of this material in constructions, until now [7, 8, 10]. In the past, severe damage occurred in some RC structures with EAFS: slag swelling caused concrete cracking, with consequent serviceability limit state failures, and in some cases, the need for a complete demolishing of the structure. However, in most cases, the causes of the expansion were not linked directly to the EAFS. Slag expansion is due mainly to two processes, the hydroxylation of free CaO and subsequent carbonation, in presence of moisture, and the hydroxylation and carbonation of free MgO. However, further processes may occur and are discussed in [20], where the authors analysed the effects of stabilization protocols to control the volumetric stability of EAFS particles. A limit of 0.5 % dilation is generally assumed as acceptable for concrete applications, and it is mostly satisfied by EAFS, when properly treated and managed.

Table 12.1 Main physical properties of EAFS (range of values)

Property	Unit	European Standard	EAFS
Bulk density	kg/m³	EN 1097-7	3200–4300
Water absorption	%	EN 1097-6	0.5–4.0
Los Angeles coefficient	%	EN 1097-2	12–25
Porosity	%	ISO 15901-1:2016	1.0–8.0

Concerning chemical composition, the main constituents are iron oxides (FeO, Fe_2O_3), lime (CaO), silica (SiO_2), magnesia (MgO), and alumina (Al_2O_3), with minor amounts of chromium, manganese, and phosphorus oxides. Overall, the proportions among EAFS constituents and the presence in trace of other metals depend on the type of fluxes and iron scrap composition. For instance, Cr presence is often linked to the use of high-alloy iron scraps. High amounts of iron oxides promote high density, mechanical strength, and crushing resistance of EAFS. In fact, the specific gravities of CaO, SiO_2, and Al_2O_3 range from 2.65 to 3.50, whereas for ferrite and manganese oxides these values increase up to 5.70. Therefore, the density of EAFS increases when the content of ferrite and manganese oxides is high [21]. Some minor amounts of metallic iron can be detected, generally less than 3 % in volume, that can cause some volumetric expansion phenomena. Lime is also one of the principal constituents, and it is mostly present in chemically bound form; only some small percentages are in the form of free CaO. Lastly, the content of magnesia is generally small, consisting of a major fraction that is combined with other components and a small part that can remain undissolved. This fraction is named periclase, and it can lead to late-expansion phenomena.

To characterise the slag, often the basicity indexes are calculated that represent the ratio of basic oxides to acidic ones. Such indexes are fundamental to evaluate important metal-slag balances, such as the oxidising power of one slag, desulfurization, and dephosphoration balance. The simplest basicity index is the $BI_2 = CaO/SiO_2$, whereas $BI_4 = (CaO + MgO)/(SiO_2 + Al_2O_3)$ is more representative of the whole components of a slag. This latter index, BI_4, is typically used to identify if a slag can display hydraulic reaction when it is above the unity, thanks to the possibility of developing a vitreous phase that shelves the heavy metals and prevent them from leaching [22].

Table 12.2 summarises the most common chemical compositions found in EAFS from Italy and Spain, together with the two basicity indexes. According to these compositions, it is clear that the hydraulic activity of EAFS is very limited.

Table 12.2 Chemical composition of EAFS from Italy and Spain [23]

(wt. %)	Fe$_x$O$_y$	CaO	SiO$_2$	Al$_2$O$_3$	MgO	MnO	SO$_2$	Cr$_2$O$_3$	P$_2$O$_5$	TiO$_2$	BI$_2$	BI$_4$
Italy	33.3	30.3	14.6	10.2	2.97	4.34	-	2.67	-	-	2.07	1.24
Spain	22.3	32.9	20.3	12.2	3.00	5.10	0.42	2.00	0.5	0.8	1.6	1.10

EAFS have typically a crystalline structure, whose complexity depends on the type of cooling method and slag composition. The phases that are most often identified are solid solutions of FeO (wustite) and the RO phase (a solid solution of CaO-FeO-MnO-MgO), gehlenite ($Ca_2Al(AlSi)O_7$), merwinite ($3CaO \cdot MgO \cdot 2SiO_2$), magnetite ($Fe_3O_4$), free lime ($CaO$) and kirstenite ($CaFeSiO_4$). An exhaustive review of the phases identified in different types of EAFS can be found in [24].

12.3.2 Leaching Potential of Electric Arc Furnace Slag

Other than its swelling potential, leaching is considered as one main risk related to the use of EAFS. It is worth recalling that the main risk refers to the application as a granular material, in an unbonded form; however, when EAFS are included in a blended form with cement, the risk for a contamination of water bodies is mitigated by the low hydraulic conductivity of the hardened concrete. However, as confirmed by the chemical compositions shown before, EAFS contain some small amounts of toxic materials, so the quality of the water that comes into contact with the slag has to be controlled. Thus, most of the current regulations in each country prescribe maximum concentration limits (CLs) of some components in the leachate. The values assumed by the CLs vary country by country, as well as the standard reference used for carrying out the leaching test. The main variables refer to the liquid/solid ratio, the contact time between the particles and the solution, the maximum diameter of the particles, the crushing method, and the pH correction [25].

Some works observed that the most critical elements in the leachate of EAFS are chromium and vanadium [26], whereas the leaching of other heavy metals is mostly avoided because they are bound in the crystalline network. Most of the EAFS analysed in literature display little or no risk for the environment, regardless of the leaching test method used and the pH of the contact solution.

12.3.3 Electric Arc Furnace Slag Concrete: Workability, Mechanical Properties

The most well-known and promising use of EAFS is as replacement of fine or coarse aggregates in concrete production. This substitution has some clear effects on the performance of the concrete, with some advantages and disadvantages.

The first impact detectable is the increase of density of EAFS concrete, compared to a reference mix. The increase in density is proportional to the replacement ratio of the natural aggregates. This entails a higher self-weight of the RC elements, i.e., higher transportation costs, and higher dead loads. It also means that higher inertial loads due to seismic actions apply to the structure. However, such higher density does not directly cause a worse performance under earthquake, because also the stiffness of the structure would change thanks to the variation of EAFS concrete elastic properties [27]. Instead, higher density implies a better behaviour for gravity structures, e.g., retaining walls, dams, marine blocks, etc.

In terms of workability, the use of crushed particles decreases this property regardless of the nature of the aggregates. Also, the higher porosity and water absorption of EAFS may result in a complex water control inside the mix, especially when slag is used replacing the natural sand [28]. The use of water-reducing and viscosity admixtures are useful to avoid slump losses. In the experience of the authors, generally the increase of the dosage of the admixtures (referring to water-reducing additives and plasticizers) is in the range of 10–20 % to ensure a sufficient workability. However, in many cases, a modification of the aggregates grading curve is also necessary, because the content in the (0–1) mm fraction is generally lower than in natural aggregates [29].

All the mechanical properties of EAFS concrete are typically improved compared to those of natural mixes. Strength improvements are as high as +30 % for both compressive and tensile strength. This is due to different co-causes: the higher strength of the EAFS particles than the natural aggregates; the shape of EAFS, which is angular, and promotes a good bond with the cementing matrix; a slight hydraulic activity of EAFS, that promotes the reduction of the interphase transition zone (ITZ) thickness, thanks to the formation of new hydration products in time. Hence, some improvements occur in terms of compressive strength, indirect tensile strength, and elastic modulus. However, in some particular cases, when the slag is very porous, some strength reductions were also observed. Strength increase depends on aggregates replacement ratio, and it is higher for concretes where only the coarse fraction of the aggregates was substituted.

The above considerations on mechanical strength were obtained through the compilation of a wide dataset, which includes about 172 experimental observations of EAFS concrete mechanical tests [30]. According to these data, it was possible to identify three homogeneous concrete classes containing EAFS (full or partial aggregate replacement) and a class of reference concrete (REF), which are listed in Table 12.3.

Table 12.3   Properties of three homogeneous EAFS and reference concrete categories [30]

	REF	EAFS1	EAFS2	EAFS3
EAFS aggregates size (mm)	none	4–31	4–31	0–31
w/c ratio (-)	0.35–0.70	0.40–0.50	0.51–0.67	0.35–0.70
Density (kg/m³)	2260–2510	2630–3001	2380–3045	2500–3180
f_c (MPa)	22.0–65.0	31.6–70.3	20.3–55.7	25.1–77.9
f_{ct} (MPa)	2.20–5.21	3.45–5.65	1.80–4.38	3.56–5.89
E_c (GPa)	24.0–42.7	42.2–49.5	23.6–40.5	37.4–48.3

From the collected results, it is possible to calculate the following indexes:

$$CSR = f_{c,EAFS}/f_{c,REF} \tag{12.1}$$

$$TSR = f_{ct,EAFS}/f_{ct,REF} \tag{12.2}$$

$$SWR = \rho_{EAFS}/\rho_{REF} \tag{12.3}$$

$$EMR = E_{c,EAFS}/E_{c,REF} \tag{12.4}$$

$$F = CSR/SWR \tag{12.5}$$

where CSR is the Compressive Strength Ratio, TSR is the Tensile Strength Ratio, SWR is the Self Weight Ratio, EMR is the Elastic Modulus Ratio, and lastly, F is a summarising index. In the above formulations, $f_{c,EAFS}$ and $f_{c,REF}$ are the compressive strength values of the EAFS concrete and its reference (i.e., made with the same mix design, with natural aggregates), $f_{c,EAFS}$ and $f_{c,REF}$ are the tensile strength values of the EAFS concrete and its reference, evaluated with the indirect tensile test (Brazilian test), ρ_{EAFS} and ρ_{REF} are the density of the EAFS concrete and its reference, and lastly, $E_{c,EAFS}$ and $E_{c,REF}$ are the secant elastic modulus values of the EAFS concrete and its reference.

The indexes *CSR* and *TSR* allow to evaluate the compressive and tensile strength gain when a mix of EAFS concrete is compared to a reference with only natural aggregates, respectively. Instead, the index *SWR* indicates its increase in self-weight, and *EMR* indicates the variation in terms of elastic modulus. The index *F* evaluates when the increase in compressive strength is sufficient to counterbalance the self-weight increase due to the addition of EAFS in the mix. When *F* assumes values above the unity, this means that the mix might display enough strength to not be penalised by the higher EAFS density. Table 12.4 shows the average value of the above parameters assumed within each EAFS concrete category: note particularly that, for the class EAFS3, the *F* index is the only one that assumes a value below 1, this meaning that when the fine fraction of the slag is used, concrete density increases at a higher rate than the strength. In all the other cases, strength gains are high, more for the compressive than for the tensile strength; the elastic modulus increases in EAFS mixes, too.

Table 12.4   Average performance indexes for the EAFS concrete

	EAFS1	EAFS2	EAFS3
CSR (-)	1.395	1.404	0.915
TSR (-)	1.280	1.100	1.080
SWR (-)	1.166	1.166	1.040
EMR (-)	1.330	1.100	1.154
F (-)	1.196	1.204	0.879

The higher elastic modulus is often associated with higher drying shrinkage, observed more in concretes with a high aggregates' replacement ratio and with high water absorption. This phenomenon makes the EAFS concrete more prone to cracks formation at early ages, if not properly cured. A solution can be to pre-soak the EAFS, which also enhances the water control in the mix [31]. So far, few works in literature dealt with the characterisation of the Poisson coefficient, however, those who analysed the transverse dilation of EAFS concrete observed slightly lower deformability than in reference mixes [32].

12.3.4 Electric Arc Furnace Slag Concrete: Probabilistic Models

Probabilistic models were developed to characterise the compressive strength f_c and the density ρ_c of the EAFS concrete [30]. These models represent a practical tool to design RC members with EAFS: in fact, it is possible to use the same formulations for the reference concrete, just considering the variations in strength and density, and the higher variability of the results of EAFS concrete. The main principle is to modify, e.g., the resistance model for the reference concrete present in EN 1990 or JCSS 2001 Probabilistic Model Code, applying the specific coefficients obtained from the experimental observations on EAFS concrete collected in literature. In the present case, these coefficients were derived fitting the experimental observations of f_c and ρ_c through normal probability density functions (*pdfs*). This was done separately for each EAFS and reference concrete category. Then, the main parameters that characterise the *pdfs* are obtained, i.e., the average value and the coefficient of variation (C.O.V. = standard deviation / average value). Afterwards, the ratios between the parameters characterising the strength and density *pdfs* of EAFS and reference concrete were calculated (see Table 12.5).

Table 12.5 Coefficients to be applied for the probabilistic models based on the calibrated distributions for each EAFS concrete category

	EAFS1	EAFS2	EAFS3
$\mu^*_{fc\_EAFS}$	1.349	1.340	0.932
$COV^*_{fc\_EAFS}$	1.065	1.910	1.015
$\mu^*_{\rho c\_EAFS}$	1.178	1.194	1.159
$COV^*_{\rho c\_EAFS}$	1.551	3.373	1.825

Let us assume to design a RC member made with EAFS concrete realised with 100 % EAFS replacing coarse natural aggregates, no substitution in the fine fraction, and a w/c ratio equal to 0.45. In this case, the EAFS concrete belongs to the class defined as EAFS1. The reference concrete mix, made with natural aggregates only, corresponds to a strength class C30/37 (according to EN 1992). Its f_c and ρ_c are characterised by two normal *pdfs*. It is now possible to obtain the parameters characterising the distribution of f_c and ρ_c for the EAFS concrete made with the same mix of the C30/37 reference, just applying Eqs. 12.6–12.9, using the coefficients proposed in Table 12.5:

$$\mu_{fc\_EAFS} = \mu_{fc\_REF} \cdot \mu^*_{fc\_EAFS} \tag{12.6}$$

$$COV_{fc\_EAFS} = COV_{fc\_REF} \cdot COV^*_{fc\_EAFS} \tag{12.7}$$

$$\mu_{\rho c\_EAFS} = \mu_{\rho c\_REF} \cdot \mu^*_{\rho c\_EAFS} \tag{12.8}$$

$$COV_{\rho c\_EAFS} = COV_{\rho c\_REF} \cdot COV^*_{\rho c\_EAFS} \tag{12.9}$$

Here, μ_{fc_EAFS} and COV_{fc_EAFS} are the main parameters to characterise the distribution of f_c for the EAFS concrete that we need to design; $\mu_{\rho c\_EAFS}$ and $COV_{\rho c\_EAFS}$ are the main parameters of the distribution of ρ_c for the EAFS concrete that we need to design; μ_{fc_REF} and COV_{fc_REF} are the main parameters of the distribution of f_c for the reference concrete (C30/37); $\mu_{\rho c\_REF}$ and $COV_{\rho c\_REF}$ are the main parameters of the distribution of ρ_c for the reference concrete (C30/37); lastly, $\mu^*_{fc\_EAFS}$ and $COV^*_{fc\_EAFS}$ are the coefficients for modifying the parameters of the compressive strength *pdf*, taken from the EAFS1 concrete, i.e., the most similar EAFS concrete class to the one we need to design; $\mu^*_{\rho c\_EAFS}$ and $COV^*_{\rho c\_EAFS}$ are the coefficients for modifying the parameters of the density *pdf*, taken from the EAFS1 concrete.

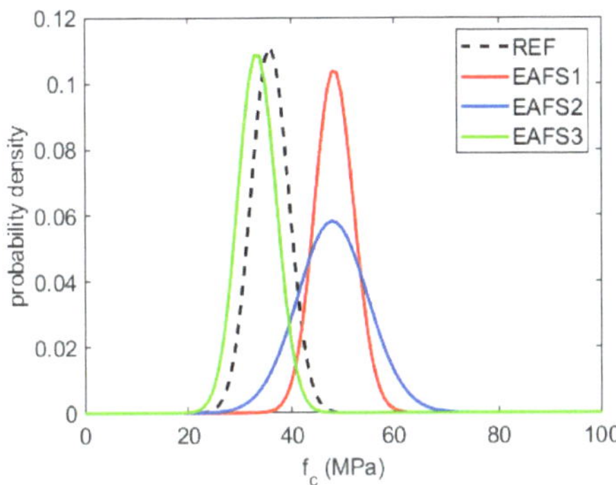

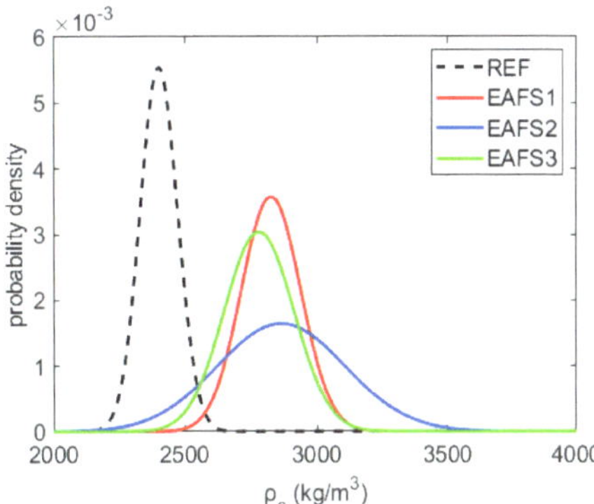

Fig. 12.2 Probability density functions for the compressive strength of EAFS concrete cylinders starting from C30/37 reference mix

Fig. 12.3 Probability density functions for the density of EAFS concrete cylinders starting from C30/37 reference mix

An example of the above procedure is shown in Figs. 12.2 and 12.3, showing the compressive strength of concrete cylinders and density *pdfs* of three EAFS concrete categories, respectively, starting from a C30/37 reference (dashed line). The normal *pdf* for the f_c of the reference concrete is characterised by a mean value of 35.9 MPa and a C.O.V. of 0.1; the normal *pdf* for the ρ_c of the reference concrete is characterised by a mean value of 2400 kg/m³ and a C.O.V. of 0.03. Note that the compressive strength *pdfs* are shifted to the right from the *pdf* of the reference mix, with the exception of the EAFS3 mixture. Furthermore, the standard deviation of the EAF curves is higher than for the reference concrete, this representing the higher variability of the observed strengths.

12.3.5 EAFS Concrete: Durability Properties

When dealing with innovative materials, it is fundamental to assess their long-term behaviour to ensure adequate performances over time. For this purpose, some research works evaluated various durability-related properties of EAFS concrete. Some contrasting results were observed due to the heterogeneity of the analysed EAFS. Slags with different porosity may display completely different durability performances. Indeed, there are mainly two types of EAFS identified in the literature, one that is constituted by higher iron oxides content, low porosity, and higher density (up to 3800–3900 kg/m³), and one with fewer iron oxides, higher porosity, and relatively lower density (up to 3400–3500 kg/m³). If this difference has a limited impact on EAFS concrete strength development, this is not true for concrete durability.

For instance, similar or less carbonation depths to natural concretes were recorded using the same accelerated carbonation tests: when the heavy-weight EAFS were used, experimental evidence demonstrates that steel slag concrete shows an excellent response against carbonation [33–34]. Conversely, when the lightweight EAFS are mixed into the concrete, carbonation depths were found to be almost the same to that of ordinary concrete [35]. Only in one study in literature [36] higher carbonation depths were recorded, but in that case, concretes were cast mixing EAFS with recycled aggregates from construction and demolition waste. Thus, it was not possible to correlate directly the variation of the carbonation resistance with one single mix component. This evidence seems to support the existence of an inverse relationship between the content of EAFS in concrete mixes and the observed carbonation depths.

Other properties affected by the types of EAFS used are: the water penetration under pressure, resistance against chlorides penetration and corrosion, and the behaviour under wetting/drying and freezing/thawing cycles. A review of the main results obtained in literature are shown in Table 12.6.

EAFS concretes are often used in harbour walls, seawalls, breakwaters, and other coastal structures, thanks to their high density. Thus, it is worth to investigate EAFS concrete behaviour in a tidal environment and in the presence of chloride ions. Several tests were carried out to evaluate corrosion resistance of steel reinforcements immersed in EAFS concrete under natural exposure in the marine environment (e.g., specimens were placed on metal grids, suspended a few centimetres from the surface of the sea, leaving the specimens exposed to tidal changes for up to 10 months in [37]). Results confirmed that chloride penetration was very low, and the direct exposure to seawater did not negatively affect the embedded steel bars, even in case of a very small concrete cover (only 10 mm).

Sulphate attacks are also common in a marine environment, leading to concrete cracking and corrosion of steel reinforcement bars. Thus, this durability-related property was analysed for EAFS concrete, too. Results presented in literature agree in showing a high dependency on EAFS porosity and density: if a high density EAFS is used, sulphate resistance improves; conversely, it decreases when low density EAFS is added to the mix [38]. The behaviour can worsen in case if combined with steel dispersed fibres; oppositely, it improves when ground granulated BFS is added to the binder [39].

Table 12.6 Durability performance of EAFS concrete compared to the reference mixes (↑: improved durability; ↓ decreased durability; ↔ no relevant variations).

Property	EAFS ρ (kg/m³)	Result	Reference
Carbonation	> 3500	↑	[33, 34]
	< 3500	↔	[35]
Chloride penetration	> 3500	↑	[37, 40]
	< 3500	↓	[41]
Sulphate attack	> 3500	↑	[38]
	< 3500	↓	[39]
Water penetration under pressure	> 3500	↑	[19]
	< 3500	↔ ↓	[42, 43]
Wetting/drying cycles	> 3500	↔ ↑	[28]
	< 3500	↓	[43]
Freezing/thawing cycles	> 3500	↑	[28]
	< 3500	↔	[43]
High-temperature behaviour	> 3500	↑	[44]
Alkali-silica reaction	> 3500	↓	[45]
	< 3500	↑	[45]

In conclusion, it can be stated that when EAFS are characterised by high micro-porosity, their use in concrete facilitates the penetration of moisture, which drives most of the physical and chemical phenomena responsible for concrete deterioration. As a consequence, concrete capillarity increases, and thus the penetration of water solutions increases, making the material more vulnerable to durability-related issues. However, the latent hydraulic activity of EAFS leads to the formation of new hydration products in the long-term [45], thus limiting a severe deterioration of EAFS concrete. Conversely, when low-porosity aggregates are used, generally EAFS concrete durability is enhanced.

12.3.6 EAFS Concrete: Environmental Impacts

From a regulatory point of view, EAFS can be classified either as a steelmaking by-product or as waste. In both cases, for a sustainable reuse in concrete, EAFS should comply with some general prescriptions in terms of technical and environmental performance. On this last aspect, the requirements follow the End-of-Waste criteria stated in the Waste Framework Directive, i.e.:

– the substance or object is commonly used for specific purposes;
– there is an existing market or demand for the substance or object;
– the use is lawful (substance or object fulfils the technical requirements for the specific purposes and meets the existing legislation and standards applicable to products);
– the use will not lead to overall adverse environmental or human health impacts.

If the first two requirements can be easily fulfilled and the third is linked to the technical performances, the environmental impacts should instead be quantified to evaluate if there is a convenience to recycle the waste or not. To this scope, and more generally speaking, to evaluate the environmental impacts of a product or a process, a Life Cycle Assessment (LCA) study [46] should be carried out.

Three main LCA studies were carried out on EAFS concrete, which are worth to be cited: one deals with the Italian context [47], one was developed in Slovenia [48], and the latest in Brazil [49]. All these works show that CO_2 emissions are the same or less in EAFS concrete than in reference mixes with natural aggregates only, even when transportation emissions are considered up to a certain distance (about 100 km). The carbon footprint is not the sole indicator which is improved by the EAFS replacement of natural aggregates: benefits were observed practically in all the analysed performance indicators. Table 12.7 lists the main results obtained from these works, summarising also the main hypotheses made by each author and the source of the information.

The first study [47] investigated both the impacts related to the production of 1 t of EAFS and 1 m³ of ready mix EAFS concrete, thus focusing both on the productive chain of the aggregates alone and of the concrete, considering a cradle-to-gate approach. Substantial reductions in terms of impacts were obtained in the first case, due to the possibility of neglecting the impacts carried by the steel industry for producing the slag ("zero-burden" assumption). When the functional unit (FU) is the whole concrete, instead, similar results were obtained for ordinary and EAFS mixes, because the impacts due to cement use constitute more than 90 % of those of the whole concrete. The second study [48] obtains almost the same results than [47], whereas the last one [49] was able to catch great environmental emission reductions thanks to the choice of considering not only mid-point but also end-point impact categories. In this way, the authors were able to include in the analysis the recycling process initially implemented to recover the metal scrap, in order to save final disposal costs.

Table 12.7 Environmental emissions from LCA studies [47–49]

Functional unit	Analysis	Inventory	Results	Ref.
1 t of EAFS	Cradle-to-gate CML 2001	Local data collected at a real plant placed in Italy + existing literature	Reductions up to 60 % for CO_2 and more than 30 % for all other categories	[47]
1 m³ of ready mix EAFS concrete with replacement ratio 100 % in the coarse fraction	Cradle-to-gate CML 2001	Local data collected at a real plant placed in Italy + existing literature	Limited differences (cement impacts are much higher than all other constituents)	[47]
1 m³ of ready mix EAFS concrete with replacement ratio 10 %	Cradle-to-gate CML 2001 Energy Consumption	GaBi datasets	No effects on CO_2 emission, great reduction for the Eutrophication category	[48]
1 m² Paving blocks with replacement ratio 50 %	Cradle-to-gate IMPACT 2002	Ecoinvent v.3.01 (2014) and US LCI (2003) databases + local data on energy, paving blocks industry and quarrying	Total impacts reduced by 50 %	[49]

It is worth recalling that LCA results cannot be directly extended to different scenarios and geographical contexts, since the inventory of the data is often site-specific. Furthermore, some peculiar environmental indicators can be considered too, such as the abiotic depletion index (that helps to reply to the question "How can the decreasing availability of a given resource be determined?"), and the land transformation due to quarrying activity (for natural aggregates) or slag treatment plants.

12.3.7 Electric Arc Furnace Slag Concrete: Special Applications

In this section, three special applications are shown: EAFS self-compacting concrete, EAFS fibre reinforced concrete, and EAFS radiation attenuation concrete.

Self-compacting concrete (SCC) is a kind of concrete mix characterised by high filling and consistent flowability that needs no vibration to be placed, with clear advantages in terms of costs and energy savings. To realize these mixtures, typically plasticizer admixtures, ultrafine aggregates such as limestone filler and a low dosage of coarse aggregates are used. SCCE Project Group recommendations [50] are generally adopted when designing SCC mixes, as they provide guidance concerning slump flow, viscosity, passing ability, segregation resistance, etc. The use of high-weight, sharp aggregates such as the EAFS is particularly challenging for this application due to different reasons: first, heavy-weight particles require a "bearing paste", able to transport EAFS with high density, without displaying segregation and bleeding risks. Then, roundish particles allow the mix to flow better; conversely, sharp edges create links inside the mix, increasing the cohesiveness of the fresh concrete. In practice, this means an increase in the rheological parameters that describe the flow curve of a SCC, e.g., the Bingham parameters yield strength and plastic viscosity. Despite the above limitations, successful results were obtained in different works, where authors demonstrate the technical feasibility of realizing EAFS-SCC mixes through the adoption of a proper mix design and chemical admixtures [29]. Particularly, the increase of the fines content in the mix (that may be as high as over 30 %) was found to be very effective to obtain a visible enhancement of the concrete flowability. Only in this way the lubrication provided by the limestone fines allows the yield stress to decrease and at the same time to maintain good values of viscosity, which slightly increases due to the enlarged amount of bearing paste. It should be recalled that these dosage modifications to the concrete mixes are not always effective to obtain the desirable fresh behaviour: in fact, there are evidences from literature of the low compatibility of some cement types with EAFS, particularly with cement CEM IV/B [29].

In fibre reinforced concrete (FRC), randomly distributed small fibres are added to a mix to overcome the high brittleness, poor tensile strength and resistance to impact strength, fatigue, and low ductility of ordinary concrete. A range of studies reports, also in this case, about the fresh behaviour of this material: in fact, fibre additions cause problems with mixing and workability due to their tendency to clump together, forming balls, or even to distribute in non-uniform ways, with the final result of altering the properties of the composite. Such problem is not trivial when high-weight, sharp aggregates such as the EAFS replace natural aggregates. However, recent studies demonstrated successful results to produce EAFS-FRC mixes: it was shown that the major effect due to the fibre addition was a relevant improvement of the post-peak behaviour of the concrete, both in tension and in compression [51].

Heavy-weight concrete is typically used in medical and nuclear energy facilities, to prevent the leakage from radioactive containing structures. The presence of both high- and low-density nuclei in this kind of concrete make it particularly attractive to attenuate different radiation types, e.g., gamma ray and neutron radiation. Magnetite, limonite, hematite, and barite are some natural aggregates that can be used for this application. However, some of them display poor mechanical properties (e.g., barite has a relatively low elastic modulus and it is very brittle), their cost is higher than that of conventional aggregates, and their availability is quite limited. Thus, EAFS can represent a sustainable alternative to such aggregates: in experiments it was demonstrated that the attenuation capacity of the EAFS concrete is comparable to that of barite-concrete, and significantly higher than that of dolomite-concrete. The half-value layer, i.e., the depth of the medium which reduces the intensity of radiation to half, is from 10 to 50 % lower than that of alternative solutions with reference materials [52–53].

12.4 Structural Applications

In this section, a case study where EAFS was used in structural concrete is presented. It is worth noting that there are by now some real projects where this material was used, such as the Kubik building foundation and basement walls (owned by Tecnalia, Spain), or the concrete blocks placed at the Punta Lucero and Punta Sollana docks, both at Bilbao Harbour (Spain). In Italy, EAFS concretes are used in different applications, mainly for mass structures, e.g., walls and foundations, but some precast plants have manufactured different structural elements too, e.g., concrete blocks, pipes, etc. Fig. 12.4 shows an example of an EAFS concrete manhole made by an Italian producer.

The proposed case study includes two concrete mixtures that were used to build RC external beam-column joints (Fig. 12.5). Table 12.8 lists the mix design of the two EAFS concretes and their reference. The real-scale specimens were realized with a beam of 2.5 m length (rectangular section of 30 cm width and 50 cm height), intersecting a column of about 4 m height (square cross-section with 30 cm side length). All the beam-column joints have the same geometry and reinforcement details. To fabricate each sample, about 800 litres of pumpable EAFS concrete were mixed in a single batch. Once cast, the specimens were further tested under a design axial load and a cyclic lateral load, both applied at the column top. The loading scheme aimed to simulate both the gravity and seismic actions in a RC frame. For this application, the shear strength of the RC panel joints was evaluated for each specimen; additionally, load-displacement curves, ductility, dissipated energy, stiffness deterioration, and the local behaviour of the joint, i.e., the strain development during loading in the reinforcement and in the concrete, were calculated. Results indicate an improvement of all the above properties; the observed improvements of the specimens' global behaviour can be attributed to the substantial enhancement of the mechanical properties of the EAFS concrete [54–55].

Table 12.8 Mix design (for 1 m³) and properties of EAFS concrete: RC beam-column joints

	EAFS concrete 1	EAFS concrete 2	REF concrete
Cement II/A-LL 42.5R (kg/m³)	400	320	400
Water (kg/m³)	180	160	180
Water / Cement ratio	0.45	0.50	0.45
EAFS 4–8 mm (kg/m³)	384	411	-
EAFS 8–12 mm (kg/m³)	127	136	-
EAFS 8–16 mm (kg/m³)	631	675	-
Natural 4–16 mm (kg/m³)	-	-	992
Natural 0–4 mm (kg/m³)	992	1071	812
Super plasticizer (kg/m³)	4.80	3.84	4.00
f_c (MPa)	60.60	54.90	48.70
$f_{c,t}$ (MPa)	3.59	4.25	3.45
E_c (GPa)	42.55	42.19	32.67

Fig. 12.4   Concrete manholes made with EAFS concrete

Fig. 12.5   Operations to cast the RC beam column joints with EAFS concrete

12.5 Conclusions

There are multiple strategies to improve the sustainability of the construction industry, and one of these is to recycle industrial waste that cannot be used alternatively except for landfilling. At the same time, however, it is necessary to create a solid knowledge on recycled materials as construction products. Designers and end-users should be helped to take decisions on which materials can be used as aggregates for concrete, which benefits can be achieved, and which rather represent the limits of their application. Furthermore, the effect of substituting a traditional material (such as a natural stone, dolomite, or limestone) with a new one (such as the EAFS), needs to be clear, as its introduction in the mix design implies variations in the target mechanical properties, in the durability, and in the environmental performances. As seen from the derivation of the probabilistic models for the compressive strength of EAFS concrete, it should also be recalled, when designing a reinforced concrete member, that the variability of material properties is generally larger than that of the ordinary concrete. In addition, there are still some challenges that need to be solved to allow a widespread use of this material in concrete, and they relate to the application of reliable models for the prediction of concrete properties, especially in the long-term. Durability-related properties are still too little investigated, thus making the estimation of the service life-time of structures made with EAFS concrete difficult. This chapter aimed to provide some useful references and indications for practitioners wanting to further improve the design of reinforced concrete structures made with EAFS.

12.6 Acknowledgements

The authors acknowledge the part played by colleagues, project co-collaborators, and friends in stimulating their thinking on sustainable development. A particular thanks goes to the colleagues of the groups "Sustainable Construction Research Group" from the University of Burgos, "Minimización de los efectos del cambio climático en el edificio y su entorno: desde la eficiencia en el uso de la energía a la generación mediante fuentes renovables" from the University of Basque Country, and "Design of Structures" from the University of Padova.

12.7 References

[1]   United Nations: Sustainable development goals, online, 2017. http://www.un.org/sustainabledevelopment/sustainable-development-goals.

[2]   European Parliament: Council Directive 2008/98/EC of the European Parliament and of the Council of 19 November 2008 on waste and repealing certain Directives.

[3]   World Steel Association: Steel Statistical Yearbook, online, 2021. https://worldsteel.org/steel-by-topic/statistics/steel-statistical-yearbook/

[4]   Bhaskar, A., Assadi, M., & Nikpey Somehsaraei, H. (2020). Decarbonization of the iron and steel industry with direct reduction of iron ore with green hydrogen. Energies, 13(3), 758. https://doi.org/10.3390/en13030758

[5]   Kuramochi, T.: Assessment of midterm CO_2 emissions reduction potential in the iron and steel industry: a case of Japan. Journal of Cleaner Production, 132, 2016, pp. 81–97. https://doi.org/10.1016/j.jclepro.2015.02.055

[6]   Özbay, E., Erdemir, M., & Durmuş, H. İ.: Utilization and efficiency of ground granulated blast furnace slag on concrete properties-A review. Construction and Building Materials, 105, 2016, pp. 423–434. https://doi.org/10.1016/j.conbuildmat.2015.12.153

[7]   Brand, A. S., & Fanijo, E. O.: A review of the influence of steel furnace slag type on the properties of cementitious composites. Applied Sciences, 10(22), 2020, paper no. 8210. https://doi.org/10.3390/app10228210

[8]   Skaf, M., Manso, J. M., Aragón, Á., Fuente-Alonso, J. A., & Ortega-López, V.: EAF slag in asphalt mixes: A brief review of its possible re-use. Resources, Conservation and Recycling, 120, 2017, pp. 176–185. https://doi.org/10.1016/j.resconrec.2016.12.009

[9]   Etxebarria, X. B.: Altos hornos de Vizcaya: análisis crítico del cierre y testimonios vitales. Universidad del País Vasco, Servicio Editorial Euskal Herriko Unibertsitatea, Argitalpen Zerbitzua, 2013.

[10]   Santamaría, A., Ortega-López, V., Skaf, M., Faleschini, F., Orbe, A., & San-José, J. T.: Ladle furnace slags for construction and civil works: A promising reality. In Waste and Byproducts in Cement-Based Materials, Woodhead Publishing, 2021, pp. 659–679. https://doi.org/10.1016/b978-0-12-820549-5.00023-1

[11]   Herrero, T., Vegas, I. J., Santamaría, A., San-José, J. T., & Skaf, M.: Effect of high-alumina ladle furnace slag as cement substitution in masonry mortars. Construction and Building Materials, 123, 2016, pp. 404–413. https://doi.org/10.1016/j.conbuildmat.2016.07.014

[12]   Xing, W., Tam, V. W., Le, K. N., Hao, J. L., & Wang, J.: Life cycle assessment of recycled aggregate concrete on its environmental impacts: A critical review. Construction and Building Materials, 317, 2022, article no. 125950. https://doi.org/10.1016/j.conbuildmat.2021.125950

[13]   Wang, B., Yan, L., Fu, Q., & Kasal, B.: A comprehensive review on recycled aggregate and recycled aggregate concrete. Resources, Conservation and Recycling, 171, 2021, article no. 105565. https://doi.org/10.1016/j.resconrec.2021.105565

[14]   De Brito, J., & Saikia, N.: Recycled aggregate in concrete: use of industrial, construction and demolition waste. Springer Science & Business Media, 2012. https://doi.org/10.1007/978-1-4471-4540-0

[15]   Brand, A. S., & Roesler, J. R.: Interfacial transition zone of cement composites with steel furnace slag aggregates. Cement and Concrete Composites, 86, 2018, pp. 117–129. https://doi.org/10.1016/j.cemconcomp.2017.11.012

[16]   Teo, P. T., Zakaria, S. K., Salleh, S. Z., Taib, M. A. A., Mohd Sharif, N., Abu Seman, A., & Mamat, S.: Assessment of electric arc furnace (EAF) steel slag waste's recycling options into value added green products: A review. Metals, 10(10), 2020, article no. 1347. https://doi.org/10.3390/met10101347

[17]   Abu-Eishah, S. I., El-Dieb, A. S., & Bedir, M. S.: Performance of concrete mixtures made with electric arc furnace (EAF) steel slag aggregate produced in the Arabian Gulf region. Construction and Building Materials, 34, 2012, pp. 249–256. https://doi.org/10.1016/j.conbuildmat.2012.02.012

[18]   Beaucour, A. L., Pliya, P., Faleschini, F., Njinwoua, R., Pellegrino, C., & Noumowé, A.: Influence of elevated temperature on properties of radiation shielding concrete with electric arc furnace slag as coarse aggregate. Construction and Building Materials, 256, 2020, article no. 119385. https://doi.org/10.1016/j.conbuildmat.2020.119385

[19]   Roy, S., Miura, T., Nakamura, H., & Yamamoto, Y.: Investigation on material stability of spherical shaped EAF slag fine aggregate concrete for pavement during thermal change. Construction and Building Materials, 215, 2019, pp. 862–874. https://doi.org/10.1016/j.conbuildmat.2019.04.228

[20]   Santamaria, A., Faleschini, F., Giacomello, G., Brunelli, K., San José, J. T., Pellegrino, C., & Pasetto, M.: Dimensional stability of electric arc furnace slag in civil engineering applications. Journal of Cleaner Production, 205, 2018, pp. 599–609. https://doi.org/10.1016/j.jclepro.2018.09.122

[21] Dong, Q., Wang, G., Chen, X., Tan, J., & Gu, X.: Recycling of steel slag aggregate in portland cement concrete: An overview. Journal of Cleaner Production, 282, 2021, article no. 124447. https://doi.org/10.1016/j.jclepro.2020.124447

[22] Gobetti, A., Cornacchia, G., & Ramorino, G.: Innovative Reuse of Electric Arc Furnace Slag as Filler for Different Polymer Matrixes. Minerals, 11(8), 2021, article no. 832. https://doi.org/10.3390/min11080832

[23] Santamaria, A., San José, J.T., Gonzalez, J.J., Faleschini, F., & Pellegrino, C.: Analyse comparative et propriétés de deux laitiers d'aciérie de four électrique en vue de leur utilisations dans des bétons hydrauliques. Laitiers sidérurgiques, 109, June 2018, pp. 19–23.

[24] Yildirim, I. Z., & Prezzi, M.: Chemical, mineralogical, and morphological properties of steel slag. Advances in Civil Engineering, 2011. https://doi.org/10.1155/2011/463638

[25] Singh, S. K., Vashistha, P., Chandra, R., & Rai, A. K.: Study on leaching of electric arc furnace (EAF) slag for its sustainable applications as construction material. Process Safety and Environmental Protection, 148, 2021, pp. 1315–1326. https://doi.org/10.1016/j.psep.2021.01.039

[26] Neuhold, S., van Zomeren, A., Dijkstra, J. J., van der Sloot, H. A., Drissen, P., Algermissen, D., & Vollprecht, D.: Investigation of possible leaching control mechanisms for chromium and vanadium in electric arc furnace (EAF) slags using combined experimental and modeling approaches. Minerals, 9(9), 2019, article no. 525. https://doi.org/10.3390/min9090525

[27] Faleschini, F., Zanini, M. A., & Toska, K.: Seismic reliability assessment of code-conforming reinforced concrete buildings made with electric arc furnace slag aggregates. Engineering Structures, 195, 2019, pp. 324–339. https://doi.org/10.1016/j.engstruct.2019.05.083

[28] Pellegrino, C., Cavagnis, P., Faleschini, F., & Brunelli, K.: Properties of concretes with black/oxidizing electric arc furnace slag aggregate. Cement and Concrete Composites, 37, 2013, pp. 232–240. https://doi.org/10.1016/j.cemconcomp.2012.09.001

[29] Santamaría, A., Ortega-López, V., Skaf, M., Chica, J. A., & Manso, J. M.: The study of properties and behavior of self compacting concrete containing Electric Arc Furnace Slag (EAFS) as aggregate. Ain Shams Engineering Journal, 11(1), 2020, pp. 231–243. https://doi.org/10.1016/j.asej.2019.10.001

[30] Zanini, M. A.: Structural reliability of bridges realized with reinforced concretes containing electric arc furnace slag aggregates. Engineering Structures, 188, 2019, pp. 305–319. https://doi.org/10.1016/j.engstruct.2019.02.052

[31] Piemonti, A., Conforti, A., Cominoli, L., Sorlini, S., Luciano, A., & Plizzari, G.: Use of iron and steel slags in concrete: State of the art and future perspectives. Sustainability, 13(2), 2021, article no. 556. https://doi.org/10.3390/su13020556

[32] Santamaría, A., Romera, J. M., Marcos, I., Revilla-Cuesta, V., & Ortega-López, V.: Shear strength assessment of reinforced concrete components containing EAF steel slag aggregates. Journal of Building Engineering, 46, 2022, article no. 103730. https://doi.org/10.1016/j.jobe.2021.103730

[33] Sosa, I., Thomas, C., Polanco, J. A., Setién, J., Sainz-Aja, J. A., & Tamayo, P.: Durability of high-performance self-compacted concrete using electric arc furnace slag

aggregate and cupola slag powder. Cement and Concrete Composites, 2022, article no. 104399. https://doi.org/10.1016/j.cemconcomp.2021.104399

[34]   Andrade, H. D., de Carvalho, J. M. F., Costa, L. C. B., da Fonseca Elói, F. P., e Silva, K. D. D. C., & Peixoto, R. A. F.: Mechanical performance and resistance to carbonation of steel slag reinforced concrete. Construction and Building Materials, 298, 2021, article no. 123910. https://doi.org/10.1016/j.conbuildmat.2021.123910

[35]   Ortega-López, V., Faleschini, F., Pellegrino, C., Revilla-Cuesta, V., & Manso, J. M. (2022). Validation of slag-binder fiber-reinforced self-compacting concrete with slag aggregate under field conditions: Durability and real strength development. *Construction and Building Materials*, 320, 126280. https://doi.org/10.1016/j. conbuildmat.2021.126280

[36]   Tamayo, P., Pacheco, J., Thomas, C., de Brito, J., & Rico, J.: Mechanical and durability properties of concrete with coarse recycled aggregate produced with electric arc furnace slag concrete. Applied Sciences, 10(1), 2019, article id. 216. https://doi.org/10.3390/app10010216

[37]   Sosa, I., Tamayo, P., Sainz-Aja, J. A., Thomas, C., Setién, J., & Polanco, J. A.: Durability aspects in self-compacting siderurgical aggregate concrete. Journal of Building Engineering, 39, 2021, article no. 102268. https://doi.org/10.1016/j. jobe.2021.102268

[38]   Park, M. S., & Kim, Y. S.: A Study on the Sulfate Attack Resistance of Concrete Using EAF Slag as Fine Aggregate. Journal of the Korea Institute of Building Construction, 9(1), 2009, pp. 81–87. https://doi.org/10.5345/jkic.2009.9.1.081

[39]   Ortega-Lopez, V., Fuente-Alonso, J. A., Santamaria, A., San-José, J. T., & Aragon, A.: Durability studies on fiber-reinforced EAF slag concrete for pavements. Construction and Building Materials, 163, 2018, pp. 471–481. https://doi.org/10.1016/j. conbuildmat.2017.12.121

[40]   Faleschini, F., Fernández-Ruíz, M. A., Zanini, M. A., Brunelli, K., Pellegrino, C., & Hernández-Montes, E.: High performance concrete with electric arc furnace slag as aggregate: Mechanical and durability properties. Construction and Building Materials, 101, 2015, pp. 113–121. https://doi.org/10.1016/j.conbuildmat.2015.10.022

[41]   Malathy, R., Arivoli, M., Chung, I. M., & Prabakaran, M.: Effect of surface-treated energy optimized furnace steel slag as coarse aggregate in the performance of concrete under corrosive environment. Construction and Building Materials, 284, 2021, article no. 122840. https://doi.org/10.1016/j.conbuildmat.2021.122840

[42]   San-José, J. T., Vegas, I., Arribas, I., & Marcos, I.: The performance of steel-making slag concretes in the hardened state. Materials & Design, 60, 2014, pp. 612–619. https://doi.org/10.1016/j.matdes.2014.04.030

[43]   González-Ortega, M. A., Cavalaro, S. H. P., de Sensale, G. R., & Aguado, A.: Durability of concrete with electric arc furnace slag aggregate. Construction and Building Materials, 217, 2019, pp. 543–556. https://doi.org/10.1016/j.cemconcomp.2021.104399

[44]   Beaucour, A. L., Pliya, P., Faleschini, F., Njinwoua, R., Pellegrino, C., & Noumowé, A.: Influence of elevated temperature on properties of radiation shielding concrete with electric arc furnace slag as coarse aggregate. Construction and Building Materials, 256, 2020, article no. 119385. https://doi.org/10.1016/j.conbuildmat.2020.119385

[45] Arribas, I., Vegas, I., San-Jose, J. T., & Manso, J. M.: Durability studies on steelmaking slag concretes. Materials & design, 63, 2014, pp. 168–176. https://doi.org/10.1016/j.matdes.2014.06.002

[46] EN ISO 14040: Environmental management – Life cycle assessment – Principles and framework, 2021.

[47] Faleschini, F., De Marzi, P., & Pellegrino, C.: Recycled concrete containing EAF slag: environmental assessment through LCA. European Journal of Environmental and Civil Engineering, 18(9), 2014, pp. 1009–1024. https://doi.org/10.1080/19648189.2014.922505

[48] Turk, J., Cotič, Z., Mladenovič, A., & Šajna, A.: Environmental evaluation of green concretes versus conventional concrete by means of LCA. Waste management, 45, 2015, pp. 194–205. https://doi.org/10.1016/j.wasman.2015.06.035

[49] Evangelista, B. L., Rosado, L. P., & Penteado, C. S. G.: Life cycle assessment of concrete paving blocks using electric arc furnace slag as natural coarse aggregate substitute. Journal of Cleaner Production, 178, 2018, pp. 176–185. https://doi.org/10.1016/j.jclepro.2018.01.007

[50] SCCE Project Group: The European Guidelines for Self-Compacting Concrete, The Self-Compacting Concrete European Project Group, 2005.

[51] Garcia-Llona, A., Ortega-Lopez, V., Piñero, I., Santamaría, A., & Aguirre, M.: Effects of fiber material in concrete manufactured with electric arc furnace slag: Experimental and numerical study. Construction and Building Materials, 316, 2022, article no. 125553. https://doi.org/10.1016/j.conbuildmat.2021.125553

[52] Tyagi, G., Singhal, A., Routroy, S., Bhunia, D., & Lahoti, M.: Radiation Shielding Concrete with alternate constituents: An approach to address multiple hazards. Journal of Hazardous Materials, 404, 2021, article no.+ 124201. https://doi.org/10.1016/j.jhazmat.2020.124201

[53] Pomaro, B., Gramegna, F., Cherubini, R., De Nadal, V., Salomoni, V., & Faleschini, F.: Gamma-ray shielding properties of heavyweight concrete with Electric Arc Furnace slag as aggregate: an experimental and numerical study. Construction and Building Materials, 200, 2019, pp. 188–197. https://doi.org/10.1016/j.conbuildmat.2018.12.098

[54] Faleschini, F., Hofer L., Zanini M.A., Pellegrino C., Dalla Benetta M.: Experimental behavior of beam-column joints made with EAF concrete under cyclic loading. Engineering Structures, 139, 2017, pp. 81–95. https://doi.org/10.1016/j.engstruct.2017.02.038

[55] Faleschini, F., Bragolusi P., Zanini M.A., Zampieri P., Pellegrino C.: Experimental and numerical investigation of the cyclic behavior of RC beam column joints with EAF slag concrete. Engineering Structures, 152, 2017, pp. 335–347. https://doi.org/10.1016/j.engstruct.2017.09.022

Chapter 13

Alkali-Activated Concrete Performance

Francisca Puertas[1], Ruby Mejía de Gutiérrez[2]

1 Eduardo Torroja Institute for Construction Science (IETcc-CSIC), Madrid, Spain
2 Universidad del Valle (GMC-CENM), Cali, Colombia

13.1 Introduction

Portland cement concrete is the construction material most widely used and the commodity ranking second only to water in terms of overall human consumption. The worldwide yearly output amounts to over 3 Gt. Concrete has improved and continues to improve quality of life while contributing to societal development with the construction of infrastructures such as roads, bridges, and dams, as well as buildings for housing and other uses. The modern world would not be what it is without it [1]. Ordinary Portland Cement (OPC) concrete is characterised by many excellent features: nearly universal availability of the raw materials used in its manufacture, versatility for adaptability to a wide variety of civil works and building designs, high mechanical strength, and long durability.

The question is, then: *Why is there a need to study and develop alternatives to Portland cement concrete?* There are a number of very sound reasons. The United Nations (UN) forecasts that by 2060 the global human population will be four-fold greater than in 1950 (Fig. 13.1). The inference is that many more and better infrastructures, dwellings, and other applications will be required to meet that population's needs, which will in turn translate into a demand for immense volumes of building materials. Along those same lines, further to data furnished by the Bill and Melinda Gates Foundation [2], over the coming 40 years the planet's inhabitants will build a city the size of New York every 30 days. This demand should be mainly in countries as India, China, and other developing countries. Given its massive output and consumption, Portland cement concrete is associated with the consumption of vast amounts of raw materials, energy- and water-intensive, greenhouse gas-emitting manufacture, and the generation of waste at the end of the service life of the structures and other products bearing these materials. Monteiro et al. [3] estimated that concrete manufacture-related CO_2 emissions will rise by 85 Gt to 105 Gt between 2017 and 2050, a figure equal to the total greenhouse gases (GHGs) emitted by all industrial sectors, worldwide, in 2009 and 2010.

In other words, whilst Portland cement concrete is a construction material with many excellent technical features (which continue to be studied and improved), its vast produc-

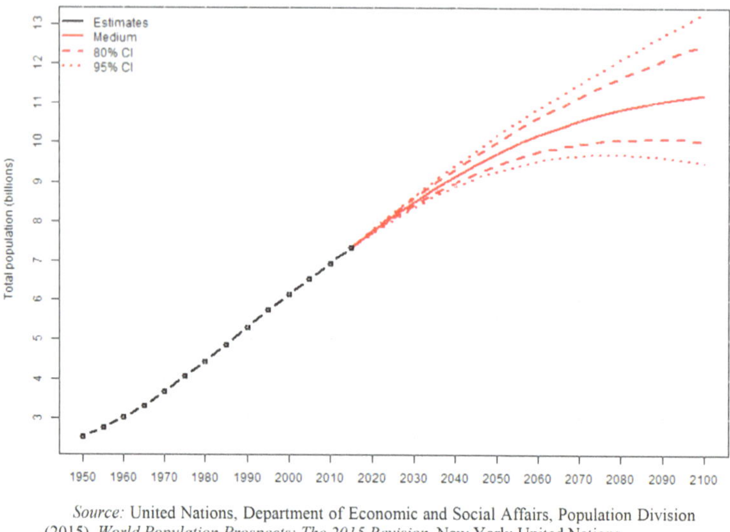

Source: United Nations, Department of Economic and Social Affairs, Population Division (2015). *World Population Prospects: The 2015 Revision.* New York: United Nations.

Fig. 13.1 Population growth forecast (1950–2100) (source: ONU, 2015)

tion and consumption, expected to grow substantially in the decades to come, have very adverse environmental impacts, some of which are associated with the manufacture of Portland cement, its primary component [4, 5]. OPC production calls for enormous quantities of natural raw materials (such as limestone and clay) to prepare the raw mix (around 1.7 t of such materials per t of cement) and energy-intensive milling to make clinker (850 kcal/kg Clinker). It also emits gas, especially GHG such as CO_2, associated with the consumption of fossil fuels in clinker furnaces, raw material, and end product transport and above all the decarbonation of the limestone used as a raw material. Around 8 % to 10 % of anthropogenic CO_2 emitted worldwide is attributable to the cement industry [6].

The answer to the question posed in the second paragraph of this section lies in the foregoing facts and figures. The world must develop and manufacture and commit to consuming concrete requiring less energy and associated with less environmental harm but with the same properties and performance as the traditional OPC material. One of the primary objectives of the cement-based construction materials industry is consequently to lower the environmental impact of concrete production without compromising society's building and infrastructure needs. The end product sought is known as 'green concrete'. That idea ties in with very topical concerns around 'sustainable construction' and 'circular economy', notions directly relevant to the Sustainable Development Goals (SDGs) defined by the UN in 2015 and set out in Agenda 2030 [7, 8]. The ultimate aim is to build a more sustainable world able to meet the needs of today's population without compromising those of tomorrow's population.

One of the scientifically and technologically viable means of reaching the construction industry's sustainability targets is to reuse or valorise industrial waste/by-products of widely differing origin and composition to manufacture new concrete with a low or nil Portland cement content. Alkali-activated concrete (AAC, also known as alkaline concrete or geopolymer) is a product closely aligned with sustainability principles [9]. Alkaline concrete or geopolymer use and the market for the product have been growing worldwide in recent years, particularly in emerging economies [10]. The sections that follow define and describe alkaline cements and concrete, their main components, fresh and hardened state properties, chemical stability, resistance to different types of aggressive agents, standardisation and worldwide output at present, as well as their most prominent applications.

13.2 What are Alkali-Activated Cements and Concretes (ACC)?

13.2.1 Definition and Brief History

Alkali-activated cements, a family of hydraulic binders that differ from Portland cements, consist essentially in two components: an aluminosilicate precursor and an alkaline activator to ensure high alkalinity in the system. The nature of the precursors and activators in the AACs developed to date varies substantially. The sub-sections below describe how these precursors and alkaline activators, together with the conditions under which such cements and concrete are prepared, affect the characteristics of the end product. Particular attention is paid to the properties of the main reaction products forming, which are determined by the chemical interaction between precursors and activators.

Alkaline cements and concrete have not appeared as recently as might be thought. The earliest patents claiming that combining vitreous blast furnace slag with alkaline solutions resulted in binders somewhat similar to Portland cements were published in 1895 and 1908. In 1940, Purdon [11] pursued that line of research further. In the nineteen sixties and seventies Prof. Glukhovsky's team at the University of Kyiv built structures that are still standing with concrete consisting of steel slag alkali-activated with sodium carbonate solutions they had developed [12]. In the early nineteen eighties Joseph Davidovits [13] patented new formulas and materials for alkali-activated burnt clay (essentially metakaolin). He called the new concrete 'geopolymers', a term subsequently applied to alkali-activated materials (AAMs) in general. Beginning in the nineteen nineties, many research groups the world over began to explore this field. Study and development have intensified in the decades since and the area has become a priority line of research for nearly all groups engaging in cement- and fired clay-based construction materials. Today, the aim is to optimise the starting materials, their dosage, properties, and the characteristics and durability of the resulting alkaline pastes, mortars, and concrete, as well as to establish the sustainability and long service lives of these new alternative construction materials.

13.2.2 Precursors and Alkaline Activators

Precursors, which act as skeletons in the initial structure of alkali-activated systems, dissolve wholly or partially in the highly alkaline environment subsequently generated by the activator. The dissolved species then coagulate, prompting the precipitation of the cohesive reaction (or activation) products to which the end product owes its strength and durability.

The nature of precursors may vary widely, with regard to both their (natural or artificial) origin and their composition (essentially aluminosilicates in the $CaO-SiO_2-Al_2O_3$ system) [14–23] (Fig. 13.2). They may be clays heated to a temperature level at which alkaline activation is technologically feasible (such as metakaolin) [24], although in most alkaline cements or geopolymers the precursor consists in waste or by-products such as glassy blast furnace slag (GBFS), (ASTM type F) aluminosilicate fly ash (FA), or a blend of the two [25]. Other materials shown to be apt as precursors include rice husk ash [26], waste glass [27], urban solid (including construction and demolition) waste [28], spent Fluid Catalytic Cracking (FCC) catalysts [29], and fired clay industry waste [30]. Nonetheless, thermally or mechanically treated clay, abundant the world over and consequently readily available, constitutes a very appropriate precursor for preparing alkaline cements and concrete [9]. Synthetic glass, whose composition resembles that of fly ash and glassy blast furnace slag, may also be used as a precursor [31]. Such glass can be produced on an industrial scale, thereby ensuring uniformity and reproducibility of the end product. In addition to those differences among precursors, two or more may also be blended. With the resulting variety in alkali-activated material composition and performance, these systems are often referred to as a 'family of binders'. The key requirements for clay- as well as for waste-based precursors are a high amorphous and/or glassy phase content and a large fraction of reactive SiO_2 and Al_2O_3, preferably at Si/Al ratios of >1.5.

Their production calls for a solid or liquid alkaline activator to ensure the medium has a very basic pH (over 13) able to dissolve the precursor/s and induce condensation, coagulation, and the precipitation of two- and/or three-dimensional, very compact, high strength reaction products. Waste may also be used to prepare the alkaline activators conventionally used: NaOH, KOH and waterglass. Waste glass waste [32, 33, 34] and biomass ash [35, 36] have also proven to be highly effective alkaline activators.

The consensus is to group alkali-activated binders under three major headings, based on their chemistry, mineralogy, and the presence or absence of Portland cement in their composition, as follows:

1. Group A): OPC-free alkali-activated cements bearing precursors with a high CaO content;

2. Group B): OPC-free alkali-activated cements bearing precursors with a low CaO content;

3. Group C): hybrid cements containing up to 30 % OPC and precursors with either a high or a low CaO content.

Two major types of alkaline activators can be distinguished, depending on their liquid or solid form [37].

1. If liquid, the activator also serves as the mixing liquid instead of the water used in Portland cement systems, generating what may be termed 'double-part alkali-activated' pastes and concrete;

2. If the activator is solid, water is used as the mixing fluid as in Portland cement systems, in which case the pastes and concrete are called 'one-part alkali-activated' materials.

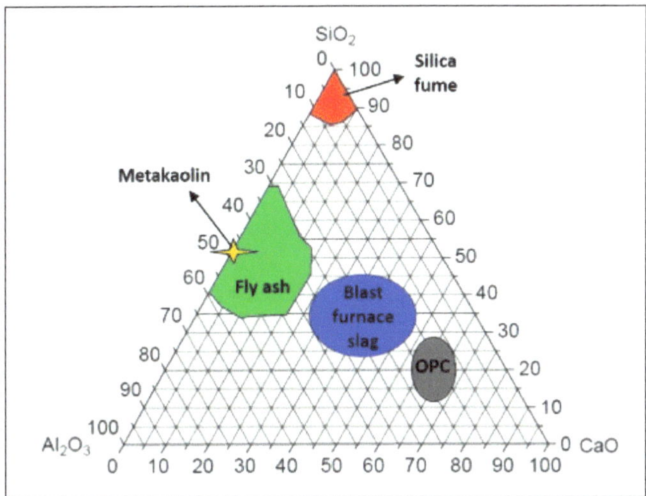

Fig. 13.2   CaO-SiO$_2$-Al$_2$O$_3$ diagram for conventionally used precursors.

Precursor and alkaline activator properties condition activation kinetics and mechanisms as well as the nature of the primary reaction products and consequently the performance of the alkaline pastes, mortars, and concrete generated. Activation processes and the nature of the main reaction products formed during alkali activation of the different groups of materials are described briefly in the following sub-section.

13.2.3 Alkaline Cement and Concrete Preparation: Nature of the Main Activation Products

Precursor-alkaline activator reaction kinetics are closely related to precursor composition. The most prominent factors include the presence or otherwise of CaO, high reactive SiO$_2$ and Al$_2$O$_3$ content, and the proportion of amorphous and/or glassy phases. Kinetics are also related to the nature of the activator (presence of Si, for instance), along with its concentration, pH, and liquid or solid state.

When the precursor has a high CaO content (Group A, typically glassy blast furnace slag), the chemical interaction between the precursor and activator solution and associated reaction kinetics are comparable to those observed in Portland cement-water systems, although the mechanisms differ [38, 39, 40]. Under such conditions, setting, hardening and strength development may take place at ambient temperature with curing likewise similar to what is found in Portland cement systems. Nevertheless, behaviour varies greatly depending on activator nature and concentration [41]. Setting is very rapid and initial strength high when the activator has an alkaline silicate base such as in water-glass, whilst the reaction rate is slower when it comprises alkaline hydroxides, and even slower when alkaline carbonates are involved, with strength developing primarily at later ages. In such systems, the optimal alkaline activator concentration ranges from 2 M to 4 M NaOH (equivalent to 3 % to 5 % NaOH of precursor weight).

If, on the contrary, the precursor has a low CaO content (Group B), such as in type F fly ash or metakaolin-like burnt clay, the reaction between precursor and activator is characterised by very slow kinetics at ambient temperature and therefore requires thermal curing at >45 °C for the process to be technologically viable [42]. The precursor-activator chemical reaction may take place in a moist or dry medium, although the former is more favourable in terms of material soundness. Activator composition is less relevant in these than in the preceding systems, although much higher concentrations, at around 8 M to 10 M, are required [43].

The conditions required in hybrid (Group C) systems with up to a content of 30 % Portland cement in the cement depend on the cement content in the blend, for in some cases thermal curing is recommended whereas others require high humidity and ambient temperature. Activator concentration also varies depending on precursor CaO content.

Different reaction products form in Group A, B, and C alkali-activated binders. The main reaction product in Group A materials (high CaO content) is a scantly crystalline two-dimensional C-A-S-H gel [44, 45, 46], depicted schematically in Fig. 13.3 [44]. The nature of the activator (presence of Si or otherwise) induces change in the two-dimensional structure of this product. In addition to C-A-S-H gel to which these alkali-activated systems owe their strength and durability, hydrotalcite-like phases may also be present, impacting their carbonation resistance [46].

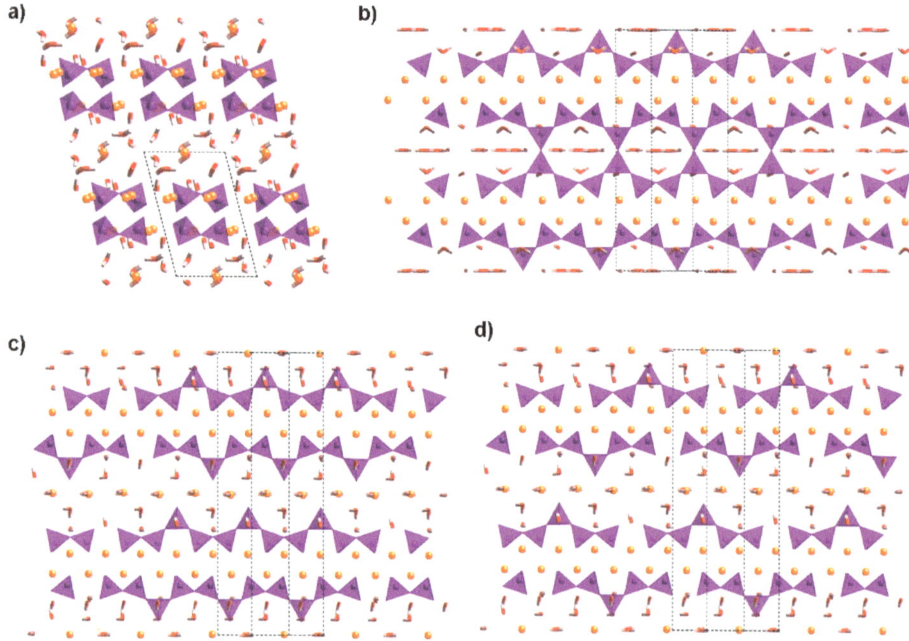

Fig. 13.3   Two-dimensional C-A-S-H gels forming during activation of high CaO content precursors [44]: a) J2 ; b) T11 (14) ; c) T14 (11) ; d) T14(5). (J=Jennite ; T=Tobermorite 11 and 14)

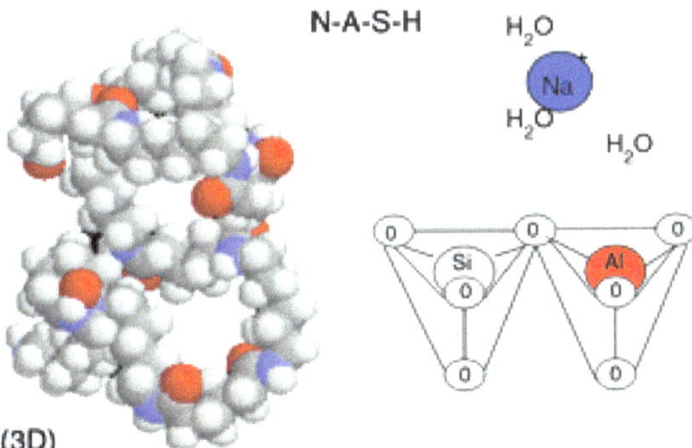

Fig. 13.4 Three-dimensional N-A-S-H gel forming during low CaO content precursor activation [18]

The primary reaction product in Group B binders (low CaO content precursors) is a family of three-dimensional materials whose general formula is M-A-S-H (where 'M' is an alkaline cation, normally Na or K) [47, 48, 49]. The structure of the N-A-S-H gels forming in such alkaline systems is illustrated in Fig. 13.4 [18]. As in the Group A materials, Group B strength and resistance to chemical and physical agents are attributable to these three-dimensional gels.

C-A-S-H and M-A-S-H are normally the majority or even the sole reaction products when the alkaline system has a single precursor, be it a glassy blast furnace slag (prompting C-A-S-H formation, possibly together with hydrotalcites as noted earlier) or a type F fly ash (associated with N-A-S-H gel formation). In the latter case, sodalite-, zeolite A- and faujasite-like zeolites may also appear, depending on the activation conditions. The two phases may co-exist when a blend of precursors with low and high CaO contents is used [25, 50, 51]. Under those circumstances mixed phases such as (N, C)-A-S-H may also form [52]. Identical phases may also appear in Group C (hybrid cement) materials, for the OPC would in all certainty raise the Ca content in the system [53].

All these reaction products generate normally low porosity cementitious pastes, and are therefore characterised by high early and late age strength as well as high resistance to aggressive media. The details of paste, mortar, and concrete performance are described in Section 13.3 of this chapter.

Generally speaking, alkaline cement concrete is prepared in much the same way as the analogous Portland cement materials. Studies conducted on mortars and concrete have shown the matrix-aggregate bond to be stronger in alkaline than Portland cement concrete [54]. Unlike the interfacial transition zone (ITZ) in concrete based on OPC only, which is normally porous, in alkali-activated concrete the ITZ is dense and uniform, as shown in the micrographs in Fig. 13.5 of concrete prepared with fly ash and blast furnace slag [55, 56].

Here also, the nature of the precursor and the nature and concentration of the activator, along with the nature, percentage, and particle size distribution of the aggregates define the end characteristics and properties of the respective alkaline concrete. The same very different kinds of aggregate (siliceous, calcareous…) as used in OPC concrete, including recycled concrete aggregate (RCA) [57, 58, 59], have been shown to be technologically apt for preparing alkaline concrete.

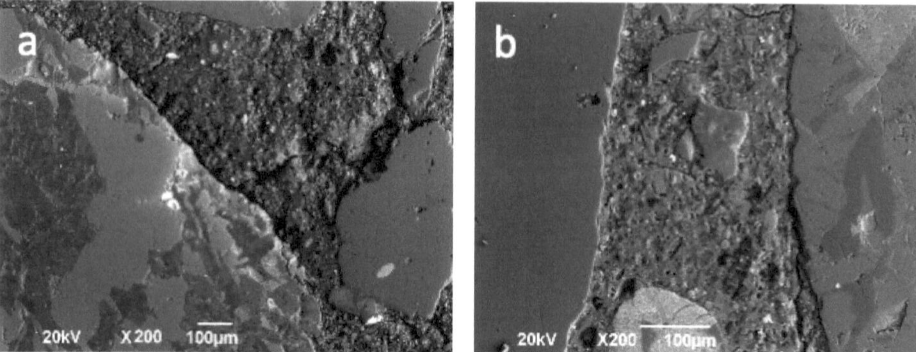

Fig. 13.5   SEM (Scanning Electron Microscopy) micrographs of AAC interfacial transition zone in concrete (28 days): (left) 100 % FA and (right) an FA/GBFS blend [56].

13.3 Fresh and Hardened Alkali-Activated Concrete (AAC) Performance

Fresh state performance in alkali-activated paste, mortar, and concrete may differ from the behaviour observed in their OPC counterparts. Such differences are associated primarily with alkaline cement composition (Groups A, B, or C), preparation, and curing conditions, along with aggregate batching and particle size distribution. That explains the intensification of research in recent years on alkaline system rheology [60–64], as understanding and controlling that factor is essential for alkaline cement, mortar, and concrete with on-site placement and standardisation.

Rheology is the science that studies matter flowability and deformation. Cement paste, mortar, and concrete rheological properties determine these materials' consistency, workability, and consequently ease of on-site placement. Cementitious system rheology also affects material macrostructure and consequently early and late age mechanical strength and durability.

Alkali-activated cement paste research has shown that the nature of the precursor and activator is a determinant in its consistency, workability, and setting times. That effect is more prominent in Group A binders, where the precursor is a CaO-high aluminosilicate, the most characteristic example being glassy blast furnace slag. The nature and composition of the activator are also key factors in the fresh state behaviour of alkali-activated slag (AAS) [61]. When the activator is a waterglass (alkaline silicate hydrate) solution, consistency and workability decline more rapidly and in a less orderly fashion, whilst setting times are shorter than when NaOH, KOH, or alkaline carbonate solutions are used. Puertas et al. [61] showed the importance of limiting the waterglass silica modulus (SiO_2/Na_2O) to 1.5 and lengthening mixing times when that activator is used in AAS systems. Such conditions contribute to good workability and consistency, as well as initial and final setting times technologically suitable for the on-site use of pastes [61], mortars [65] and concrete [64]. The effect of lengthening mixing time from 15 min to 60 min on waterglass-activated slag concrete's slump, viscosity, threshold shear strength, setting times, and mechanical strength is illustrated in Figs. 13.6 and 13.7 [64].

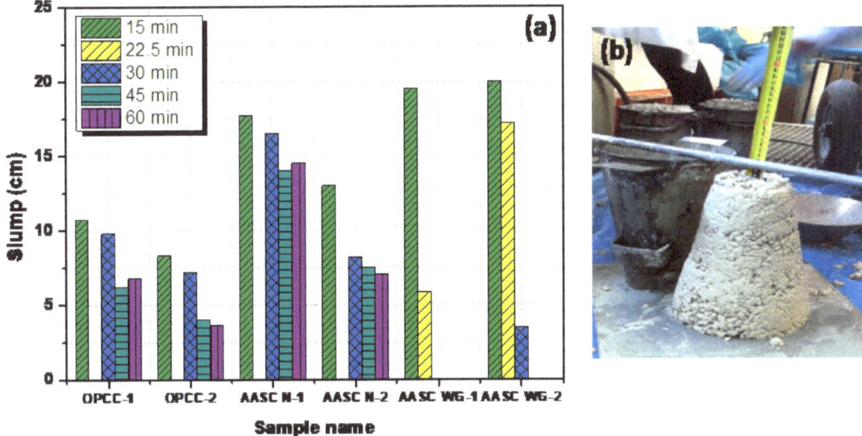

Fig. 13.6 (a) Slump dimensions for concrete of different ages; and (b) 22.5 min slump for waterglass-activated slag concrete prepared [64]

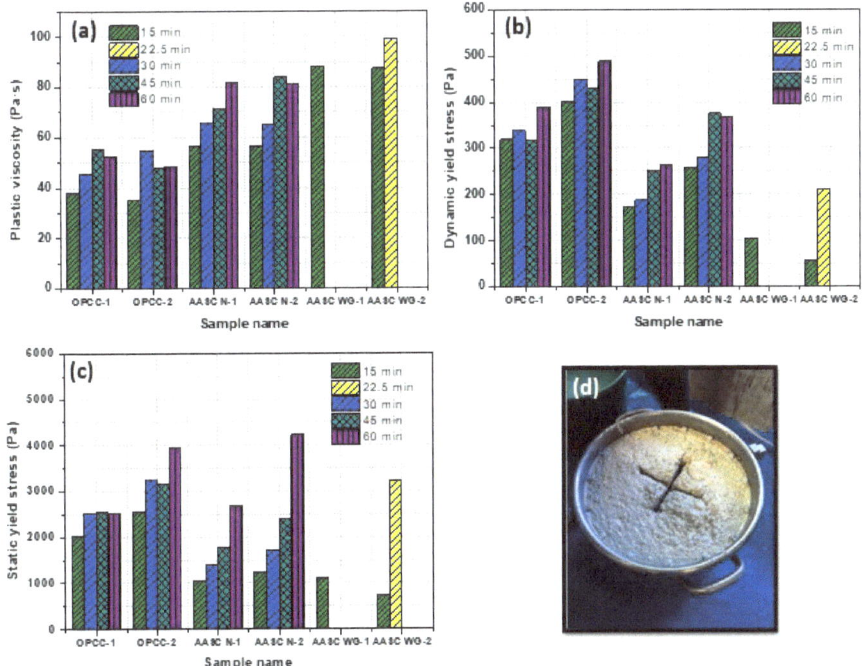

Fig. 13.7 (a) Concrete plastic viscosity values; (b) dynamic yield stress values; (c) static yield stress values; and (d) 30 min waterglass-activated slag concrete at age = 30 min [64]

The rheological behaviour (workability, consistency, setting times) of Groups B and C alkali-activated pastes, mortars, and concrete is less problematic [66], for which performance is closer to the behaviour observed in OPC materials. That notwithstanding, activator batching and curing conditions (these systems may require thermal curing) must always be closely monitored.

Alonso et al. [65] studied the role of aggregates in alkali-activated system rheology. Their findings showed that activated slag (AAS) and fly ash (AAFA) workability was more sensitive to changes in both the liquid/solid ratio and aggregate batching than OPC mortars. The role and nature of aggregate particle size distribution in alkali-activated mortar (AAM) rheology were studied by Gismera et al. [62] (Fig. 13.8). Testing siliceous, calcareous, and recycled aggregates, these authors found that AAMs required higher liquid/solid ratios than OPC mortars to attain similar plastic consistency. Furthermore, the nature of the aggregate had a heavier impact on the rheological properties of sodium silicate-AAS mortars than on NaOH-AAFA and OPC mortars. The use of recycled aggregates raised the water demand due to the high water absorption of these aggregates because of their high porosity, although 20% replacement of siliceous by recycled aggregates did not alter the amount of water required to reach plastic consistency in either OPC or AAM mortars [65].

Such research attests to the need to conduct workability, consistency, and setting trials in alkali-activated concrete prior to their on-site placement to ensure optimal performance.

The graph in Fig. 13.9 was drawn from a study [56] comparing the 1440 days compressive strength in alkali-activated concrete prepared with 100 % FA to binary FA/GBFS and hybrid FA/OPC concrete strength. According to the findings, whereas the 7 days compressive strength in the FA/GBFS material was 35 MPa with a 75 % rise after 360 days, the FA/OPC hybrid exhibited 41 MPa strength in the same timeframe. At longer curing times (1440 days), however, FA/GBFS was the higher performer, with a compressive strength of 93 MPa. The 28 days tensile strength in the FA/GBFS concrete was 3.57 MPa, 143 % greater than the value recorded for the concrete made with 100 % FA. The steeper rise in mechanical performance in the systems containing GBFS or (20 %) OPC compared to the 100 % FA material is related to the calcium content present in those blends, which alters gel microstructure relative to the systems prepared with FA only [56, 67].

Other studies by the same researchers corroborated the importance of the nature of the precursor in determining hardened alkali-activated concrete mechanical performance. Fig. 13.10 graphs the compressive strength found for two concrete mixes, a binary concrete (author-labelled AABC) comprising a 70/30 blend of a natural volcanic pozzolan (NP) and GBFS and a hybrid concrete (AAHC) prepared with 90 % alkali-activated construction and demolition waste (CDW) and 10 % OPC. In the latter CDW was used in both the hybrid cement as binder and the fine and coarse aggregates. That experiment reused 100 % of the raw materials (ground to between 25.4 mm and 0.075 mm) and by-products (powder ≤ 0,075 mm) generated during CDW crushing and processing, even in scenarios where the various types of raw waste could not be segregated (mixed waste) [68, 19]. The alkaline activator used was a blend of sodium hydroxide (NaOH) and sodium silicate ($Na_2SiO_3 \cdot 5H_2O$), whilst both were batched to 400 kg/m^3 of precursor and cured at ambient temperature (25 °C). Although early age (7 day) alkaline concrete AABC exhibited somewhat lower strength than OPC, after 28 days the reverse was observed [19]. The twenty-eight-day compressive strength of 31.5 MPa was attained by adjusting the precursor content (from 300 kg/m^3 to 600 kg/m^3). Those findings, in conjunction with

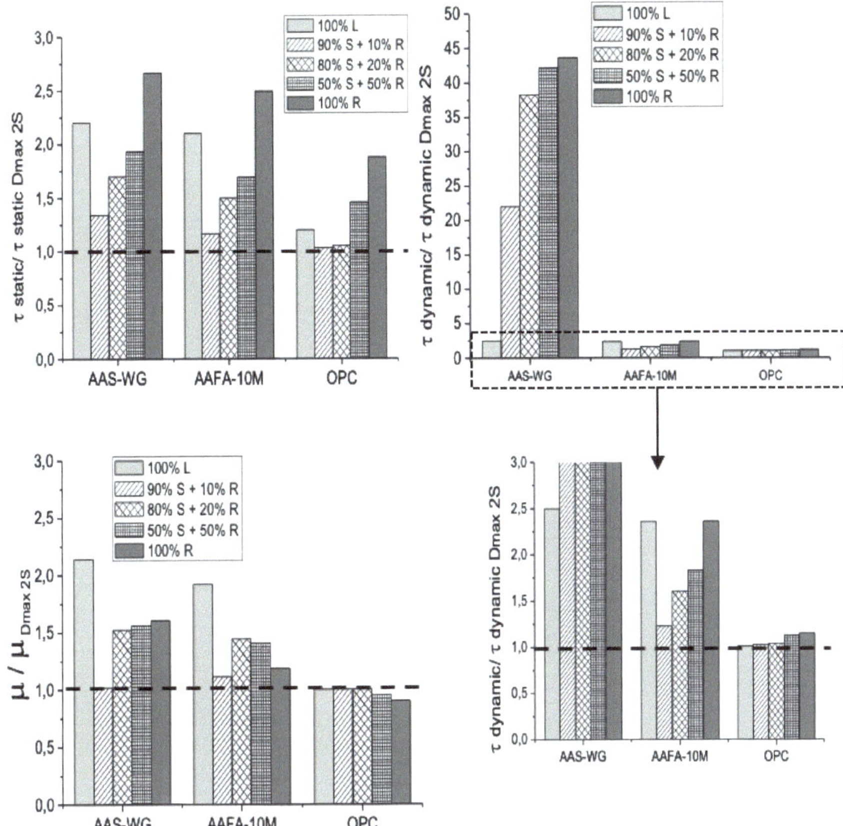

Fig. 13.8 Variation in static (t_{static}) and dynamic ($t_{dynamic}$) shear and plastic viscosity (μ) when siliceous (*S*) aggregate was replaced with 100 % limestone (*L*) or different percentages of recycled (*R*) aggregate (values normalized to mortars with 100 % siliceous aggregate and a D_{max} = 2 mm) [62]

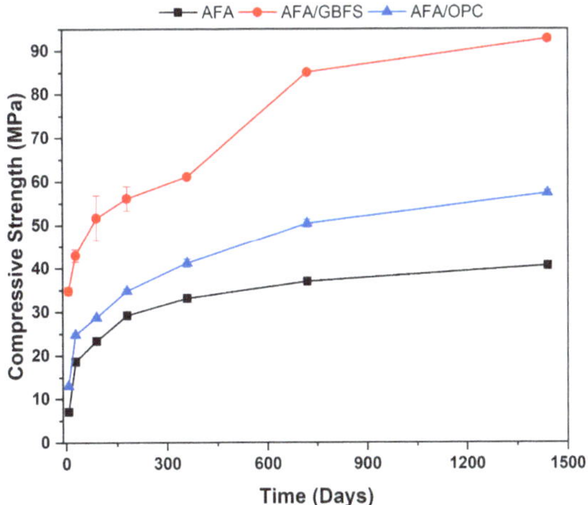

Fig. 13.9 Compressive strength in three alkali-activated concrete [56]

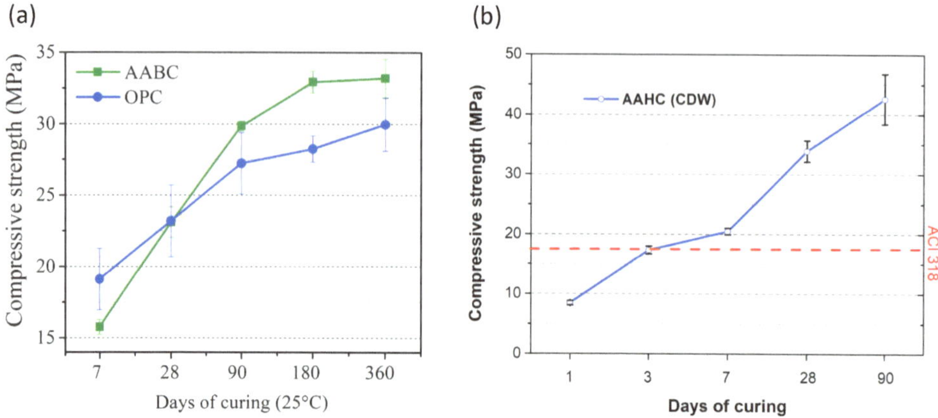

Fig. 13.10   Concrete compressive strength: a) comparing AABC (NP/GBFS) to OPC; and b) for AAHC prepared with CDW [19, 68]

a 4.2 % higher modulus of elasticity (29.14 GPa) in AABC than in OPC and comparable flexural strength in both (modulus of rupture = 4.7 MPa to 5.8 MPa), endorsed the use of AABC as a feasible alternative to OPC concrete in low-to-medium strength applications. Based on the 17.5 MPa limit strength defined in ACI 318 and further to the 1 d (8.5 MPa), 3 day (17.4 MPa), 7 day (20.5 MPa), 28 day (33.9 MPa), and 90 day (42.6 MPa) compressive strength in the CDW-based hybrid concrete, AAHC was standard-compliant from day 7 onward [68].

Generally speaking, alkaline concrete has a lower micro- and meso-porosity than OPC materials prepared under similar conditions, although the values may differ depending on the type of alkaline binder (Group A, B, or C). Such lower total porosity and better pore distribution would largely explain the good mechanical performance of this alkaline concrete and its high chemical and physical resistance to aggressive media.

13.4 Alkali-Activated Concrete Durability

To be apt for use in civil construction and ensure durability throughout a long service life, alkali-activated concrete (AAC) must meet a series of mechanical strength and workability requirements. AAC durability is directly related to the nature of the reaction products formed and the resulting microstructure, which governs fluid transport and matrix resistance to aggressive agents. As noted, its behaviour in that regard is closely associated with the nature of the precursor, the nature and concentration of the activator, the design parameters, and the curing age of the material. As in OPC concrete, aggregate nature, shape, and size also play a significant role in AAC durability.

The behavior of AAC against deterioration processes due to alkali-aggregate reactions has been studied and, in general terms, it is comparable or superior to that of OPC [69, 70] concrete. In relation to physical attack by freezing-thawing, it has been demonstrated that alkali-activated slag concretes have much better behaviour than those based on the activation of fly ash [71]. Alkali-activated concrete behaviour when exposed to aggressive agents is described and compared to OPC concrete performance in the sub-sections that follow.

Steel corrosion, identified as the key problem in reinforced concrete, is associated with chemical decay in the concrete itself such as sulfate attack, acid-induced corrosion, carbonation, and/or the presence of chlorides [72, 73]. Decay may also be induced by physical processes, most prominently high temperature, although alkaline materials often exhibit higher heat performance than OPC-based binders. AAC resistance to aggressive agents is generally assessed using the same protocols as OPC concrete, though recommendations should be specifically developed for the former.

13.4.1 Sulphate Resistance

Concrete is attacked by sulphate when exposed to underground water, lakes, wells or soils containing sulphate ions [72, 74]. These ions contribute to gypsum formation by reacting with OPC hydration products, in particular calcium hydroxide (in portlandite form), any remaining anhydrous tricalcium aluminate, ettringite or C-S-H gels, as well as secondary ettringite and even thaumasite. The outcome is concrete expansion, cracking, and disintegration [74]. In other words, sulphate attack is a physical-chemical process in which external or internal sulphates raise matrix porosity by interacting with the components of the gels or crystals resulting from binder hydration [75].

The higher performance generally observed in alkali-activated concrete in that respect is attributed to the compositional and microstructural properties of their systems. Here also, however, differences in precursor nature (Group A, B, or C) likewise inform differences in resistance to a given source of sulphates. Alkaline concrete based on type A binders, i.e., with a high calcium content, resists sulphate attack better than OPC materials despite the similarity between some of their reaction products such as C-(A)-S-H gel [73, 75, 76]. The presence of that gel is offset by the near total absence in the alkaline binder of portlandite and other sources of aluminates liable to react with sulphates. Type B (low calcium) binder-based alkaline concrete hydration generates reaction products such as N-A-S-H gels that differ widely from those found in OPC materials. The higher resistance of alkaline than OPC concrete to sulphate attack has been corroborated by studies conducted in several countries [77–81].

Nonetheless, the severity of the attack depends primarily on ion sulphate concentration in the surrounding soil or water as well as the cation (sodium or magnesium) involved, for magnesium sulphate is generally deemed more aggressive [72, 82]. The mechanical strength of alkaline concrete based on fly ash / slag blends has been reported to rise in the presence of sodium sulphate [83]. Exposure to Na_2SO_4 appears to favour the structural growth of the geopolymer-based gel, whilst $MgSO_4$ enhances phase decomposition in systems with a high calcium content, specifically by inducing C-(A)-S-H gel decalcification and gypsum formation while lowering medium alkalinity, resulting in a less stable gel. Valencia-Saavedra et al. [83] observed 0.04 % expansion and a 33 % decline in mechanical strength in 80/20 FA/GBFS concrete exposed to 5 % $MgSO_4$ for 360 days, although the specimens exhibited no dimensional change during the first 6 months of

Fig. 13.11   FA/GBFS concrete exposed to 5% Na_2SO_4 and $MgSO_4$ for 270 days [83]

exposure. In the presence of Na_2SO_4, however, strength rose in that same concrete. Visual inspection revealed greater salt deposits on the geopolymer than on OPC concrete, although the latter exhibited spalling around the edges and strength declines of up to 48% (Fig. 13.11). Duan et al. [84] observed that adding metakaolin raised sulphate resistance in Group B (fly ash-based) concrete.

Bondar et al. [85], studying natural pozzolan NP-activated systems, observed strength declines of 8% after 2 years of exposure to 2.5% Na_2SO_4 and of 19.5% where they were exposed to 2.5% $MgSO_4$. Concrete made with 70/30 NP/GBFS (author-labelled AABC) activated with a mix of sodium hydroxide and sodium silicate expanded by 0.0985% and declined a mere 0.58% in strength after soaking for 730 days in 5% Na_2SO_4 [86]. In contrast, when soaked in 5% $MgSO_4$ they expanded by 0.1471% and strength was observed to drop by 20.85%. That still compared favourably to OPC concrete performance, where a 60% decline in strength and 0.7566% expansion after soaking in Na_2SO_4 and 0.2192% after exposure to $MgSO_4$ were recorded (Fig. 13.12) [86]. Photographs of the specimens after 90 days and 360 days of exposure are reproduced in Fig. 13.13. Using XRD and SEM, the authors also corroborated the presence of gypsum in the specimens exposed to $MgSO_4$ [86].

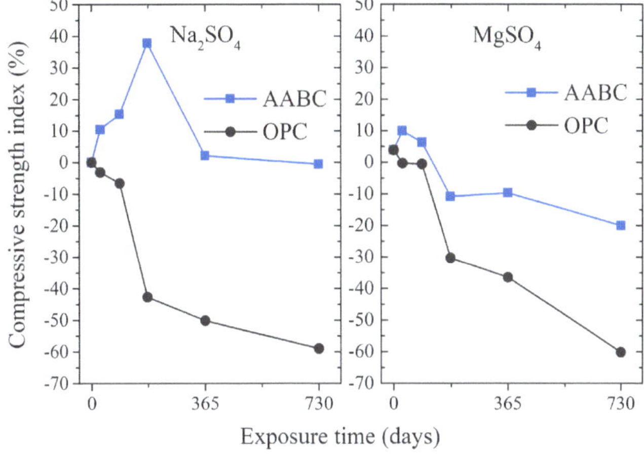

Fig. 13.12 Na₂SO₄ and MgSO₄ sulfate attack: comparison of effect on compressive strength in AABC (70/30 NP/GBFS) and OPC concrete [86]

AABC		OPC	
Na₂SO₄ 90 days	Na₂SO₄ 360 days	Na₂SO₄ 90 days	Na₂SO₄ 360 days
MgSO₄ 90 days	MgSO₄ 360 days	MgSO₄ 90 days	MgSO₄ 360 days

Fig. 13.13 AABC and OPC concrete specimens after 90- and 360-days exposure to Na₂SO₄ and MgSO₄ [86]

13.4.2 Resistance to Acid Media

Due to the high pH of their aqueous base, concrete is susceptible to both organic and inorganic acid attacks. In OPC concrete, acids may react with the calcium hydroxide (in the form of portlandite), ettringite, and C-S-H gel in the paste, generating soluble or insoluble calcium salts. Such salts leach, thereby raising porosity, whereas their insoluble counterparts may locate in pores, inducing cracks. The effects of both developments are lower mechanical strength and greater concrete permeability.

Alkaline concrete has been shown to resist acid media more effectively, thanks primarily to their lower permeability than OPC, although that again varies with the type of alkaline binder involved (Group A, B, or C) and therefore with the gel or gels generated. Other significant factors to be considered include the nature and concentration of the acid and the solubility of the resulting salt.

High acid resistance has been reported in alkaline concrete where the primary reaction products are the N-A-S-H gels that form in Group B binders (such as those based on metakaolin) [87–90]. In contrast, concrete with high calcium precursors (Group A) has been observed to form calcium sulphate (gypsum) in sulfuric acid media, which induces expansion-mediated decay. In alkali-activated GBFS mortars exposed to lactic, nitric, sulphuric, or hydrochloric acid solutions, severe attack was recorded only in the case of sulphuric acid [91], with mass losses of under 7 % in all other cases, compared to 95 % loss observed in OPC concrete. Studying OPC and activated GBFS concrete exposed to acetic acid (pH=4), Bakharev et al. [92] reported 33 % strength loss in the latter and 47% in the former, findings attributed to the higher Ca content in the OPC materials.

Zhang et al. [93] assessed 30/70 FA/GBFS alkaline concrete, observing that despite the pore structure refinement and lower acid penetration induced by the addition of GBFS, mechanical performance declined by up to 60 % when the samples were soaked for 28 days in sulphuric acid (pH=1). Koenig et al. [94] strength-tested FA and FA/GBFS alkaline mortar and concrete after exposure to a 1.5 % acetic acid, 0.5 % propionic acid, and 3.10 % lactic acid mix. After 18 weeks of exposure to those organic acids, the authors concluded that the alkaline concrete exhibited higher resistance than OPC materials and that acid resistance generally declined with rising CaO content in the binder. Duan et al. [95] reported that FA/MK binary systems exhibited less strength and mass loss after exposure to sulphuric and hydrochloric acid than OPC materials.

In alkaline concrete, in contrast to OPC systems, acid attack is closely associated with Si-O-Si and Si-O-Al bond cleavage, depolymerisation and silicic acid formation [96, 74]. A study of exposure to 1 M sulphuric and acetic acids for 360 days revealed higher performance than in OPC in two types of binary concrete, FA/GBFS and hybrid FA/OPC, with residual strengths of 10 MPa to 20 MPa (Fig. 13.14) and under 6 % mass loss [97]. The same study found obvious proof of gypsum formation in all the specimens exposed to sulphuric acid.

As noted for sulphates, the lack of standardised methods to assess alkaline concrete in acid media hinders comparison of different research groups' findings. Most studies have been conducted in pastes and mortars at fairly short exposure times and variable acid concentrations.

13.4.3 Carbonation Resistance and Reinforcement Corrosion

Carbonation results from the ingress of the CO_2 present in urban and industrial environments and environmental pollution into concrete and the concomitant chemical interaction with cementitious system reaction products. The high (~13) pH in the aqueous

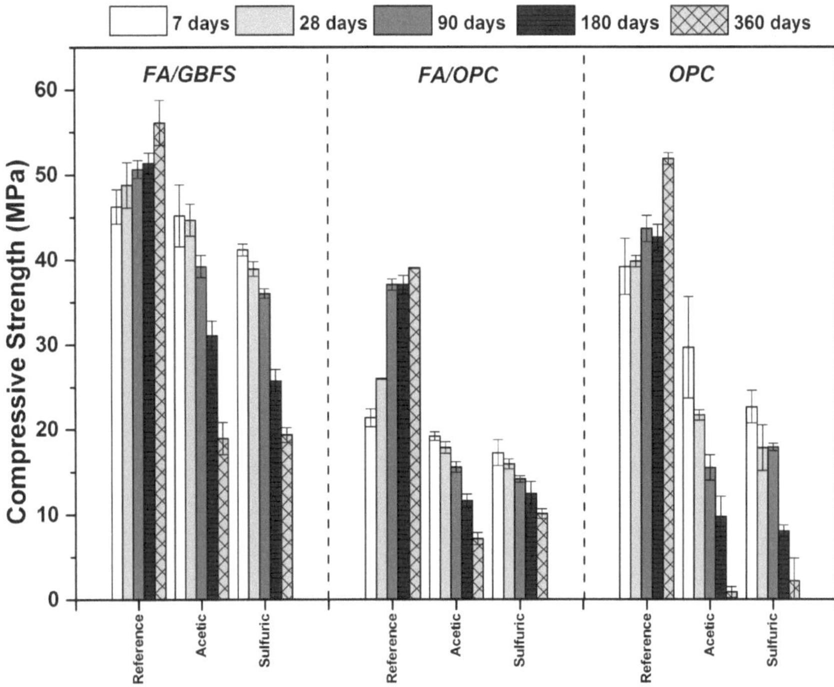

Fig. 13.14 Acid resistance in OPC and alkali-activated FA/GBFS and FA/OPC [97]

phase of concrete that affords rebar natural anti-corrosion protection is weakened by carbonation, for it lowers the pH to approximately 9, causing steel depassivation. In OPC concrete, carbonation depends on the contact between carbon dioxide and water, and alkaline components present in the cement matrix phases, such as portlandite, C-S-H, and aluminates, are altered by CO_2 ingress. When carbonation begins, the calcium carbonate formed on the surface acts as a barrier for gas diffusion, subsequently advancing inward to the reinforcing steel. The mechanism differs in alkaline systems, however. In calcium high (Group A) systems the CO_2 reacts with and decalcifies C-A-S-H gel, generating calcium carbonates [98, 99]. In systems with low calcium content (Group B), the reaction with N-A-S-H gel gives rise to sodium carbonates and/or bicarbonates [74]. Accelerated exposure methods have been developed by research to explore this development in light of the slow CO_2 reaction rate [82, 100]

The studies conducted to date differ in terms of the exposure media parameters addressed, with some based on natural exposure and others on the use of carbonation chambers at concentrations ranging from 1 % to 100 % CO_2; the assessment methods deployed also vary from study to study (measuring effects with phenolphthalein indicators, electrochemically, or with mineralogical characterisation). With such differences, comparison is a challenge and contradictory reports are common. A single performance parameter cannot be defined for alkaline concrete, however, for it depends on precursor composition, type and quantity of the activator, and relative humidity, among others. Bernal et al. [101] reported that the N-A-S-H gel resulting from FA alkali activation remained stable during carbonation; however, when FA and GBFS were mixed, the resistance to carbonation decreased due to the presence of calcium. The presence of MgO in GBFS would appear to be relevant to Group A alkaline systems, for hydrotalcite formation inhibits or lowers carbonation rates [102]. Several authors have noted that, although carbonation resistance is lower in AAMs than in OPC, the reduction in pH in the former is not overly severe and unlikely to cross the depassivation threshold for reinforcing steel [74, 103, 104]. Mechanical strength is consequently not significantly affected.

In one of the studies cited above, linear polarisation resistance (LPR) and Tafel slopes were used to assess carbonation advance at the same time as steel corrosion in NP/GBFS concrete exposed to 1% CO_2 in a climate chamber at 25 °C and 65% RH [104]. The findings confirmed that the reinforcing steel in alkaline concrete was not at risk of corrosion after 360 days of exposure to CO_2, for the pH prevailing was high enough to ensure passive film stability [104]. That contradicted Sufian Badar et al.'s findings [105] for FA-based alkaline concrete carbonation, according to which pH declined substantially to the significant detriment of mechanical strength. Valencia-Saavedra [106] also reported up to 40% drops in strength after 1 year of exposure to CO_2 in FA/GBFS and FA/OPC concrete. Nonetheless, in reinforced concrete with such matrices, the reinforcing steel showed no visible signs of corrosion after 20 months of CO_2 exposure. In the same study, the authors found Tafel curve constant B to be 18 mV for concrete FA/GBFS and 15 mV for FA/OPC [106], whilst other research on NP/GFBS delivered a B value of 18±2 mV [104]. The inference of such data is that the values traditionally suggested for Portland concrete-embedded steel (26 mV for active and 52 mV for passive steel) differ from those for AAC, for which specific criteria need to be defined.

13.4.4 Chloride Resistance and Reinforcement Corrosion

Chlorides may be present in the concrete mix (contaminated aggregate, sea- or polluted water, additives with a high chloride content) or in de-icing salts, sea spray, or chemicals that penetrate the binder from the exterior. Although chlorides are not aggressive to concrete, they can destroy the passive layer protecting rebar, generating local corrosion (pitting), concrete detachment due to corrosion product expansion, and ultimately structural failure. The greater density in alkaline than in OPC concrete determines lower chloride permeability in the former. Chloride permeability has nonetheless been observed to vary widely in alkaline systems depending on the type of precursor (that determines whether the alkaline binder lies in Group A, B, or C). Some authors [107] have reported chloride diffusion to be approximately half as intense in alkaline as in OPC concrete. In addition, GBFS-based Group A alkaline concrete has been found to exhibit greater penetration resistance than those bearing FA [108]. At the same time, as in OPC concrete, some of the phases present in alkaline systems have been observed to bond to chloride ions, either through physical adsorption or forming Friedel's salt [109–112], retarding inward diffusion.

Alkaline concrete based on natural pozzolans (NP) has been found to be fairly susceptible to chloride diffusion [113], with penetration nonetheless mitigated with the addition of GBFS [114]. Aguirre-Guerrero et al. [115] reported up to 70% lower chloride permeability in a 70/30 NP/GBFS than in OPC concrete. GBFS content (ranging from 25 % to 75 %) in alkali-activated FA mortars [116, 117] has been found to be inversely related to the chloride ion diffusion coefficient and mixes containing over 50% of the slag to perform better than a 100% OPC mortar.

Studies conducted in reinforced mortars made from alkali-activated slag exposed to chlorides have reported high corrosion current densities, although after 9 months of exposure no sign of steel corrosion was observed [118]. The constant B values (used to calculate current density, i_{corr}) for rebar embedded in NP/GBFS concrete subjected to accelerated corrosion testing, wet-dry cycles, and soaking in 3.5 % NaCl have been found to be 17 mV when in the active and 20 mV when in the passive state [115]. Other authors have reported passive state B values for binary FA/GBFS concrete of 13 mV to 20 mV and active state values of 45 mV to 58 mV [116]. Such figures deviate from the 26 mV and 52 mV typically used for the active and passive states, respectively, in Portland cement concrete. Tennakoon et al. [119] noted that in the presence of chloride ions the corrosion resistance of reinforcing steel was greater in a 50/50 FA-GBFS than in an OPC concrete. That was consistent with Prusty and Pradhan's [120] reports of lower current density values for the former than observed for 100 % FA concrete. For the above reasons, some authors have contended that electrochemical measurement values fail to reliably measure actual corrosion in alkali-activated systems [115–120].

Inasmuch as chloride ion transport properties, like those of other aggressive agents, depend on many factors, alkali-activated concrete performance in that regard cannot be generalised. Nonetheless, most corrosion studies on steel embedded in alkaline concrete have delivered satisfactory results, as noted. What should be stressed is that since the criteria traditionally used for OPC concrete are not necessarily the most appropriate for alkaline concrete, specific standards and parameters should be defined, as contended in the sub-section on carbonation.

13.4.5 Heat and Fire Resistance

OPC concrete exhibits low resistance to high temperatures, and more specifically to temperatures of over 300 °C, above which their hydrates begin to disintegrate. Hence, concrete in a structure continuously exposed to intense fire detaches from the rebar, causing structural damage, affecting steel stability, and possibly inducing collapse. In contrast, alkali-activated materials perform generally well under such conditions [74] and some even exhibit significant rises in mechanical strength after exposure to >1200 °C. Such behaviour is attributed to the formation of new, more stable ceramic phases such as leucite, nepheline, kalsilite, mullite, and similar [74, 121, 122, 123]. Residual strength of 15 MPa has been reported for FA/GBFS and of 5.5 MPa for FA/OPC alkaline concrete exposed to temperatures of 1100 °C, whereas under the same conditions OPC concrete was observed to retain 0 % of its bearing capacity.

The crystalline phases identified by the authors of the respective study included sodalite, nepheline, albite, and akermanite [121]. Leucite has been detected at 1100 °C in metakaolin-based geopolymers activated with potassium alkaline solutions, although the material was observed to lose ~8 % to 9 % of its mass and potentially to shrink linearly by 7 % to 9 % at that temperature [122]. The effect, a decline in flexural strength, was observed to be partially offset by the addition of (ceramic, carbon, or similar) reinforcing particles and fibres [124, 125, 126]. Under such conditions 80 MPa to 160 MPa strength was recorded even after exposure to temperatures of 1200 °C [127]. Diffractograms for MK-based simple and compound geopolymers before and after exposure to 1200 °C are reproduced in Fig. 13.15 [124]. According to He et al. [128], structural densification at high temperatures and subsequent formation of leucite-like minerals may induce the development of high-performance ceramics with fairly high flexural strength (>100 MPa). That has led to proposals for their use as thermal, refractory, and fire insulation, particularly in structures where fire hazard is high [129–130] and even as cladding for steel exposed to high temperatures [131].

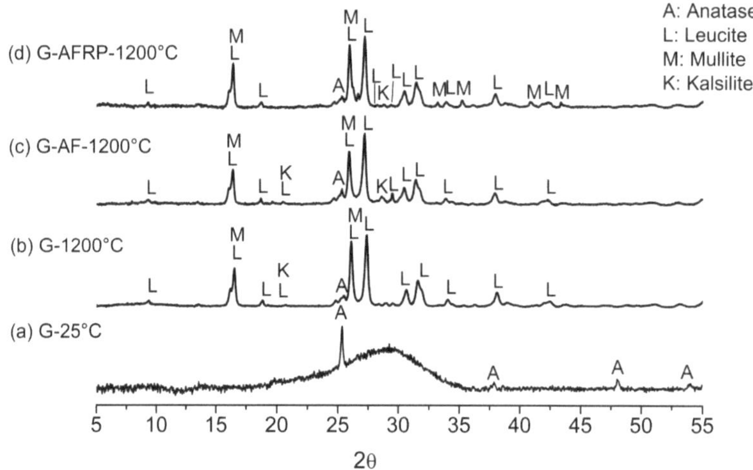

Fig. 13.15   XRD (X-Ray Diffraction) scans before and after exposure to 1200 °C (G: geopolymer activated with potassium solution; AF: alumina fibre; AFRP: alumina fibre and refractory particles) [124]

13.5 Final remarks

The data furnished in this chapter attest to the sustainability and viability of alkaline concrete as an alternative to OPC concrete. To date, however, such alkaline concrete has not proved apt for structural applications. Further study and certification are needed before such use can be recommended. They are nonetheless suitable for a wide range of non-structural applications (walkways, pavements, and similar), whilst their strength, insulating capacity, and durability ensure their versatility.

Before alkaline cements and concrete can be unconditionally adopted by the construction industry, certain important technical and legislative issues must be addressed. Firstly, specific standards must be implemented for the study, characterisation, and validation of the alkaline cement and concrete family of materials. The standards routinely applied to Portland cement and concrete have been shown to be non-valid. RILEM TC 247-DTA on durability testing of alkali-activated materials (2012–2019) has been working to identify and validate specific tests to generate reliable and reproducible data on the alkaline concrete [100].

Likewise imperative for the ultimate implementation of these new sustainable construction materials is the enactment of specific legislation and codes of good practice. Whilst some countries (Ukraine, China, and United Kingdom, among others) have such regulations in place, broader legislative coverage is required for the enthusiastic adoption of these alkaline systems.

The paradigm for such systems today is the pursuit of precursors and activators widely available worldwide to ensure that the materials at issue can be produced anywhere on the planet.

13.6 References

[1]   F. Puertas, J.A. Suárez-Navarro, M.M. Alonso, C. Gascó. (2021). *NORM wastes, cements and concretes. A Review.* Materiales de Construcción, vol. 71, nº 344, e259. https://doi.org/10.3989/mc.2021.13520

[2]   B. Gates. (2019). How to make concrete green. Available at: https://www.youtube. com/watch?v=0pAH-6R5J2A

[3]   P.J.M. Monteiro, S.A. Miller, A. Horvath. *Towards Sustainble Concrete.* (2017). Nature Materials, vol. 16, 698–699. https://doi.org/10.1038/nmat4930

[4]   M. Schneider. *The Cement Industry on the Way to a Low Carbon Future.* (2019). Cement and Concrete Research, vol. 124, 105792. https://doi.org/10.1016/j. cemconres.2019.105792

[5]   Cembureau, Ecofys Report Shows Potential for Further Uptake of Alternative Fuels in the Cement Industry. (2017) Available at: https://cembureau.eu/news-views/blog/ecofys-report-shows-potential-for-further-uptake-of-alternative-fuels-in-thecement-industry/

[6]   F. Belaïd. (2022). *How does concrete and cement industry transformation contribute to mitigating climate change challenges?* Resources, Conservation & Recycling Advances, vol. 15, article 200084. https://doi.org/10.1016/j.rcradv.2022.200084

[7]   S. Mindess (2019). Sustainability of Concrete. Chapter 1. Developments in the Formulation and Reinforcement of Concrete (Second Edition). Woodhead Publishing Series in Civil and Structural Engineering, pp. 3–17

[8]   Circular Economy. UE. 2016. Available at: https://ec.europa.eu/commission/priorities/jobs-growth-and-investment/towards-circular-economy_es.

[9]   A. Palomo, O. Maltseva, I. García-Lodeiro, A. Fernández-Jiménez. (2021). *Portland versus alkaline cement: Continuity or clean break: "A key decision for global sustainability".* Frontiers in Chemistry, vol. 9, article 705475. https://doi.org/ 10.3389/ fchem.2021.705475

[10] N. Shehata, O.A. Mohamed, E.T. Sayed, M.A. Abdelkareem, A.G. Olabi. (2022). *Geopolymer concrete as green building materials: Recent applications, sustainable development and circular economy potentials.* Science of the Total Environment, vol. 836, article 155577, https://doi.org/10.1016/j.scitotenv.2022.155577

[11] A.O. Purdon. (1940). *The action of alkalis on blast-furnace slag.* Journal of the Society of Chemical Industry 59, 191–202.

[12] V.D. Glukhovsky (1959). *Gruntosilikaty (Soil Silicates).* Grosstroyizdat, Kiev

[13] J. Davidovits (1982). *Mineral polymers and methods of making them.* US Patent 4,349,386

[14] F. Puertas (1985). *Cementos de escorias activadas alcalinamente. Situación actual y perspectivas de futuro.* Materiales de Construcción, vol. 45, nº 239, 53–64. https://doi.org/10.3989/mc.1995.v45.i239.553

[15] R. Robayo-Salazar, R. Mejía de Gutiérrez, F. Puertas. (2016). *Effect of metakaolin on natural volcanic pozzolan-based geopolymer cement.* Applied Clay Sciences, 132–133, 491–497. https://doi.org/10.1016/j.clay.2016.07.020

[16] J.S.J. Van Deventer (2017). Chapter 10. *Progress in the adoption of geopolymer cement.* J.S.J. Van Deventer: Handbook of Low Carbon Concrete, 2017, Butterworth Heinemann, 217–262. https://doi.org/10.1016/B978-0-12-804524-4.00010-5

[17] J. Provis, A. Palomo, Shi, C. (2015). *Advances in understanding alkali-activated materials.* Cement and Concrete Research, 78, 110–125. https://doi.org/10.1016/j.cemconres.2015.04.013

[18] A. Palomo, P. Krivenko, E. Kavalerova, O. Maltseva (2014). *A review on alkaline activation: new analytical perspectives.* Materiales de Construcción, vol. 64, nº 315, e022. https://doi.org/10.3989/mc.2014.00314

[19] R. A. Robayo-Salazar, R. Mejía de Gutiérrez, F. Puertas. (2019). *Alkali-activated binary concrete based on a natural pozzolan and granulated blast furnace slag: physical, mechanical and microstructural characterization.* Materiales de Construcción, vol. 69, nº 335, e191. https://doi.org/10.3989/mc.2019.06618

[20] F. Pacheco-Torgal, J.A. Labrincha, C. Leonelli, A. Palomo, A., Chindaprasirt, R. (Eds.). (2015). Handbook of Alkali-activated cements, mortars and concretes. Woodhead Publishing Series in Civil and Structural Engineering. https://doi.org/10.1016/C2013-0-16511-7

[21] J.L. Provis and J.S.J. van Deventer (Eds). (2014). Alkali Activated Materials. State of the Art Report, ILEM TC 224-AAM. Springer. https://doi.org/10.1007/978-94-007-7672-2

[22] C. Shi, P. Krivenko, D. Roy. (2006). Alkali-Activated Cements and Concretes. Taylor and Francis, London and New York. https://doi.org/10.4324/9780203390672

[23] C. Shi, A. Jiménez-Fernández, A. Palomo. (2011). *New cements for the 21st century: The pursuit of an alternative to Portland.* Cement and Concrete Research, 41, 750–763. http://doi:10.1016/j.cemconres.2011.03.016

[24] P. Pérez-Cortes, J. I. Escalante-García. (2020). *Alkali activated metakaolin with high limestone contents-Statistical modeling of strength and environmental and cost analyses.* Cement and Concrete Composites 106. 103450. https://doi.org/10.1016/j.cemconcomp.2019.103450

[25] F. Puertas, S. Martínez-Ramírez, A. Alonso, T. Vázquez. (2000). *Alkali-activated fly ash/slag cement. Strength behaviour and hydration products.* Cement and Concrete Research, vol. 30, pp. 1625–1632. https://doi.org/10.1680/adcr.1999.11.4.189

[26] J.M. Mejía, R. Mejía de Gutiérrez, F. Puertas. (2013). *Rice husk ash as a source of silica in alkali-activated fly ash and granulated blast furnace slag systems.* Materiales de Construcción, 63, 361–375. https://doi.org/10.3989/mc.2013.04712

[27] M. Torres-Carrasco, F. Puertas. (2017). *Waste glass as a precursor in alkaline activation: chemical process and hydration products.* Construction and Building Materials, 139, 342–351. https://doi.org/10.1016/j.conbuildmat.2017.02.071

[28] J. Payá, F. Agrela, J. Rosales, M.M. Morales, M.V. Borrachero. (2018). *Application of alkali-activated industrial waste.* New Trends in Eco-efficient and Recycled Concrete, 357–424. https://doi.org/10.1016/b978-0-08-102480-5.00013-0

[29] M. M. Mas, M. M. Tashima, J. Payá, M.V. Borrachero, L. Soriano, J. M. Monzó. (2016). *A binder from alkali activation of FCC waste: Use in roof tiles fabrication.* Key Engineering Materials, vol. 668, 411–418. https://doi.org/10.4028/www.scientific.net/kem.668.411

[30] F. Puertas, A. Barba, M.F. Gazulla, M.P. Gómez, M. Palacios, S. Martínez-Ramírez. (2006). *Ceramic wastes as raw materials in pórtland cement clinker fabrication: characterization and alkaline activation.* Materiales de Construcción, vol. 56 (281), pp. 71–83. https://doi.org/10.3989/mc.2006.v56.i281.94

[31] I. García Lodeiro, E. Aparicio-Rebollo, A. Fernández-Jiménez, A. Palomo. (2016). *Effect of calcium on the alklaline activation of aluminosilicate glass.* Ceramics International, 42, 7697–7707. https://doi.org/10.1016/j.ceramint.2016.01.184

[32] F. Puertas, M. Torres-Carrasco. (2014). *Use of glass waste as an activator in the preparation of alkali-activated slag cements. Mechanical strength and paste characterisation.* Cement and Concrete Research, vol. 57, 95–104. https://doi.org/10.1016/j.cemconres.2013.12.005

[33] M. Torres-Carrasco, F. Puertas. (2015). *Waste glass in the geopolymer preparation. Mechanical and microstructural characterisation.* Journal of Cleaner Production; 90, 397–408. https://doi.org/10.1016/j.jclepro.2014.11.074

[34] O. Burciaga-Díaz, M. Durón-Sifuentes, J.A. Díaz-Guillén, J.I. Escalante-García. (2020). *Effect of waste glass incorporation on the properties of geopolymers formulated with low purity metakaolin.* Cement and Concrete Composites, Vol 107, 103492. https://doi.org/10.1016/j.cemconcomp.2019.103492

[35] M.M. Alonso, C. Gascó, M. Martín Morales, J.A. Suárez-Navarro, M. Zamorano, F. Puertas. (2019). *Olive Biomas ash as an alternative activator in geopolymer formation. Strenth, radiological and leaching behaviour.* Cement and Concrete Composites, vol. 104. https://doi.org/10.1016/j.cemconcomp.2019.103384S

[36] M. de Moraes Pinheiro, A. Font, L. Soriano, M.M. Tashima, M.J. Monzó, M.V. Borrachero, J. Payá. (2018). *Olive-stone biomass ash (OBA): An alternative alkaline source for the blast furnace slag activation.* Construction and Building Materials, vol. 178, 30, 327–338. https://doi.org/10.1016/j.conbuildmat.2018.05.157

[37] T. Luukkonena, Z. Abdollahnejada, J. Yliniemia, P. Kinnunena, M. Illikainena. (2018). *One-part activated Materials: A review.* Cement and Concrete Research, 103, 21–34. doi.https://doi.org/10.1016/j.cemconres.2017.10.001

[38] A. Fernández-Jimenez, J.G. Palomo, F. Puertas. (1999). *Alkali-activated slag mortars: mechanical strength behaviour.* Cement and Concrete Research, 29, 1313–1321

[39]   S. A. Bernal, J. L. Provis, A. Fernández-Jiménez, P. V. Krivenko, E. Kavalerova, M. Palacios, C. Shi. (2014). *Chapter 3. Binder Chemistr. High- Calcium Alkali-Activated Materials.* Alkali-Activated Materials. State-of-the-Art Reports, RILEM TC 224-AAM. Ed. J.L. Provisand J.S.J. van Deventer. Springer

[40]   F. Puertas, A. Fernández-Jiménez, M.T. Blanco-Varela. (2004). *Pore solution in alkali-activated slag cement pastes. Relation to the composition and structure of calcium silicate hydrate.* Cement and Concrete Research, 34, 139–148. https://doi.org/10.1016/s0008-8846(03)00254-0

[41]   A. Fernández-Jiménez, F. Puertas. (2003). *Effect of activator mix on the hydration and strenght behaviour of alkali-activated slag cement.* Advances in Cement Research, 15 (3), 129–136. https://doi.org/10.1680/adcr.2003.15.3.129

[42]   J.L. Provis, A. Fernández-Jiménez, E. Kamsey, C. Leonelli, A. Palomo. (2014). *Chapter 4. Binder Chemistry. Low-Calcium Alkali-Activated Materials.* State-of-the-Art Reports, RILEM TC 224-AAM. Ed. J.L. Provisand J.S.J. van Deventer. Springer

[43]   A. Fernández-Jiménez, F. Puertas, J. Sanz, I. Sobrados. (2001). *Structure of calcium silicate hydrate formed in alkaline activated slag. Influence of the type of alkaline activator.* Journal of American Ceramic Society, 86 (8), 1389–94. https://doi.org/10.1111/j.1151-2916.2003.tb03481.x

[44]   F. Puertas, M. Palacios, H. Manzano, J.S. Dolado, A. Rico, J. Rodríguez (2011). *A model for the C-A-S-H gel formed in alkali-activated slag cements.* Journal of European Ceramic Society, 31, 2043–2056. https://doi.org/10.1016/j.jeurceramsoc2011.04.036

[45]   J. Schneider, M.A. Cincotto, H. Panepucci. (2001). *^{29}Si and ^{27}Al High-Resolution NMR characterization of calcium silicate hydrate phases in activated blast-furnace slag pastes.* Cement and Concrete Research, 31, 993–1001. https://doi.org/10.1016/s0008-8846(01)00530-0

[46]   S.A. Bernal, R. San Nicolas, R.J. Myers, R. Mejía de Gutiérrez, F. Puertas, J.S.J. van Deventer, J.L .Provis. (2014). *Slag chemistry controls phase evolution and structural changes induced by accelerated carbonation in alkali-activated binders.* Cement and Concrete Research, 57, 33–43. https://doi.org/10.1016/j.cemconres.2013.12.003

[47]   A. Fernández-Jiménez, F. Palomo. (2005). *Composition and microstructure of alkali-activated fly ash binder: Effect of the activator.* Cement and Concrete Research, 35, 1984–1992. https://doi.org/10.1016/j.cemconres.2005.03.0

[48]   I. Garcia-Lodeiro et al. (2015). *An overview of the chemistry of alkali-activated cement-based binders*, in Handbook of Alkali-activated Cements, Mortars and Concretes, Edited by F. Pacheco-Torgal, J. A. Labrincha, C. Leonelli, A. Palomo and P. Chindaprasirt, Woodhead Publishing, Oxford. https://doi.org/10.1533/9781782422884.1.19

[49]   A. Palomo, P. Krivenko, I. García-lodeiro, E. Kavalerona, O. Maltseva, A. Fernández-Jiménez. (2014). *A review on alkaline activation: New analitycal perspectives.* Materiales de Construcción, 64, 315, e022. https://doi.org/10.3989/mc.201.00314

[50]   F. Puertas, A. Fernández-Jiménez. (2003). *Mineralogical and microstructural characterization of alkali-activated fly ash/slag pastes.* Cement and Concrete Composites, 25, 287–292. https://doi.org/10.1016/s0958-9465(02)00059-8

[51] P. Pérez-Cortes, J. V. Escalante-García. (2020). *Gel composition and molecular structure of alkali-activated metakaolin-limestone cements.* Cement and Concrete Research, 137, 106211. https://doi.org/10.1016/j.cemconres.2020.106211

[52] I. García-Lodeiro, A. Fernández-Jiménez, A. Palomo, D. Macphee. (2010). *Effect of calcium additions on N-A-S-H cementitious gels.* Journal of American Ceramic Society, 93 (7), 1934–1940. https://doi.org/10.1111/j.1551.2916.2010.03668.x

[53] A. Palomo, P. Monteiro, P. Martauz, V. Bilek, A. Fernández-Jiménez (2019). *Hybrid binders: A journey from the past to a sustainable future (opus caementicium futurum).* Cement and Concrete Research, 124, 105829. https://doi.org/10.1016/j.cemconces.2019.105829

[54] V. Scarpini Cândido, A. Clay Rios da Silva, N. Tonini Simonassi, E. Sousa Lima, F. Santos da Luz, S. Neves Monteiro. (2018). *Mechanical and microstructural characterization of geopolymeric concrete subjected to fatigue.* Journal of Materials Research and Technology, 7(4), 566–570. https://doi.org/10.1016/j.jmrt.2018.07.011

[55] F. Pacheco T, J. Castro-Gomes, S. Jalali. (2007). *Investigations about the effect of aggregates on strength and microstructure of geopolymeric mine waste mud binders, Cement and Concrete Research,,* 37(6), 933–941. https://doi.org/10.1016/j.cemconres.2007.02.006

[56] W.G. Valencia-Saavedra, R.A. Robayo-Salazar, R. Mejía de Gutiérrez. (2021). *Propiedades de ingeniería de concretos híbridos activados alcalinamente basados en altos contenidos de ceniza volante: un análisis a largas edades.* Rev. UIS Ing. 20, 1–18. https://doi.org/10.18273/revuin.v20n3-2021001

[57] F. Puertas, R. Santos, M.M. Alonso, M. de. Rio. (2015). *Alkali-activated cement mortars containing recycled clay-based construction and demolition wastes.* Ceramics – Silikáty, 59(3), 202–210

[58] C. Arenas, Y. Luna-Galiano, C. Leiva, L.F. Vilches, F. Arroyo, R. Villegas, C. Fernández-Pereira. (2017). *Development of a fly ash-based geopolymeric concrete with construction and demolition wastes as aggregates in acoustic barriers.* Construction and Building Materials, 134, 433–442. https://doi.org/10.1016/j.conbuildmat.2016.12.119

[59] Z. Abdollahnejad, M. Mastali, M. Falah, T. Luukkonen, M. Mazari, M. Illikainen. (2019). *Construction and Demolition waste as recycled aggregates in alkali-activated concretes. Review.* Materials, 12, 4016. https://doi.org/10.3390/ma12234016

[60] C. Lu, Z. Zhang, C. Shi, N. Li, D. Jiao, Q. Yuan. (2021). *Rheology of alkali-activated material. A Review.* Cement and Concrete Composites, 121, 104061. https://doi.org/10.1016/j.cemconcomp.2021.104061

[61] F. Puertas, C. Varga, M.M. Alonso. (2014). *Rheology of alkali-activated slag pastes. Effect of the nature and concentration of the activating solution.* Cement and Concrete Composites. 53, 279–288. https://doi.org/10.1016/j.cemconcomp.2014.07.012

[62] S. Gismera, M. M. Alonso, M. Palacios, F. Puertas. (2020). *Rheology of alkali-activated mortars: Influence of particle size and nature of aggregates.* Minerals. 10, 726. https://doi.org/10.3390/min10080726

[63] M. Palacios, P.F.G. Banfill, F. Puertas. (2008). *Rheology and setting of alkali-activated slag pastes and mortars. Effect of organic admixtures.* ACI Materials Journal, 105, 140–148. https://doi.org/10.14359/19754

[64] F. Puertas, B. Gonzalez-Fonteboa, I. Gonzalez-Taboada, M.M. Alonso, M. Torres-Carrasco, G. Rojo, F. Martínez-Abella. (2018). *Alkali-activated slag concrete: Fresh and hardened behaviour.* Cement and Concrete Composites, 85, 22–31. https://doi.org/10.1016/j.cemconcomp.2017.10.003

[65] M.M. Alonso, S. Gismera, M.T. Blanco, M. Lanzón, F. Puertas. (2017). *Alkali-activated mortars: Workability and rheological behaviour.* Construction and Building Materials, 145, 576–587. https://doi.org/10.1016/j.conbuildmat.2017.04.020

[66] M. Palacios, M.M. Alonso, C. Varga, F. Puertas. (2019). *Influence of the alkaline solution and temperatura on the rheology and reactivity of alkali-activated fly ash pastes.* Cement and Concrete Composites, 95, 277–284. https://doi.org/10.1016/j.cemconcomp.2018.08.010

[67] G. Fang, W. K. Ho, W. Tu, M. Zhang. (2018). *Workability and mechanical properties of alkali-activated fly ash-slag concrete cured at ambient temperature* Construction and Building Materials, 172, 476–487. https://doi.org/10.1016/j.conbuildmat.2018.04.008

[68] R.A. Robayo-Salazar, W. Valencia-Saavedra, R. Mejía de Gutiérrez. (2020). *Construction and Demolition Waste (CDW) Recycling – as Both Binder and Aggregates – in alkali-activated Materials: A novel Re-use Concept.* Sustainability, 12, 5775. https://doi.org(10.3390/su12145775

[69] I. G. Lodeiro. A. Palomo, A. Fernández-Jiménez. (2007). *Alkali-aggregate reaction in activated fly ash systems.* Cement and Concrete Research, vol. 37, 175–183. https://doi.org/10.1016/j.cemconres.2006.11.002

[70] F. Puertas, M. Palacios, A. Gil-Maroto, T. Vázquez. (2009). *Alkali-aggregate behaviour of alkali-activated slag mortars: Effect of aggregate type.* Cement and Concrete Composites, vol. 31, 277–284. https://doi.org/10.1016/j.cemconcomp.2009.02.008

[71] M. Torres-Carrasco, T.M. Tognonvi, A. Tagnit-Hamou, F. Puertas. (2015). *Durability of alkali-activated slag concretes prepared using waste glass as alternative activator.* ACI Materials Journal, vol. 112 (6), 791–800. https://doi.org/10.14359/51687903

[72] A.M. Aguirre, R. Mejía de Gutiérrez. (2013) *Durability of reinforced concrete exposed to aggressive conditions.* Materiales de Construcción, 63, 309, 7–38. https://doi.org/10.3989/mc.2013.00313

[73] A. Wang, Y. Zheng, Z. Zhang, et al. (2020). *The Durability of Alkali-Activated Materials in Comparison with Ordinary Portland Cements and Concretes: A Review,* Engineering, 6(6), 695–706. https://doi.org/10.1016/j.eng.2019.08.019

[74] K. Chen, D. Wu, L. Xia, Q. Cai, Z. Zhang. (2021). *Geopolymer concrete durability subjected to aggressive environments – A review of influence factors and comparison with ordinary Portland cement.* Construction and Building Materials, 279, 122496. https://doi.org/10.1016/j.conbuildmat.2021.122496

[75] H. A. Alcamand, PHR. Borges, FA. Silva, ACC. Trindade. (2018). *The effect of matrix composition and calcium content on the sulfate durability of metakaolin and metakaolin/slag alkali-activated mortars.* Ceramics International, 44(5), 5037–5044. https://doi.org/10.1016/j.ceramint.2017.12.102

[76] F. Puertas, R. de Gutierrez, A. Fernández-Jiménez, S. Delvasto, J. Maldonado. (2002). *Alkaline cement mortars. Chemical resistance to sulfate and seawater attack.*

Materiales de Construcción, 52(267), 55–71. https://doi.org/10.3989/mc.2002.v52.i267.326

[77] S. Thokchom, P. Ghosh, S. Ghosh. (2010). *Performance of fly ash based geopolymer mortars in sulphate solution.* Journal of Engineering Science and Technology Review, 3(1), 36–38. https://doi.org/10.25103/jestr.031.07

[78] P. Duan, C. Yan, W. Zhou. (2016). *Influence of partial replacement of fly ash by metakaolin on mechanical properties and microstructure of fly ash geopolymer paste exposed to sulfate attack.* Ceramics International, 42(2), 3504–3517. https://doi.org/10.1016/j.ceramint.2015.10.154

[79] D. Ren, C. Yan, P. Duan, Z. Zhang, L. Li, Z. Yan. (2017). *Durability performances of wollastonite, tremolite and basalt fiber-reinforced metakaolin geopolymer composites under sulfate and chloride attack.* Construction and Building Materials, 134, 56–66. https://doi.org/10.1016/j.conbuildmat.2016.12.103

[80] P. Chindaprasirt, P. Paisitsrisawat, U. Rattanasak. (2014). *Strength and resistance to sulfate and sulfuric acid of* ground *fluidized bed combustion fly ash-silica fume alkali-activated composite.* Advanced Powder Technology, 25(3),1087–1093. https://doi.org/10.1016/j.apt.2014.02.007

[81] J. Zhang, CJ. Shi, ZH. Zhang, ZH. Ou. (2017). *Durability of alkali-activated materials in aggressive environments: a review on recent studies.* Construction and Building Materials, 152, 598–613. https://doi.org/10.1016/j.conbuildmat.2017.07.027

[82] S. Bernal, J.L. Provis. (2014) *Durability of alkali-activated materials: Progress and perspectives.* Journal of the American Ceramic Society, 97(4), 997–1008. https://doi.org/10.1111/jace.12831

[83] W. Valencia-Saavedra, D.E. Angulo, R. Mejía de Gutiérrez. (2016). *FlyAsh-Slag Geopolymer Concrete: Resistance to* Sodium *and Magnesium Sulfate Attack.* Journal of Materials in Civil Engineering (04016148-1/9). https://doi.org/10.1061/(ASCE)MT.1943-5533.0001618

[84] P. Duan, C.J. Yan, W. Zhou. (2016). *Influence of partial replacement of fly ash by metakaolin on mechanical* properties *and microstructure of fly ash geopolymer paste exposed to sulfate attack.* Ceramic International, 42(2), 3504–3517. https://doi.org/10.1016/j.ceramint.2015.10.154

[85] D. Bondar, C.J. Lynsdale, N.B. Milestone, N. Hassani. (2015). *Sulfate Resistance of Alkali Activated Pozzolans.* International Journal of Concrete Structures and Matererials, 9, 145–158. https://doi.org/10.1007/s40069-014-0093-0

[86] P. Aragón, R.A. Robayo-Salazar, R. Mejía de Gutiérrez. (2020). *Alkali-Activated Concrete Based on Natural* Volcanic Pozzolan: *Chemical Resistance to Sulfate Attack.* Journal of Materials in Civil Engineering, 32(5). https://doi.org/10.1061/(asce)mt.1943-5533.0003161

[87] K. Bouguermouh, N. Bouzidi, L. Mahtout, L. Pérez-Villarejo, ML. Martínez-Cartas. (2017). *Effect of acid attack on microstructure and composition of metakaolin-based geopolymers: the role of alkaline activator.* Journal of Non-Crystalline Solids, 463, 128–137. https://doi.org/10.1016/j.jnoncrysol.2017.03.011

[88] JR. Zheng, LN. Liu, LX. Xie. (2009). *Properties of the mortar and concrete of alkali-activated fly-ash cementing materials.* Concrete, 5, 77–79

[89]　XH. Zhao, CY. Liu, LM. Zuo, YZ. Pang, YF. Liu. (2018). *Experimental research on durability of new grouting materials with soda residue and fly ash matrix*. Industrial Construction, 48(3), 31–36 (in Chinese)

[90]　XJ. Song, M. Marosszeky, M. Brungs, R. Munn. (2015). *Durability of fly ash based Geopolymer concrete against sulphuric acid attack*. In 10DBMC International Conference On Durability of Building Materials and Components, Lyon (France), 17–20 April 2005. https://www.irbnet.de/daten/iconda/06059017363.pdf

[91]　W. *Jiang*, MR. Silsbee, E. Breval, MD. Roy, (1997). *Alkali-activated cementitious materials in chemically aggressive environments*. In: Young JF, editor. Mechanisms of chemically degradation of cement-based systems. E&FN SPON (1997) London, 289–296. https://doi.org/10.1201/9781482294958-44

[92]　T. Bakharev, JG. Sanjayan, YB. Cheng. (2003). *Resistance of alkali-activated slag concrete to acid attack*. Cement and Concrete Research, 33, 1607–1611. https://doi.org/10.1016/s0008-8846(03)00125-x

[93]　J. Zhang, C. Shi, N. Li, Z. Zhang, N. Farzadnia. (2018). 12 – *Carbon dioxide sequestration by alkali-activated materials*, in: F. Pacheco-Torgal, C. Shi, A.P. Sanchez (Eds.), Carbon Dioxide Sequestration in Cementitious Construction Materials, Woodhead Publishing, pp. 279–298. https://doi.org/10.1016/B978-0-08-102444-7.00012-5

[94]　A. Koenig, A. Herrmann, S. Overmann, F. Dehn. (2017). *Resistance of alkali-activated binders to organic acid attack: Assessment of evaluation criteria and damage mechanisms*, Construction and Building Materials, Construction and Building Materials151, 405–413. https://doi.org/10.1016/j.conbuildmat.2017.06.117

[95]　P. Duan, C. Yan, W. Zhou, W. Luo, C. Shen. (2015). *An investigation of the microstructure and durability of a fluidized bed fly ash-metakaolin geopolymer after heat and acid exposure*, Materials & Design, 74, 125–137 https://doi.org/10.1016/j.matdes.2015.03.009

[96]　A. *Mehta*, R. Siddique. (2017). *Sulfuric acid resistance of fly ash based geopolymer concrete*. Construction and Building Materials, 146, 136–143. https://doi.org/10.1016/j.conbuildmat.2017.04.077

[97]　W. Valencia, Ruby Mejía de Gutiérrez, F. Puertas. (2020). *Performance of FA-based geopolymer concretes exposed to acetic and sulfuric acids*. Construction and Building Materials, 257, 119503, 1–12. https://doi.org/10.1016/j.conbuildmat.2020.119503

[98]　F. Puertas, M. Palacios, T. Vázquez. (2006). *Carbonation process of alkali-activated slag mortars.* Journal of Materials Science, 41(10), 3071–3082. https://doi.org/10.1007/s10853-005-1821-2

[99]　M. Palacios, F. Puertas. (2006). *Effect of carbonation on alkali-activated slag pastes. Journal* of American Ceramic Society, 89(10), 3211–3221. https://doi.org/10.1111/j.1551-2916.2006.01214.x

[100]　G.J.G. Gluth, K. Arbi, S.A. Bernal, D. Bondar, A. Castel, S. Chithiraputhiran, A. Dehghan, K. Dombrowski-Daube, A. Dubey, V. Ducman, K. Peterson, P. Pipilikaki, S. L. A. Valcke, G. Ye, Y. Zuo, J. L. Provis. (2020). *RILEM TC 247-DTA round robin test: carbonation and chloride penetration testing of alkali-activated concretes*. Materials and Structures, 53(21). https://doi.org/10.1617/s11527-020-1449-3

[101] S. Bernal, J. L. Provis, B. Walkley, R.S. Nicolas, J.D. Gehman, DG. Brice, et al. (2013). *Gel nanostructure in alkali-activated binders based on slag and fly ash, and effects of accelerated carbonation.* Cement and Concrete Research, 53(2), 127–144. https://doi.org/10.1016/j.cemconres.2013.06.007

[102] S. Bernal S., R. San Nicolas, R.J. Myers, R. Mejía de Gutiérrez, F. Puertas, J.S. van Deventer, J. Provis J. (2014). *MgO content of slag controls phase evolution and structural changes induced by accelerated carbonation in alkali-activated binders.* Cement and Concrete Research, 57, 33–43. https://doi.org/10.1016/j.cemconres.2013.12.003

[103] R. Pouhet, M. Cyr, (2016). *Carbonation in the pore solution of metakaolin-based geopolymer.* Cement and Concrete Research, 88, 227–235. https://doi.org/10.1016/j.cemconres.2016.05.008

[104] R. A. Robayo-Salazar, A. M. Aguirre-Guerrero, R. Mejía de Gutiérrez. (2020). *Carbonation-induced corrosion of alkali-activated binary concrete based on natural volcanic pozzolan.* Construction and Building Materials, 232, 117189. https://doi.org/10.1016/j.conbuildmat.2019.117189

[105] M. Sufian Badar, K. Kupwade-Patil, SA. Bernal, JL. Provis, EN. Allouche. (2014). *Corrosion of* steel *bars induced by accelerated carbonation in low and high calcium fly ash geopolymer concretes.* Construction and Building Materials, 61, 79–89. https://doi.org/10.1016/j.conbuildmat.2014.03.015

[106] W.G. Valencia-Saavedra, A. María Aguirre-Guerrero, Ruby Mejía de Gutiérrez. (2020). *Alkali-activated* concretes *based on high unburned carbon content fly ash: carbonation and corrosion performance.* European Journal of Environmental and Civil Engineering, 26(8), 3292–3312. https://doi.org/10.1080/19648189.2020.1785948

[107] DM. Roy, W. Jiang, MR. Silsbee. (2000). *Chloride diffusion in ordinary blended and alkali-activated cement pastes and its relation to other properties.* Cement and Concrete Research, 30, 1879–1884. https://doi.org/10.1016/s0008-8846(00)00406-3

[108] RJ. Thomas, E. Ariyachandra, D. Lezama, S. Peethamparan. (2018). *Comparison of chloride* permeability *methods for alkali-activated concrete.* Construction and Building Materials, 165, 104–111. https://doi.org/10.1016/j.conbuildmat.2018.01.016

[109] JJ. Shi, CH. Deng, YM. Zhang. (2016). *Early corrosion behavior of rebars* embedded *in the alkali-activated slag mortar.* Journal of Building Materials, 19(6), 969–975 (in Chinese). http://doi.org/10.3969/j.issn.1007-9629.2016.06.003

[110] MSH. Khan, O. Kayali, U. Troitzsch, (2016). *Chloride binding capacity of hydrotalcite and the* competition *with carbonates in ground granulated blast furnace slag concrete.* Materials and Structures, 49(11), 4609–4619. https://doi.org/10.1617/s11527-016-0810-z

[111] XY. Ke, SA. Bernal, JL. Provis. (2017). *Uptake of chloride and carbonate by Mg-Al and Ca-Al layered double hydroxides in simulated pore solutions of alkali- activated slag cement.* Cement and Concrete Research, 100, 1–13. https://doi.org/10.1016/j.cemconres.2017.05.015

[112] Zhang, J., Shi, C., Zhang, Z. (2019). *Chloride binding of alkali-activated slag/fly ash cements.* Construction and Building Materials, 226, 21–31. https://doi.org/10.1016/j.conbuildmat.2019.07.281

[113] C.J.L. Dali Bondar Neil B. Milestone, and Nemat Hassani. (2012). Oxygen and Chloride Permeability of Alkali-Activated Natural Pozzolan Concrete, ACI Materials Journal, 109, 53–62. https://doi.org/10.14359/51683570

[114] M. Najimi, N. Ghafoori, M. Sharbaf. (2018). *Alkali-activated natural pozzolan/slag mortars: A parametric study.* Construction and Building Materials, 164, 625–643. https://doi.org/10.1016/j.conbuildmat.2017.12.222

[115] A. Aguirre-Guerrero, R.A. Robayo-Salazar, Ruby Mejía de Gutiérrez. (2021). *Corrosion resistance of alkali-activated binary reinforced concrete based on natural volcanic pozzolan exposed to chlorides* Journal of Building Engineering, 33, 101593. https://doi.org/10.1016/j.jobe.2020.101593

[116] M. Babaee, A. Castel. (2016). *Chloride-induced corrosion of reinforcement in low-calcium fly ash-based geopolymer concrete,* Cement and Concrete Research, 88, 96–107. https://doi.org/10.1016/j.cemconres.2016.05.012

[117] M. Babaee, A. Castel. (2018). *Chloride diffusivity, chloride threshold, and corrosion initiation in reinforced alkali-activated mortars: Role of calcium, alkali, and silicate content.* Cement and Concrete Research, 111, 56–71. https://doi.org/10.1016/j.cemconres.2018.06.009

[118] M. Criado and J.L. Provis (2018) *Alkali Activated Slag Mortars Provide High Resistance to Chloride-Induced Corrosion of Steel.* Frontiers in Materials, 5, 34, https://doi.org/10.3389/fmats.2018.00034

[119] C. Tennakoon, A. Shayan, J.G. Sanjayan, A. Xu. (2017). *Chloride ingress and steel corrosion in geopolymer concrete based on long term tests.* Materials & Design, 116, 287–299. https://doi.org/10.1016/j.matdes.2016.12.030

[120] J.K. Prusty, B. Pradhan. (2020). *Effect of GGBS and chloride on compressive strength and corrosion performance of steel in fly ash-GGBS based geopolymer concrete.* Materials Today: Proceedings, 3rd International Conference on Innovative Technologies for Clean and Sustainable Development, 32, 850–855. https://doi.org/10.1016/j.matpr.2020.04.210

[121] W.G. Valencia-Saavedra, R. Mejía de Gutiérrez. (2017). *Performance of geopolymer concrete composed of fly ash after exposure to elevated temperatures.* Construction and Building Materials, 154, 229–235. https://doi.org/10.1016/j.conbuildmat.2017.07.208

[122] M. Villaquirán-Caicedo, R. Mejía-de Gutiérrez, E.D. Rodríguez (2015). *Assessment of Metakaolin Based-Geopolymers Produced with Alternative Silica Sources Exposed to High Temperatures.* Ingeniería, Investigación y Tecnología, 16, 113–122 (in Spanish)

[123] M. Villaquirán-Caicedo, R. Mejía de Gutierrez, S. Sulekar, C. Davis, J. Nino. (2015). *Thermal properties of novel binary geopolymers based on metakaolin and alternative silica sources.* Applied Clay Science, 118, 276–282. https://doi.org/10.1016/j.clay.2015.10.005

[124] M. Villaquirán-Caicedo, R. Mejía de Gutiérrez, N.C. Gallego. (2017). *A Novel MK-based Geopolymer Composite Activated with Rice Husk Ash and KOH : Performance at High Temperature.* Materiales de Construcción, 67, 1–13. https://doi.org/10.3989/mc.2017.02316

[125] S. Bernal, J. Bejarano, C. Garzón, R. Mejía de Gutierrez, S. Delvasto, E. Rodríguez. (2012) *Performance of refractory aluminosilicate particle/fiber-reinforced geopolymer composites*. Composites Part B: Engineering, 43(4), 1919–1928. http://doi.org/10.1016/j.compositesb.2012.02.027

[126] P. He, D. Jia, T. Lin, M. Wang, Y. Zhao. (2010) *Effects of high-temperature heat treatment on the mechanical proper- ties of unidirectional carbon fiber reinforced geopolymer composites*. Ceramics International, 36(4), 1447–1453. doi.https://doi.org/10.1016/j.ceramint.2010.02.012

[127] M. Villaquirán-Caicedo, R. Mejia de Gutiérrez, R. (2019). *Mechanical and microstructural analysis of geopolymer composites based on metakaolin and recycled silica.* Journal of American Ceramic Society, 102, 3653–3662. https://doi.org/10.1111/jace.16208

[128] P. He, Z. Yang, J. Yang, X. Duan, D. Jia, S. Wang, Y. Zhao, Y. Wang, P. Zhang. (2015) *Preparation of fully stabilized cubic-leucite composite through heat-treating Cs-substituted K-geopolymer composite at high temperatures.* Composites Science and Technology, 107, 44–53. https://doi. org/10.1016/j.compscitech.2014.11.009

[129] P. Krivenko, S. Guziy. (2007). *Fire resistant alkaline Portland cements. In: Alkali activated materials – research, production and utilization.* 3rd conference, Prague, Czech Republic, 333–347

[130] I. Perná, T. Hanzlicek, P. Straka, M. Steinerova. (2007). *Utilization of fluidized bed ashes in thermal resistance applications. In: Alkali activated materials – research, production and utilization.* 3rd conference, Prague, Czech Republic, 527–537

[131] J. Temuujin, W. Rickard, M. Lee, A. Van Riessen. (2011). *Preparation and thermal properties of fire* resistant *metakaolin-based geopolymer-type coatings.* Journal of Non-Crystalline Solids, 357, 1399–1404. https://doi.org/10.1016/j.jnoncrysol.2010.09.063

Chapter 14

Fibre-Reinforced Polymer Reinforcement Towards More Durable and Sustainable Concrete Structures

João P. Firmo[1], Inês C. Rosa[1], João R. Correia[2], Tommaso D'Antino[3], Wallace Souza[4]

1 PhD, CERIS, Instituto Superior Técnico, University of Lisbon, Lisboa, Portugal
2 Full Professor, CERIS, Instituto Superior Técnico, University of Lisbon, Lisboa, Portugal
3 Associate Professor, Politecnico di Milano, Milano, Italy
4 Assistant Professor, Universidade Federal do Maranhão, São Luís, Brazil

14.1 Introductory Remarks

The costs associated with the rehabilitation of civil engineering structures built with traditional materials, namely with reinforced concrete (RC), steel, and timber, have increased quite significantly during the last decades. In many cases, this is due to the durability-related anomalies experienced by structures exposed to different environmental agents or due to the lack of preventive maintenance. In the future decades, those costs are expected to increase in many countries, owing to the age of existing infrastructure and the backlog of constructions presently requiring repair. As an example, it has been recently estimated that 42 % of all bridges in the United States of America are more than 50 years old and that 7.5 % of them are structurally deficient [1].

In RC structures, the most severe durability-related anomaly is associated with the corrosion of the steel reinforcement, caused by either carbonation or penetration of chlorides, namely in marine environments or in structures where de-icing salts are frequently used. These problems have promoted changes in codes and guidelines, which now address not only the structural design, but also durability requirements, aiming at guaranteeing that the design service life is attained. At the same time, efforts have been made to develop alternative reinforcing materials for concrete structures, less prone to corrosion or even non-corrodible. Among these materials, fibre-reinforced polymer (FRP) bars are being increasingly adopted due to their several advantages over mild steel, namely the non-corrosiveness, electromagnetic transparency, lightness, and high strength, thus offering a competitive alternative to epoxy-coated steel, galvanized steel, or stainless steel [2]. On the other hand, FRP composites present a few specificities and drawbacks that need to be duly accounted for in design, namely the brittle failure, the generally low elastic modulus, and the sensitivity to elevated temperature and to some environmental agents.

This chapter presents an overview of the use of FRP bars as internal reinforcement of concrete structures. Section 14.2 summarizes the main features of FRP bars, describing their constituent materials, typical geometries, manufacturing process, main physical and mechanical properties, and adherence to concrete; the most frequent applications are also described. Section 14.3 discusses the main aspects concerning the structural behaviour of FRP-RC members, namely the concepts for bending and shear design, highlighting the main differences with respect to conventional steel-RC members; the most relevant design codes and guidelines available in different countries are also presented. Section 14.4 addresses critical behavioural aspects of FRP-RC structures due to the unique inherent characteristics of the FRP reinforcement, namely concerning the fire behaviour, seismic behaviour, and long-term durability. Section 14.5 discusses the sustainability of using FRP reinforcement in RC structures. Finally, Section 14.6 presents some brief concluding remarks.

14.2 Fibre-Reinforced Polymer Bars for Reinforced Concrete Structures

14.2.1 Constituent Materials

FRP bars are made of two main constituents: (i) the fibre reinforcement (rovings), oriented in the axial direction of the bars, and (ii) the polymeric matrix that embeds the fibres (typically composed by a thermoset resin, that may also include fillers and additives). The fibre reinforcement, which governs the tensile properties of FRP bars, presents high strength, brittle behaviour (for all types of fibres), and variable stiffness, function of the type of fibre. The polymeric matrix has much lower mechanical properties compared to the fibre reinforcement, yet it is responsible for promoting a uniform stress distribution among the fibres, protecting them from environmental agents, and keeping them in position and aligned, delaying or preventing their buckling in compression.

Most FRP bars used as internal reinforcement of concrete members are made of glass fibres (due to their lower cost with respect to other fibres), generally of type AR or E-CR, as they offer improved performance in alkaline environments, such as concrete, compared to conventional type E glass fibres; some manufacturers also provide FRP bars made of carbon or aramid fibres and, more recently, also made of basalt fibres. As for the polymeric matrix, FRP bars for civil engineering applications are often produced with vinylester and, less frequently, with epoxy thermoset resins [3].

14.2.2 Geometries and Manufacturing

FRP bars are commercially available with very similar sizes to steel rebars: their diameters usually range between 6 mm and 36 mm, and they are supplied in lengths that typically vary from 10 m to 14 m. Some manufactures also provide FRP bars with small diameters (typically lower than 10 mm) in form of coil. The surface finish, which strongly affects the bond between FRP bars and concrete, depends on the manufacturer and the following types are available (Fig. 14.1): (i) surface formed deformations or ribs; (ii) sand coating, and (iii) exterior wound fibres, often applied together with a sand coating. In terms of geometry, most manufacturers only provide straight FRP bars, which are generally manufactured by pultrusion, an automated process of continuous production in which the fibres are first pulled (by a system of synchronized reciprocating pullers) into an open bath, where they are impregnated with the polymeric matrix, and then enter a heated die (with the cross-section of the bars) where the resin cures at temperatures that typically range between 90 °C and 180 °C. After exiting the die, the surface finish is applied, which can involve mechanical grinding (ribs) or bonding the sand coating with resin. FRP bars can also be produced by braiding or weaving.

Although conventional sawing equipment can be used to cut FRP bars to the desired length, because they are made of a thermoset resin, it is not possible to bend their extremities, either mechanically or by applying heat. Therefore, a few manufacturers also provide bent FRP bars (Fig. 14.2), with different shapes, namely L, U or closed (for stirrups); in this case, the FRP bars are not produced by pultrusion, but rather by injection of the polymeric resin into a closed mould already containing the (dry) fibre reinforcement.

Fig. 14.1   FRP bars with different types of surface finishes: a) sand coating b) sand coating with exterior wound fibres ribs; c) and d) ribs.

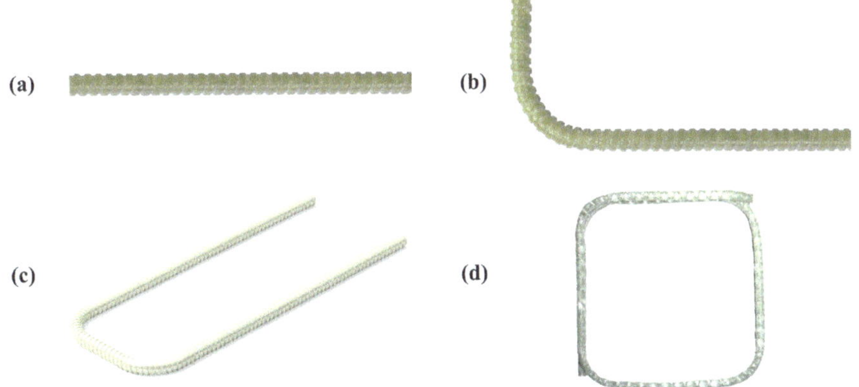

Fig. 14.2   FRP bars with different geometries: a) straight; b) L-bent; c) U-bent (courtesy of Schöck Bauteile GmbH); d) closed.

14.2.3 Mechanical and Physical Properties

Table 14.1 presents typical ranges of variation of mechanical and physical properties of FRP bars produced with different types of fibres, namely glass (GFRP, by far the most often used), carbon (CFRP), and aramid (AFRP). Like other composite components, FRP bars are orthotropic and their properties result from those of their constituent materials – the type of fibre reinforcement and polymeric resin – as well as their relative fractions, compatibility, manufacturing process, and quality control in production.

Table 14.1 Typical ranges of variation of physical and mechanical properties of FRP bars made of different types of fibres (adapted from [4]).

Property		GFRP	CFRP	AFRP
Axial tensile strength [MPa]		483–1600	600–3690	1720–2540
Axial elastic modulus [GPa]		35–60	120–580	41–125
Axial strain at failure [%]		1.2–3.1	0.5–1.7	1.9–4.4
Relative density [-]		1.25–2.10	1.50–1.60	1.25–1.40
Thermal expansion coefficient [×10^{-6}]	Axial	6.0–10.0	-9.0–0.0	-6.0 – -2.0
	Transverse	21.0–23.0	74.0–105.0	60.0–80.0

In terms of mechanical behaviour, FRP bars present a linear elastic response up to failure, which occurs for relatively high strains and, unlike steel, they do not yield – this lack of material ductility needs to be duly accounted for in design. Regardless of the type of fibres, the tensile strength of FRP bars is generally (much) higher than that of conventional steel bars and, unlike steel, it appears to be influenced by the diameter of the bars [5]. It is worth mentioning that the bent portions of FRP rebars generally have lower strength (typically about 50 %) than their straight parts, due to stress concentrations and kinking of the fibres in the bent portion [4]. The compressive strength of FRP bars can be significantly lower than their tensile strength – reductions of about 50 % have been reported for GFRP bars [6] – and this also needs to be considered when designing members or reinforcement under compression. The elastic modulus of FRP bars produced with different types of fibres can be very different. For GFRP bars, the elastic modulus is only 1/6 to 1/3 of steel and, as discussed ahead (*see* Section 14.3), this has a major influence on ultimate limit states and serviceability design, namely in what concerns cracking and deformability. Unlike steel bars, FRP bars are susceptible to creep rupture – this means that they can fail under sustained stresses, which consequently need to be limited at the design stage [7]; for GFRP bars, this limit (in service conditions) is about 20 % to 30 % of the short-term tensile strength [8, 9]. The anisotropic behaviour of FRP bars is reflected in their shear strength, which is generally very low, resulting in much lower dowel action compared to steel bars.

Regarding the physical properties of FRP bars, their relative density ranges from 1.25 to 2.10, which is about 1/6 to 1/4 of that of steel. This offers a potential advantage regarding transportation and handling on the construction site. The thermal expansion coefficient in the longitudinal direction (governed by the fibres) of GFRP bars is the same order of magnitude of concrete (CFRP and AFRP bars have negative values, which means that they contract when heated). However, the thermal expansion coefficient of GFRP bars in the transverse direction (governed by the resin) is much higher than that of concrete, and this may cause cracking under exposure to elevated temperature or even spalling of the concrete cover if the confining action of concrete is not sufficient.

14.2.4 Bond to Concrete

The bond of reinforcing bars to concrete is a crucial aspect for the structural behaviour of RC members because it influences the stress transfer between both materials. In general, there are three main mechanisms of stress transfer between reinforcing bars and concrete: (i) chemical adherence (conferred by the reaction at the surface of the bars with the cement paste); (ii) mechanical adherence (associated to the deformations of the surface of the rebars); and (iii) friction (associated to the roughness of the surface of the rebars). These mechanisms can be assessed by testing (pull-out tests and beam tests are the ones most frequently used), which can provide interaction (bond stress *vs.* slip) curves between both materials.

The bond between FRP bars and concrete depends, among other factors (e.g., those inherent to the concrete, the quality of its compaction, and the level of confinement), on the type of fibre reinforcement, the elastic modulus of the FRP bar, its surface finish, and its shape [4]. Unlike steel, in which bond stems essentially from the mechanical adherence, in FRP bars the frictional mechanism can be particularly relevant, namely in FRP bars with sand coating [10].

The test data available in the literature for FRP bars show that, depending on the type of FRP bar (and surface finish), it is possible to obtain bond strengths (often defined as the maximum value of the average shear stress along the FRP-concrete interface[1]) that are similar (or even higher) than those obtained with steel bars. It has also been shown that analytical models developed to describe the bond between steel bars and concrete, namely the model proposed by Eligehausen, Popob and Bertero [10] (BPE model), can be used to describe the FRP-concrete interaction; modified versions of the BPE model have also been proposed, the most popular being the mBPE and CMR models, proposed by Cosenza *et al.* [12, 13].

14.2.5 Applications

As mentioned, the key advantages associated to the use of FRP bars in RC structures are their non-corrodibility, electromagnetic transparency, low thermal conductivity (except for CFRP bars), lightness, and high tensile strength in the axial direction. These features explain the most frequent applications of FRP reinforcement in concrete structures (Fig. 14.3): (i) members exposed to chlorides, either in maritime environments or sub-

---

[1] The actual bond strength corresponds to the peak value of the bond stress *vs.* slip curve.

Fig. 14.3 Different applications of FRP bars: a) bridge deck; b) maritime structure; c) secant pile (courtesy of Owens Corning); and d) soft eye (courtesy of Schöck Bauteile GmbH).

jected to de-icing salts, namely bridge decks and slabs of parking buildings; (ii) chemical and other industrial plants with corrosive environments (e.g., waste water treatment plants); (iii) hospital buildings where magnetic resonance imaging (MRI) equipment is used; and (iv) buried temporary structures, namely diaphragm walls for tunnelling applications or secant piles, in which the low shear strength of FRP bars is an advantage, making them easier to be cut by tunnel-boring or pilling machines, respectively.

14.3 Structural Behaviour and Design Principles of Reinforced Concrete Members Internally Reinforced with Fibre-Reinforced Polymer Bars

The current body of knowledge about the behaviour of concrete structures reinforced with conventional steel bars (hereafter designated simply as steel-RC) resulted from more than one century of research activities and practical experience acquired from numerous structures that have been built, used, and studied worldwide. Steel-RC structures are designed based on well-established principles, typically assuming an elastic-plastic behaviour of the steel reinforcement; in fact, designers are recommended to exploit steel ductility, so that when steel yields, extensive cracking and large deformations are often possible with little or no loss of the load-carrying capacity, thus providing essential warnings of structural collapse. Additionally, steel-RC members can also be designed to present significant energy dissipation capacity, which is especially relevant for applications in seismic areas. However, when FRP reinforcement is used as an alternative to steel, its linear elastic behaviour up to failure and comparatively low stiffness (especially when

GFRP are used) have significant implications on the behaviour of FRP-RC structures and, consequently, on their design. The present section describes the fundamental behavioural aspects of FRP-RC structures, highlighting the main differences when compared to conventional steel-RC and providing the general concepts for bending (Section 14.3.1) and shear (Section 14.3.2) design of FRP-RC structural members. Finally, in Section 14.3.3, the most often used design guidelines/codes are presented.

14.3.1 Bending

The experimental data available in literature about the behaviour of FRP-RC members in bending have demonstrated that design principles similar to those used in steel-RC members can be adopted [9, 11], *i.e.*, standard section analysis techniques can be used for the determination of the flexural characteristics of FRP-RC beams and slabs. Thus, the following conventional assumptions are considered in the flexural design of FRP-RC members:

- Plane sections before loading remain plane at any stage of loading; therefore, strains in both the concrete and FRP bars are proportional to the distance from the neutral axis;
- There is perfect bond between concrete and the FRP reinforcement, ensuring strain compatibility within the cross-section.

In the above-mentioned context, the flexural strength of FRP-RC cross-sections can be determined based on strain compatibility, internal force equilibrium, considering the linear elastic behaviour up to failure of the FRP reinforcement, and the following possible failure modes: (i) FRP tensile rupture, (ii) concrete crushing, or (iii) both occurring simultaneously, in a so-called balanced-condition; it is worth mentioning that due to their relatively low compressive strength (and also due to the relatively low elastic modulus of most rebars), the contribution of the FRP rebars under compression for the flexural strength is usually disregarded.

If the flexural strength is controlled by the tensile rupture of the FRP reinforcement before concrete crushing (*i.e.*, in under-reinforced cross-sections), failure of the structural member is sudden (brittle), not exhibiting the ductile response that is typically observed for tension-controlled steel-RC flexural members [14, 15]. However, due to the high failure strain and relatively low elastic modulus of FRP rebars (especially of GFRPs, the most often used), some warnings of collapse can be perceived in the form of extensive cracking and significant (mostly elastic) deflection. If the flexural strength is controlled by concrete crushing with no tensile rupture of the FRP reinforcement (*i.e.*, in over-reinforced cross-sections), the flexural member does exhibit some inelastic behaviour prior failure, due to the non-linear response of concrete; therefore, the design of over-reinforced cross-sections is usually preferable for flexural members [16]. This inelastic response, often designated in the literature as "pseudo-ductile", can be enhanced if specific reinforcement (*i.e.*, closed stirrups) is adopted to provide additional confinement, thus increasing the deformation capacity of concrete under compression (*cf.* discussion in Section 14.4.2).

Irrespective of the failure mode, and due to the above-mentioned lack of ductility, plastic hinges are not expected to form in FRP-RC members; therefore, their moment redistribution capability is limited and should not be considered in the design of statically indetermined members. The use of FRP reinforcement in primary structural members subjected to seismic actions is not recommended by the current versions of design guidelines/codes, namely for applications in RC columns. In fact, as discussed in Section 14.4.2, the seismic behaviour of FRP-RC structures is still an open issue, for which additional research activities are needed, namely the development of technological solutions to improve their energy dissipation capacity when subjected to cyclic loadings.

To compensate the above-mentioned lack of ductility and the limited warnings of failure, most design guidelines recommend that FRP-RC flexural members should be designed with a high reserve of strength. This recommendation is usually satisfied because, as discussed next, the serviceability requirements govern the design of most FRP-RC flexural members, thus their strength is generally (much) higher than that required for the ultimate limit state in bending.

The performance under service conditions of FRP-RC flexural members is highly dependent on the stiffness of the FRP reinforcement. Due to their significantly lower cost with respect to carbon FRP rebars, GFRP rebars are the most often used; however, as mentioned in Section 14.2, their elastic modulus ranges between 1/6 to 1/3 of that of steel. The use of low stiffness reinforcement results in lower post-cracking flexural stiffness and, consequently, in higher deflections than those that would be obtained with conventional steel-RC members with similar reinforcement ratios (*e.g.*, [17]). In this context, deflections of FRP-RC members in bending under service loads can often govern the selection of the tensile reinforcement.

The comparatively low stiffness of FRP reinforcement can also result in wider cracks under service conditions [18]. However, as the electrochemical corrosion of FRP rebars is not a concern, crack widths larger than those typically observed for steel-RC members (set mostly to mitigate steel corrosion) are usually permitted on FRP-RC structures; in fact, the maximum crack widths allowed in most design guidelines (in some cases, up to 0.7 mm) are set for aesthetic reasons or to ensure a certain shear strength of the concrete. Regarding the methods that can be used to estimate the crack width in FRP-RC members, different approaches are proposed in current design guidelines/codes, most of them modifying existing equations developed for steel-RC members to account for the different stiffness and bond properties of FRP reinforcement (*e.g.*, [9]) – in this regard, as mentioned above, it is now possible to have FRP rebars with similar (or even higher) bond properties than those of standardized ribbed steel bars (*see* Section 14.2.4), due to continuous technological developments during the last decades by the composites industry.

Another important aspect that needs to be accounted for in safety verifications of FRP-RC members is the maximum stress installed in the FRP reinforcement under service conditions. As mentioned, FRP bars are susceptible to creep rupture, meaning that they can fail under relatively low sustained stresses [19, 20]. The sensitivity to this phenomenon depends on the type of fibre reinforcement, being more severe on GFRP than on CFRP rebars. In fact, current design guidelines/codes recommend that the maximum allowable stress in FRP rebars at service load levels should be limited; for GFRP reinforcement, those limits are usually within the range of 20 % to 30 % of their "instantaneous (quasi-static)" tensile strength; this serviceability verification can also influence the selection/design of the tensile reinforcement.

As highlighted in the previous paragraphs, serviceability requirements often govern the design of FRP-RC flexural members. For this reason, it is unanimously accepted that the design of FRP over-reinforced cross-sections is inevitable for most structural applications. Consequently, for the ultimate limit state in bending, the collapse of structural members will occur due to concrete crushing and for stresses in the FRP reinforcement much lower than their tensile strength. This means that for most structural applications, the high tensile strength of FRP rebars is far from being fully exploited.

14.3.2 Shear

Independently of the reinforcement material, the shear behaviour of RC structural members is a complex phenomenon that relies on the contribution of four main internal load-carrying mechanisms: (i) shear resistance provided by the uncracked compression zone; (ii) contribution of the aggregate interlock developed throughout cracked zones; (iii) dowel action provided by the longitudinal reinforcement in cracked sections; and (iv) shear resistance of the specific shear reinforcement, if any. The individual contribution of these mechanisms depends not only on the concrete characteristics, but also, reciprocally, on the mechanical and bond properties of the reinforcement. In this regard, it is worth reminding that despite the important efforts developed during the last decades to fully characterize the shear behaviour of steel-RC members, this is still subject of debate. Comparatively, the shear behaviour of FRP-RC structural members has been much less investigated (*e.g.,* [21, 22]) than that of steel-RC members. The following paragraphs describe how the brittle nature of the FRP bars, their anisotropy, and low stiffness influence the above-mentioned mechanisms, and how these effects are considered in current design guidelines/codes for FRP-RC structures.

The contribution of uncracked concrete for the shear strength is highly dependent on the depth of the neutral axial, which, in turn, is dependent on the characteristics of the longitudinal reinforcement. Since most FRP rebars have lower axial stiffness than steel, FRP-RC members exhibit smaller depths of the neutral axis than those of comparable steel-RC members. As a result, the shear resistance provided by the compressed concrete is lower in FRP-RC members; however, this conclusion cannot be generalized for all scenarios, especially because very high longitudinal reinforcing ratios are often adopted in FRP-RC members to fulfil service requirements (see previous section), thus increasing the neutral axis depth.

The contribution of the shear transfer mechanism through aggregate interlock is highly dependent on the magnitude of the crack opening. As discussed in the preceding section, FRP-RC members typically present higher deflections and wider cracks that those of comparable steel-RC members, therefore, the importance of this mechanism for the overall shear strength of the structural member is smaller when FRP reinforcement is used [23].

Within the four main internal load-carrying mechanisms that contribute to the shear strength, the dowel action is unanimously identified as the least important one [24]. It depends mainly on the bending and transverse shear properties of the reinforcing bars that cross a crack. As highlighted in Section 14.2, FRP rebars are anisotropic and present very low shear stiffness and strength; for this reason, the shear capacity conferred by dowel action can be considered negligible in FRP-RC members.

When the sum of the contributions of the above-mentioned mechanisms is lower than the shear demand, transverse reinforcement needs to be adopted. Presently, many manufacturers offer bent and/or closed form FRP rebars that can be used as shear reinforcement in concrete members. However, as mentioned in Section 14.2, the bent portions of FRP bars generally exhibit significantly lower strength compared to straight bars, with reductions of around 50 % being relatively common [25]. This reduction in the tensile strength depends on the diameter of the bar and the bending radius, with this dependence being duly considered in most design guidelines/codes.

The research activities developed during the last decades allowed for concluding that the (total) shear strength of FRP-RC members can be accurately predicted by the sum of the shear resistances of the concrete and of the shear reinforcement, provided that shear cracks are effectively controlled (*e.g.*, [26, 27]). In these circumstances, the total shear strength of the member can be calculated using modified versions of existing predictive equations based on well-established principles defined for the design of steel-RC structures, taking into account the reduced stiffness and strength (of bent portions) of the FRP reinforcement. Interested readers are referred to the guidelines listed in the next section, where the modifications introduced into different formulations for steel-RC structures are described.

14.3.3 Available Design Guidelines/Codes

Since the 1990s, a significant number of organizations around the world have developed design guidelines and/or codes for the use of FRP reinforcement in concrete structural members. As highlighted in the previous sections, most of the provisions included in these guidelines/codes are based on well-consolidated knowledge and methodologies developed for conventional steel-RC structures, with the necessary adaptations to consider the different characteristics of FRP reinforcement and behavioural aspects of FRP-RC structures. It is worth referring that due to the lower technological maturity of FRP-RC structures (especially when compared to steel-RC), most of the existing design provisions are conservative, and they are only intended to be used for selected applications; in fact, there are still some critical aspects (discussed in Section 14.4) that need to be investigated in further depth before they can be incorporated into guidelines/codes, allowing for a more optimized (and economical) design and to further expand the use of FRP reinforcement to applications still not recommended or permitted in the current versions of design guidelines/codes, such as columns or primary structural members in seismic regions.

Table 14.2 lists guidelines/codes produced by different technical committees that have been used in the design and construction of FRP-RC structures worldwide. From a relatively long list of documents produced during the last decades, the authors selected those that have been widely disseminated and were published by internationally recognized organizations, such as the *Japan Society of Civil Engineers* (JSCE*)*, the *International Federation for Structural Concrete* (FIB), the Italian *National Research Council* (CNR), the *American Concrete Institute* (ACI), and the *Canadian Standard Association* (CSA). As a consequence of the continuous developments from both the composites industry and the research community, new in-depth knowledge has been acquired in recent years and, therefore, included in updated versions of the referred guidelines/codes. In this regard, it is worth mentioning that the next generation of the Eurocode 2 (part 1.1) will include an informative Annex regarding the design of concrete structural members reinforced with FRP bars – this document is presently being prepared by the *European Committee for Standardization* (CEN).

Table 14.2 Design guidelines and codes for FRP-reinforced concrete structures.

Guideline/code	Organization/ publisher	Year*
Recommendation for Design and Construction of Concrete Structures Using Continuous Fiber Reinforcing Materials [28]	JSCE	1997
Guide for the Design and Construction of Concrete Structures Reinforced with Fiber-Reinforced Polymer Bar – CNR-DT203/2006 [29]	CNR	2007
FRP Reinforcement for RC Structures – Bulletin 40 [11]	FIB	2007
Design and Construction of Building Components with Fibre-Reinforced Polymers – CSA-S806-12 [30]	CSA	2012
Guide for the Design and Construction of Structural Concrete Reinforced with FRP Bars – ACI 440.1R-15 [4]	ACI	2015
Building Code Requirements for Structural Concrete Reinforced with Glass Fiber-Reinforced Polymer (GFRP) Bars – ACI CODE-440.11-22 [9]	ACI	2022
Informative Annex R in the next generation of EN 1992-1-1	CEN	-

*Note: year of publication of the current version of the document.

14.4 Critical Topics for FRP-Reinforced Concrete Structures

14.4.1 Fire Behaviour

The high susceptibility of GFRP reinforcement to elevated temperature is a major concern regarding its use in structural applications. However, it has been shown that GFRP-RC structural members can fulfil the fire safety requirements set for building applications and even attain comparable fire resistance to that presented by steel-RC members with an equivalent ambient temperature strength [31, 32]. However, due to noteworthy differences in the behaviour of GFRP and steel rebars at elevated temperature, the fire endurance of GFRP-RC members is usually governed by the degradation of the mechanical properties of the rebars with temperature and/or by the loss of bond between the bars and concrete.

Compared to steel, the strength of GFRP bars, as well as their bond to concrete, are more intensely reduced with the temperature increase, especially at mildly elevated temperatures, near the glass transition temperature (T_g) of the polymeric resin. GFRP bars (typically produced with vinylester or epoxy thermoset resins) generally present a T_g between 93 ºC and 120 ºC [4]. When the temperature of the bars approaches the T_g, the matrix softens and becomes less capable of efficiently transferring stresses amongst the fibres. Consequently, the properties of the bars are greatly reduced, especially those that depend primarily on the polymeric matrix. For example, the compressive strength [33] and transverse shear strength [34] of GFRP bars, which are mainly governed by the fibre-matrix interaction, are severely affected by the softening of the resin and were found to be fully degraded before the decomposition temperature of the resin (T_d) is attained, which typically varies between 250 ºC and 400 ºC for most thermoset resins [35]. However, GFRP bars are capable of retaining considerable tensile strength (*cf.* Fig. 14.4) and stiffness (*cf.* Fig. 14.5) up to very high temperatures above T_d, as these properties are more reliant on the behaviour of the fibres. The most commonly used type of glass fibres (E-glass) softens at approximately 830 ºC and melts at 1070 ºC, thus they present much higher thermal stability than the polymeric resin [35]. The tensile modulus of GFRP bars, a fibre-dominated property, typically experiences minor reductions up to the decomposition of the resin; according to literature (*cf.* Fig. 14.5), some GFRP bars can retain up to nearly 70 % of their original tensile modulus at temperatures as high as 715 ºC [36]. The tensile strength, on the other hand, relies on the integrity of the fibres but also on the fibre-matrix interaction. As a result, it can be reduced up to 40 % near the T_g (compared to ambient temperature), being still non-negligible (though can be very degraded) at 500 ºC – at this temperature (above T_d), the tensile strength can decrease by 60 % [37] up to nearly 90 % [38] compared to ambient temperature (*cf.* Fig. 14.4).

Although the softening of the polymeric resin does not necessarily jeopardize the tensile capacity of GFRP bars, it is a major concern regarding the bond between the bars and concrete; in fact, it severely affects the ability to transfer tangential stresses from the concrete to the fibres, thereby potentially compromising the reinforcement anchorage capacity, even at moderately elevated temperatures. As shown in Fig. 14.6 (also *cf.* Fig. 14.7), the bond strength of GFRP bars is severely reduced near the T_g – in more severe cases (*e.g.*, [57]), it can decrease up to 80 % compared to ambient temperature, and it is generally marginal at temperatures close to the T_d.

Many parameters affect the bond of GFRP bars to concrete at ambient temperature, yet there is little information available regarding the bond behaviour at elevated temperatures. Studies have shown that the T_g and surface finish of the bars seem to be the dominant parameters governing the bond degradation rate with temperature [56, 59]. According to Rosa *et al.* [59], most of the bond strength of sand coated bars is lost below the T_g (*cf.* Fig. 14.7a), whereas that of ribbed bars is more degraded above that temperature (*cf.* Fig. 14.7b); note that in the referred figures the bars have similar values of T_g. The differences in behaviour associated with these two types of surface finishes are related with the fact that the superficial resin that binds the sand grains is typically more sensitive to thermal degradation than the resin in which the fibres are embedded, as the surface coating is usually applied over the hardened bars and therefore cured under different conditions than those of the bar core

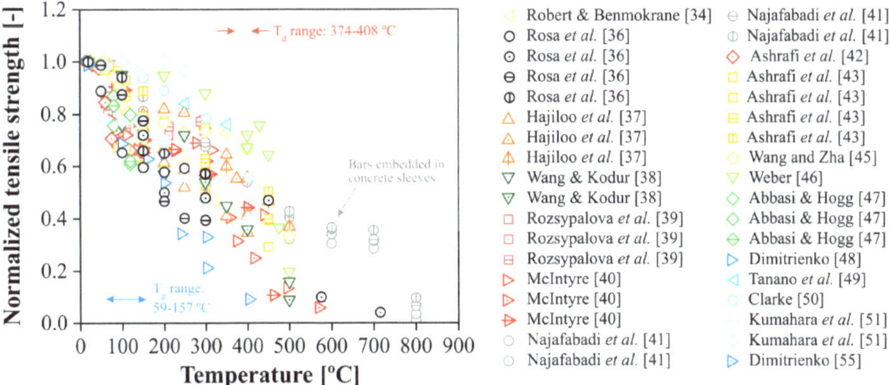

Fig. 14.4 Normalized tensile strength of GFRP bars (compared to ambient temperature) as a function of temperature – data from [34, 36–55] (adapted from [56]).

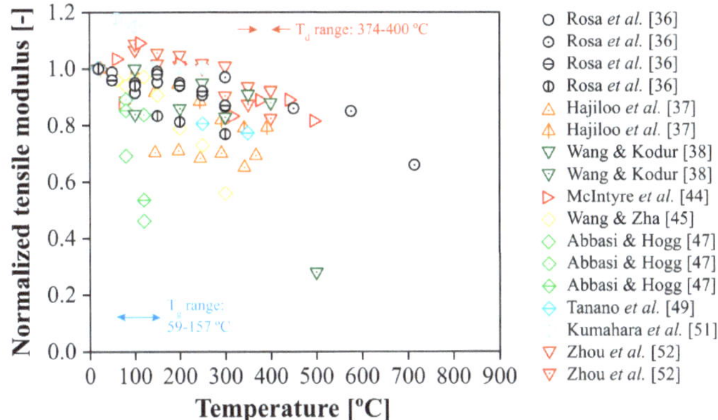

Fig. 14.5 Normalized tensile modulus of GFRP bars (compared to ambient temperature) as a function of temperature – data from [36–38, 44, 45, 47, 49, 51, 52] (adapted from [56]).

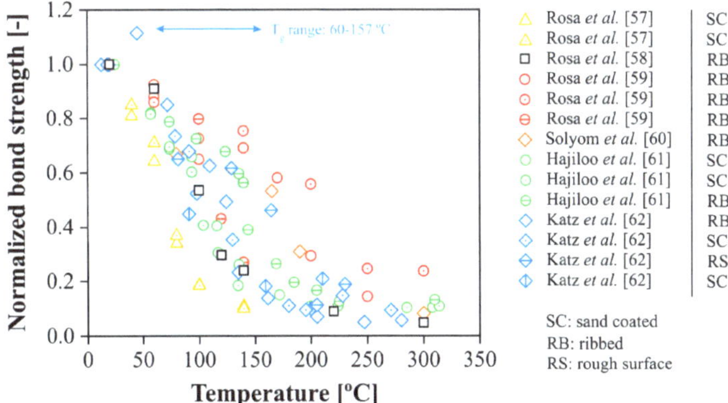

Fig. 14.6 Normalized bond strength of GFRP bars as a function of temperature – data from [57–62].

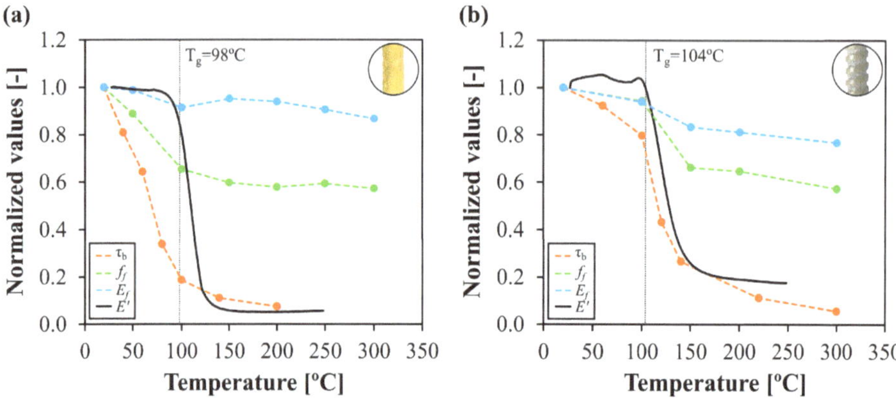

Fig. 14.7 Normalized bond strength (τ_b), tensile strength (f_f), tensile modulus (E_f) and storage modulus (E', obtained by dynamic mechanical analysis) of GFRP bars as a function of temperature – results for (a) sand coated and (b) ribbed rebars (adapted from [57, 59]).

[57]. Conversely, the ribs of ribbed rebars are ground into the hardened bars or moulded during the curing process, hence cured at the same time as the core.

The behaviour of GFRP bars at elevated temperature is complex and difficult to generalize given that the temperature effects on the degradation rate of the bar properties can present significant differences, as depicted in Fig. 14.4 to Fig. 14.6. This is explained by the fact that the degree and rate of degradation with temperature depend on various factors, namely: (i) the thermophysical properties of the bar constituent materials (matrix and fibres); (ii) geometrical features of the bars (*e.g.*, surface finish, cross-section); and (iii) extrinsic factors related with the heating and loading history (higher temperatures or longer periods of exposure to heat normally induce higher deterioration), and the exposure environment (if the rebars become directly exposed to fire as a result of a wide crack in concrete, the contact with oxygen accelerates the decomposition reaction of both the resin and fibres, leading to the premature rupture of the reinforcement) [35, 36].

The consequences of the severe degradation of the GFRP rebar mechanical and bond properties at elevated temperature on the fire resistance of RC beams and slabs have been investigated over the past two decades. Although some aspects are not yet sufficiently well understood, the knowledge about this subject has seen considerable progress in recent years, showing that it is possible to use GFRP reinforcement in load-bearing applications where fire safety is a concern. Yet, for engineering practice, the potential for replacing steel with GFRP rebars in buildings still remains greatly unexploited, mainly due to the limited availability of fire design guidance in FRP-RC codes. Presently, only the Canadian CAN/CSA S806-12 standard [30], and the American Concrete Institute's ACI 440.1R-15 guide [4] and ACI CODE-440.11-22 [9], provide informative (non-regulatory) fire design recommendations, though in general, they are limited in scope and poorly detailed comparing to those available for steel-RC fire design. The recent ACI CODE-440.11-22 [9], in particular, only allows the use of GFRP reinforcement in structures with fire resistance requirements if the required performance is demonstrated through calculations or tests.

It is well demonstrated in the literature (*e.g.*, [31, 63, 64]) that, if the extremities of the reinforcement attain the T_g, debonding (pull-out), failures are prone to occur at the anchorage zones, hence triggering premature collapses and hindering the tensile capacity of the rebars at elevated temperature to be fully exploited. However, FRP-RC standards/ guides (e.g., [9]) state that, provided that continuous reinforcement is adopted between the supports and that its extremities are kept sufficiently cold (*i.e.*, below a certain critical temperature, typically considered equal to the T_g), structural members will not be at imminent risk of collapse when the GFRP-concrete bond becomes fully deteriorated along the fire exposed span. In fact, if the above conditions are met, GFRP-RC beams and slabs are capable of enduring from 60 min to well over 180 min of fire exposure (*e.g.*, [40, 64–68]). The reason for this is that, once the resin begins to decompose (around the T_d) and bond is lost along the exposed span, the member load bearing capacity becomes fully reliant on the tensile strength of the rebars (*i.e.*, the integrity of the glass fibres) at elevated temperature, as well as on their anchorage capacity at the cold extremity zones. In other words, the main reinforcement starts behaving like cables, anchored in the cold extremities of the member where bond is less damaged. As a result, failure typically occurs when the rebars temperature is high enough to cause a meaningful degradation of the glass fibres – in fact, the most frequent failure mode of slabs with continuous rebars reported in literature is the tensile rupture of the main reinforcement at temperatures (Fig. 14.8a and Fig. 14.8b) in the range of 500 ºC to 713 ºC [40, 65, 66, 68].

The fire resistance of GFRP-RC members depends on the anchoring conditions of the rebars, on the degradation of their mechanical properties and bond to concrete at elevated temperature (*i.e.*, on the material and geometrical features of the rebars), and also on the concrete cover. Since durability issues are generally not a concern in GFRP-RC members, the concrete cover should be adequate to: (i) ensure the transmission of bond stresses; (ii) prevent heat-induced spalling resulting from the higher transverse thermal expansion coefficient of GFRP bars in comparison with that of concrete; and (iii) limit the temperature increase in the rebars to a certain critical value (currently not well-defined), preventing them from rupturing prematurely. The CAN/CSA S80612 standard [30] recommends covers of 50 to 60 mm to guarantee a 60 min fire resistance rating. However, several studies [31, 63–66] have shown that it is feasible to attain fire resistances greater than 180 min using considerably lower covers than those prescribed in [30], provided that the rebars are properly anchored in cool zones of the structure. For example, in [65], slabs with continuous rebars endured 181 min and 221 min of fire exposure with covers of 25 mm and 35 mm, respectively. It is worth noting that the fire resistance can also be affected by the concrete strength, load level, and axial restraint, among other parameters; yet, further research is still needed to comprehensively understand the influence of these parameters in the fire performance of GFRP-RC members [31, 56].

The concerns about the severe loss of the GFRP-concrete bond during fire exposure are relevant in the design of anchorage zones and also of tension lap splices. Fire resistance tests performed in beams and slabs with lap splices directly exposed to fire (*e.g.*, [40, 65, 66]) showed that when straight rebars were used and the splice lengths were designed for normal (ambient) temperature conditions, the GFRP-RC members were unable to achieve fire resistances above 60 min, due to the debonding of the spliced rebars

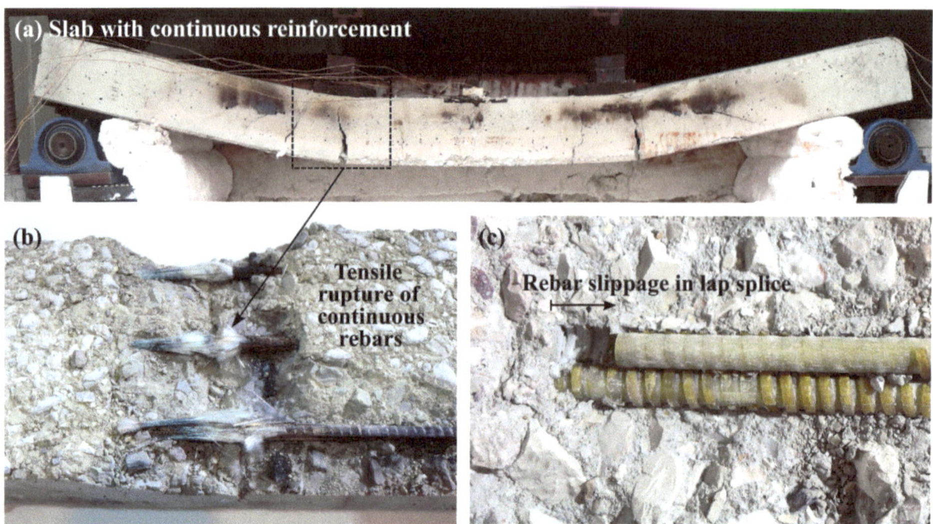

Fig. 14.8 Typical failure modes of fire-exposed GFRP-RC slabs with (a,b) continuous and (c) spliced reinforcement (adapted from [32]).

(Fig. 14.8c), unless relatively thick concrete covers (60 mm, defined according to [30]) were used – in that case, the members endured up to 90 min of fire exposure before the rebars slipped [46, 69, 70]. In more severe cases [40, 65, 66], flexural members with thinner covers (20–25 mm) collapsed after less than 20 min of fire exposure, therefore failing to comply with the minimum fire resistance period of 30 min typically required for small building applications. Regardless, even in studies where thicker covers were adopted [46, 69, 70], it was concluded that the splice lengths designed for normal temperature conditions are largely insufficient to anchor GFRP reinforcement when it is exposed to elevated temperature. It is worth mentioning that current FRP-RC codes and guidelines do not provide any design formulae for the development length of FRP rebars that consider the effects of temperature on the degradation of bond to concrete.

Based on the above, anchorage zones and lap splices should be protected from direct heat exposure to prevent the temperature of the rebars from attaining the T_g. To that end, they should be preferably located/extended into cooler areas of the structure, such as connection zones between slabs and beams, over partition walls or columns. The length and geometry of the cold anchorages and lap splices should be designed considering the influence of the T_g and surface finish of the bars on the bond degradation rate with temperature. The limited research conducted on members with bent GFRP reinforcement [58, 64, 65] demonstrated that this bar shape allows to considerably improve the rebar anchorage capacity at elevated temperature and, therefore, this geometry may be preferable over straight rebars. According to [58, 64, 65], bent reinforcement can be used to reduce the cold development lengths required to anchor GFRP bars, as it allows anchoring the rebars in cooler zones of the structure/member. It is also an effective solution to enhance the bond performance in lap splices – in [65], the fire resistance of GFRP-RC slabs with lap splices (directly exposed to fire) improved from 26 min, using straight-end rebars, to 75 min, using 90° bent-end rebars.

14.4.2 Seismic Behaviour

As mentioned in Section 14.3, the brittle nature of FRP rebars and the widely assumed lack of ductility of FRP-RC members raise well-founded concerns about their use in concrete structural members in seismic areas. In fact, due to these concerns and the limited number of studies developed on this topic, the existing guidelines/codes do not recommend the use of FRP bars as longitudinal reinforcement in columns. Moreover, they do not provide guidance regarding the seismic design of FRP-RC elements – the only exception is a simplistic comment made in the CNR design code [29], mentioning that in the case of primary structural members subjected to seismic actions, the design spectrum shall be derived from the elastic one by setting an appropriate value for the structural (behaviour) factor to account for the (mostly) elastic response of FRP-RC structural members. However, no guidance is provided about what is considered an *"appropriate value for the structural (behaviour) factor"*, neither how it can be determined. The above-mentioned concerns and the lack of guidance about seismic design have contributed to hinder a widespread adoption of FRP reinforcement in concrete structures, especially in primary structural members located in seismic regions. In fact, this lack of guidance is also explained by the relatively limited number of studies assessing: (i) the stress redistribution capabilities of statically indetermined FRP-RC members (a fundamental aspect to dissipate inelastic energy), and (ii) the structural response of FRP-RC columns and beam-to-column joints subjected to cyclic loadings.

A few experimental studies developed on continuous FRP-RC beams subjected to monotonic static loadings showed that they can exhibit semi-ductile failure modes, therefore, potentially being able to dissipate inelastic energy (*e.g.,* [71, 72]), while others have shown that the well-known beneficial effects of RC confinement with stirrups enhances the performance of "plastic" hinges also in statically indetermined FRP-RC structural members [73, 74].

Regarding the response under cyclic loading, analytical models have been proposed for the hysteretic moment-displacement relationship of FRP-RC columns under constant axial loading (*e.g.,* [73, 75]); as a consequence of the linear elastic behaviour of FRP bars, the main difference to conventional hysteretic models for steel-RC is that the unloading branches of hysteretic loops aim towards the origin of the moment-displacement curves. As discussed in Section 14.3.1, regarding pure bending, completely brittle failures can be prevented if failure occurs due to concrete crushing prior to tensile failure of the FRP bars, provided that the concrete is sufficiently well confined and that over-reinforced cross-sections are designed, as opposed to the typical design concept of under-reinforced sections used for steel-RC.

Experimental studies on FRP-RC columns under cycling lateral loading [76] showed that, if sufficient concrete confinement is provided by transverse FRP reinforcement, significant lateral drift ratios (up to 2–3 %, corresponding to the ratio between the imposed lateral displacement and the columns height) can be attained; although no ductility was observed, the elastic deformation of the specimens can be sufficient to meet seismic drift limits defined in most building codes. In light of these results, together with the typically lower stiffness of FRP-RC members (compared to steel-RC), the authors of these studies suggested that an elastic approach with sufficient deformability may be appropriate for their seismic design, as reduced spectral values associated with longer vibration periods are expected in FRP-RC structures.

Other experimental investigations on rectangular FRP-RC columns under cyclic lateral loading [77, 78] allowed for concluding that the dissipated energy was approximately half of that dissipated by steel-RC reference specimens; these studies also confirmed that adequate stirrups spacing is needed to ensure concrete confinement and thus to increase the deformation capacity of FRP-RC columns. Previous studies (*e.g.*, [79]) also pointed out that the large elastic range of the response of FRP-RC columns may provide recentring capabilities, reducing the residual deformations after a seismic event and, consequently, the repair costs.

Studies on steel-RC and GFRP-RC beam-to-column joints under cyclic loading [79–81] showed that the latter joints can present adequate deformability, but may lack ductility and energy dissipation capacity. In order to enhance such energy dissipation capacity, some authors assessed the potential beneficial effect of combining ductile reinforcement with FRP bars, *i.e.*, adopting a concept that is usually designated as hybrid reinforcement. In fact, some authors [82] confirmed that hybrid steel-FRP-RC joints can present an intermediate seismic performance between fully steel-RC and FRP-RC joints; however, the non-corrodibility of that hybrid solution is hindered by the use of mild steel. In this context, Nehdi *et al.* [83] proposed an alternative RC hybrid joint with shape memory alloy (SMA)-FRP reinforcement, which presented adequate energy dissipation capacity; however, as pointed out by the authors, this solution is presently not feasible due to the high costs of SMA.

To illustrate the differences between the overall structural response of GFRP-RC and comparable steel-RC columns, the following paragraphs summarize a recent experimental study developed by the authors, in which both types of columns were simultaneously subjected to a constant axial load and (i) monotonic or (ii) cyclic lateral loadings. It is worth mentioning that the results that are briefly presented next are part of a more comprehensive experimental and numerical study [6] that addressed the influence of various parameters on the structural response of the above-mentioned columns under different types of loading (*e.g.*, concentric, eccentric with monotonic and cyclic lateral loadings), including the use of hybrid longitudinal reinforcement, combining GFRP and stainless-steel bars.

The experimental programme of the above-mentioned study included tests on full-scale concrete columns with a rectangular cross-section of 300×300 mm^2 and height of 1800 mm. They were manufactured with steel-RC footings with dimensions of $1100 \times 1500 \times 600$ mm^3. The longitudinal reinforcement of the steel-RC (reference) and GFRP-RC columns were designed to have similar flexural strength for an applied axial load level equal to 20 % of their concentric axial capacity, leading to 4 steel and 8 GFRP rebars with 16 mm of diameter, respectively. Regarding the shear reinforcement, steel and GFRP closed stirrups were designed respectively for the steel-RC and GFRP-RC columns (i) to avoid shear failure during the tests and (ii) to ensure that both types of stirrups presented similar transversal stiffness per metre of height of the columns (*i.e.*, providing similar confinement to the concrete core of the columns). Fig. 14.9a illustrates the overall dimensions and the reinforcement adopted in both types of columns. For the reference steel-RC columns, conventional ribbed steel bars with an average yielding stress of 550 MPa were used. In the GFRP-RC columns, GFRP bars with a surface finish of he-

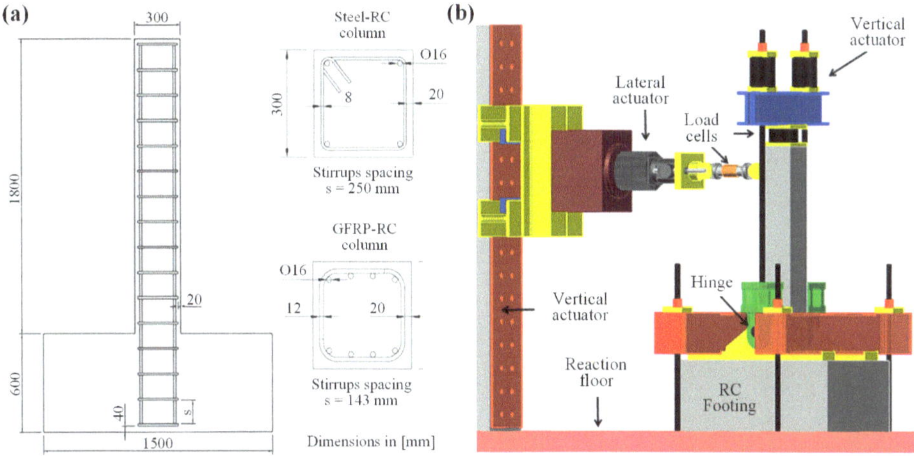

Fig. 14.9   Overall dimensions and internal reinforcement of steel-RC and GFRP-RC columns (a) and test setup (b) (adapted from [6]).

lical-wounded fibres were adopted as longitudinal and shear reinforcement, both with an average elastic modulus of 50 GPa and average tensile strengths of 900 MPa and 250 MPa, respectively, determined according to standard material characterization tests [84, 85]. Ready-mix concrete with an average compressive strength of 40 MPa (determined at the age of the column tests) was used to cast the specimens.

The test setup developed for these tests is schematically illustrated in Fig. 14.9b; it allowed the application of a constant axial load (corresponding to 20 % of the concentric axial capacity) and monotonic or cyclic lateral loadings, the latter with increasing magnitudes of imposed displacement, following a standard protocol according to the ACI 374.1 [86] recommendations.

Fig. 14.10 shows the lateral load *vs.* drift curves obtained in the tests with monotonic lateral and cyclic loadings for both types of columns. The curves from the monotonic tests confirm that both columns attained similar maximum lateral loads (≈76 kN for steel-RC *vs.* 79 kN for GFRP-RC), with the steel-RC column exhibiting a higher deformation capacity than the GFRP-RC. However, it is worth mentioning that the maximum load of the steel-RC column was attained for a considerably lower drift value than that of the GFRP-RC column – in the former column, the lateral load presented a drop (corresponding to the spalling of the compressed concrete cover, for a lateral displacement of ≈30 mm) and then remained almost constant while the deformation increased quite significantly (due to steel yielding) up to failure; whereas for the GFRP-RC column, the lateral load recovered after the concrete spalling and attained its maximum value for a very high lateral displacement (≈69 mm). This result can be explained by the more efficient concrete confinement provided by the GFRP stirrups (with lower spacing, *cf.* Fig. 14.9a) when compared to that of the steel stirrups, although both shear reinforcements presented the same transversal stiffness per metre of height. Failure of the steel-RC column occurred due to concrete crushing after steel yielding of the longitudinal bars, also involving buckling of the steel bars in compression; whereas the GFRP-RC column

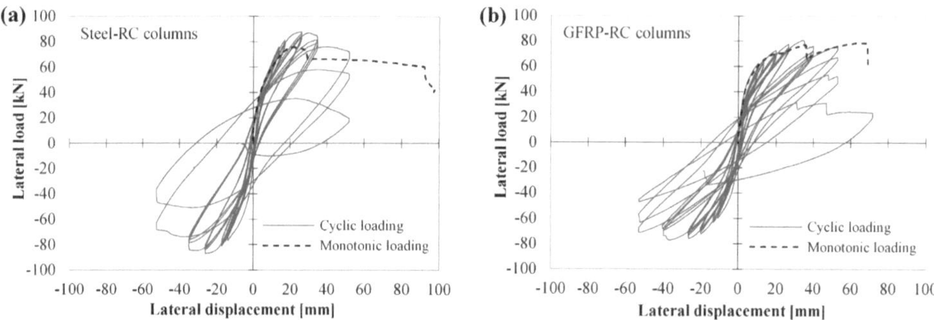

Fig. 14.10   Lateral load vs. drift curves obtained in steel-RC (a) and GFRP-RC (a) columns (adapted from [6]).

failed by crushing of the concrete and of the compressed GFRP bars, while the ones in tension remained well below their tensile capacity.

Regarding the response of the steel-RC column under cyclic loading, the hysteretic diagram depicted in Fig. 14.10 exhibits the typical behaviour of RC members when subjected to reversal actions, *i.e.*, with increasingly wider loops as the magnitude of the lateral drift increases and the elasto-plastic response of the longitudinal steel bars is further exploited. It is worth referring that this column failed prematurely due to local buckling of the longitudinal steel rebars (*cf.* Fig. 14.11a) for drift values lower than the maximum lateral deformation attained by the column subjected to monotonic lateral loading. In spite of this premature failure, the steel-RC column presented high energy dissipation capacity, as attested by the area enclosed within the hysteresis loops. On the other hand, as expected, the hysteretic diagram of the GFRP-column exhibited narrower loops due to the linear elastic behaviour of the GFRP bars. However, a non-negligible energy dissipation capacity was obtained due to the efficient concrete confinement provided by the closed-GFRP stirrups; the efficacy of the GFRP shear reinforcement was also evident by the relatively high deformation capacity of this column; in fact, it failed due to crushing of the GFRP bars (Fig. 14.11b) for a lateral drift value that exceeded 3 %, which would meet most of the drift limitations imposed for seismic design.

The results obtained in this study confirmed that GFRP-RC structural members can dissipate some inelastic energy by exploring the non-linear behaviour of confined concrete. Notwithstanding these promising results, future investigations are still needed to understand how this feature can be further explored and how FRP bars can be safely designed as internal reinforcement for concrete structural members in seismic areas.

14.4.3 Durability

As discussed in Section 14.2, FRP bars are often employed to reinforce concrete members as an alternative to traditional steel bars in particularly aggressive environments, as in the case of structures exposed to chloride attack and freeze-thaw cycles. Indeed, salt can be present in concrete due to the use of contaminated water or aggregates in the concrete mixture (which also affects the setting of concrete) or can penetrate within concrete

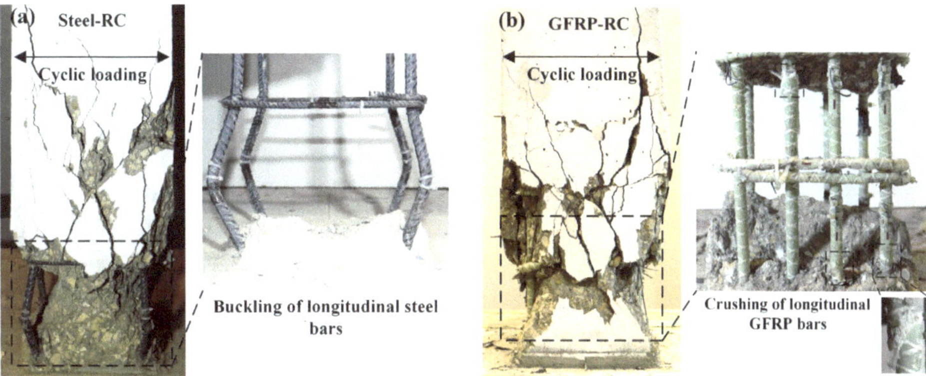

Fig. 14.11   Failure modes and details of the reinforcement after concrete removal of the columns tested under cyclic lateral loading: steel-RC (a) and GFRP-RC (b) (adapted from [6]).

members due to direct exposure to seawater or due to the use of de-icing salts, which is common in RC bridge decks located in cold climates. Salt attacks the passive layer created on the steel bar surface due to the alkalinity of concrete, thus promoting steel corrosion [87]. FRP bars are much less affected by chloride attack than steel bars and therefore represent a viable solution also for promoting the use of seawater and sea sand in concrete mixtures, which in turn would result in important savings of fresh water and river sand, among others [88, 89].

Understanding the behaviour of FRP bars with respect to typical exposure conditions of concrete members is then fundamental to assess their durability and guarantee their long-term performance. In general, available studies on the durability performance of FRP bars were focused on the effects of high alkalinity environments, which reproduce the conditions of concrete members at an early age, as well as of saline environments and freeze-thaw cycles. To increase the degradation rate and, in turn, decrease the duration of the experimental test, the conditioning solutions were heated at different temperatures and the Arrhenius law was used to predict the bar behaviour at different temperatures and periods of exposure [90]. However, high temperatures (relative to the glass transition temperature of the resin) can cause physical and chemical changes in the resin and can introduce additional degradation mechanisms (other than those directly caused by the exposure conditions), which in turn may lead to misleading predictions of the bar long-term properties [91].

The long-term behaviour of FRP bars has been typically assessed by comparing the tensile strength of control bare bars (*i.e.*, not embedded within concrete) with that of corresponding conditioned bare bars. The ratio between the tensile strength of a conditioned bar s_f and that of the control bar s_{fu} is referred to in the literature as the residual strength ratio, and it has been used to compute environmental conversion factors for different types of exposure environment, which can be used in design to take into account the reduction of the mechanical properties of the bars due to durability issues [92]. Fig. 14.12 compares the residual strength ratio of straight GFRP bars exposed to different alkaline environments as a function of the period of exposure. The figure comprises results of 210 bar specimens from 19 studies [93–110], which were divided in three sets depending on the exposure temperature. The first set comprises 70 specimens exposed to 11–25 °C, the second 75 specimens exposed to 26–53 °C, and the third 65 specimens exposed to 57–80 °C. Fig. 14.12 shows that the tensile strength of the bars was not significantly affected for exposure temperatures up to 53 °C, except for results by Chen *et al.* [101], which present considerable reductions in strength. Contrarily, the results of specimens exposed to temperatures between 57 °C and 80 °C, although more scattered than those obtained for lower temperatures, show significant decreases of the tensile strength. Neglecting the results from Chen *et al.* [101] and adopting the design by testing procedure provided in EN 1990 [111] (further details regarding the application of this procedure can be found in [92]), constant average environmental conversion factors $h_{c,m}$ equal to 0.90 and 0.85 were calibrated for specimens exposed to 11–25°C and 26–53°C, respectively. Specimens exposed to higher temperatures were not considered since these temperatures may affect the physical properties of the matrix and impair the results. It should be noted that $h_{c,m}$ is the average environmental conversion factor, while characteristic and design values of mechanical properties (and of conversion factors) depend on the scatter of results [92]. Interestingly, results from recent research, namely by Arczewska *et al.* [109] and Wu *et al.* [110], showed residual strength ratios generally higher and with less scatter than those reported in previous literature. This may indicate that basic materials, manufacturing, and testing procedures have been improving over the years, and thus new and less severe calibration of environmental conversion factors could be obtained considering new data.

The residual strength ratio of straight GFRP bars conditioned in different saline environments and exposed to 11–25°C (45 specimens), 40–50°C (26 specimens), and 60–80°C (32 specimens) are shown in Fig. 14.13 [90, 93, 94, 102, 106]. Similar to what was observed for specimens exposed to alkaline environments, bars conditioned in saline solutions up to 50°C did not show significant strength reduction, except in a few specimens. Indeed, the application of the design by testing procedure [111] did not show statistically significant degradation of strength with time. The average environmental conversion factors obtained were $h_{c,m} = 0.94$ and $h_{c,m} = 0.91$ for exposures at 11–25°C and 40–50°C, respectively. Specimens exposed to 60–80°C showed an important decrease of tensile strength after exposure, with residual strength ratios between 40 % and 60 % for some specimens. However, the high temperature employed may have altered the physical and chemical properties of the resin inducing additional degradation mechanisms and thus providing misleading results, which is the reason why no environmental conversion factor was calibrated for these specimens.

Figs. 14.12 and 14.13 confirm the good durability of GFRP bare bars exposed to high-alkalinity and saline environments. Although conditioning bare bars is convenient, it does not accurately reproduce the real bar exposure conditions. Thus, some researchers investigated the long-term behaviour of GFRP bars embedded within concrete beams and exposed to different solutions [105, 112], and generally observed slightly higher degradation of tensile strength with respect to that observed on the same bars conditioned by direct immersion in chemically aggressive solutions [92].

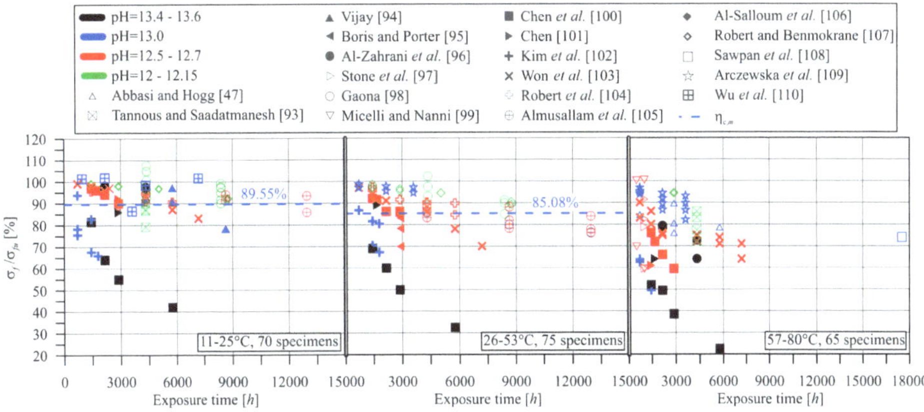

Fig. 14.12 Residual strength ratio of GFRP bars exposed to alkaline environments (results from [47, 93–110]).

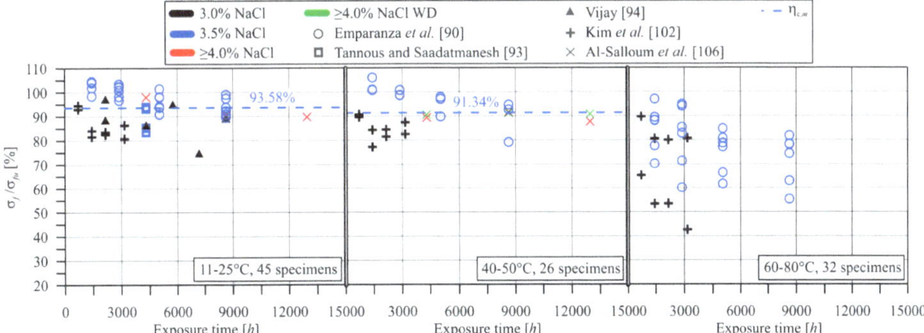

Fig. 14.13 Residual strength ratio of GFRP bars exposed to saline environments (results from [90, 93, 94, 102, 106]).

Another parameter that may affect the durability of FRP bars in real service conditions is the presence of applied loads. To investigate the synergistic effect of sustained stress on the residual strength ratio of GFRP bars, bare bar specimens were exposed to aggressive environments while subjected to a constant applied load. Fig. 14.14 shows the residual strength ratio of 52 bare bars, taken from 7 experimental campaigns [107, 109, 113–117], exposed to alkaline environments ($10.5 \leq pH \leq 13.1$) at temperatures varying from 22 °C to 65 °C and with different sustained stress ratios s_s/s_{fu}, where s_s is the stress applied to the bar during the exposure. Fig. 14.14 does not show a clear influence of the sustained stress ratio on the degradation of the bar tensile strength, as specimens exposed at relatively high temperature and relatively high s_s/s_{fu} showed high s_f/s_{fu} values [118]. It should be noted that specimens with s_s/s_{fu}=80 % [117] showed surprisingly high residual strength ratios, since sustained stresses higher than approximately 60 % of their short-term tensile strength (well above the limits defined in available codes and guidelines) were shown to induce rapid creep rupture in GFRP bars [119]. Fig. 14.14 also shows that bars made with alkali-resistant glass fibres, indicated with AR in Fig. 14.14 legend, did not show better results than their E-glass fibre counterparts tested by the same authors [113].

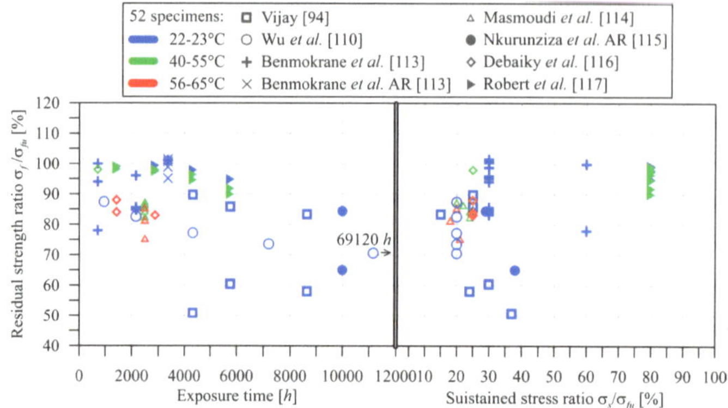

Fig. 14.14   Residual strength ratio of bars with sustained stress and exposed to alkaline environments (results from [94, 110, 113–117]).

To further investigate the effect of sustained stress on the long-term behaviour of GFRP bars, the residual strength ratio of stress-free bare bars exposed to alkaline environments at 11–25 °C (see Fig. 14.12) was compared in Fig. 14.15 with that of bare bars with sustained stress exposed to alkaline environments at 22–23 °C (see Fig. 14.14). The design by testing procedure [111] applied to Fig. 14.12 and Fig. 14.13 provided values of average $h_{c,m}$ equal to 0.90 and 0.86 for stress-free and stressed bars, respectively, which confirmed the limited influence of sustained stress on the bars' long-term behaviour. This observation shall be related only to the bar residual strength ratio, as tests described in this section were not meant to study creep and relaxation phenomena, which can impair the use of RC members in service conditions and can even induce premature (creep) rupture of the bar [119].

Results provided in this section showed that GFRP bars present a good durability with respect to real service exposure conditions of RC members, which can be particularly harsh for traditional steel reinforcement. However, the durability of concrete members reinforced with FRP bars does not depend exclusively on the durability of materials (concrete and FRP), but also on the FRP-concrete bond durability. As discussed in Section 14.2, the FRP-concrete bond behaviour has been generally investigated through pull-out tests carried out according to the ASTM D7913 [120] or ISO 10406-1 [121] standards. According to these standards, FRP bars are embedded within concrete blocks with a bonded length equal to 5 or 4 times the diameter (ϕ) of the bar, respectively, which only allows establishing a limited portion of the FRP-concrete stress transfer mechanism. The bond strength, defined as the peak load carried by the pull-out specimen divided by the FRP-concrete contact interface, was observed to generally decrease with the increase in the bar diameter, which indicates that it is not an intrinsic property of the FRP-concrete interface [122, 123]. Regardless, pull-out tests, due to their simplicity, have been the preferred methodology to investigate the bond strength of conditioned and non-conditioned specimens, and thereby allow determining the residual bond strength ratios for FRP bars. Results available in literature [124] show that, as in the case of non-conditioned specimens, the bond strength of conditioned specimens is greatly affected by the bar surface finish – sand-coated and deformed bars showed values of bond strength higher

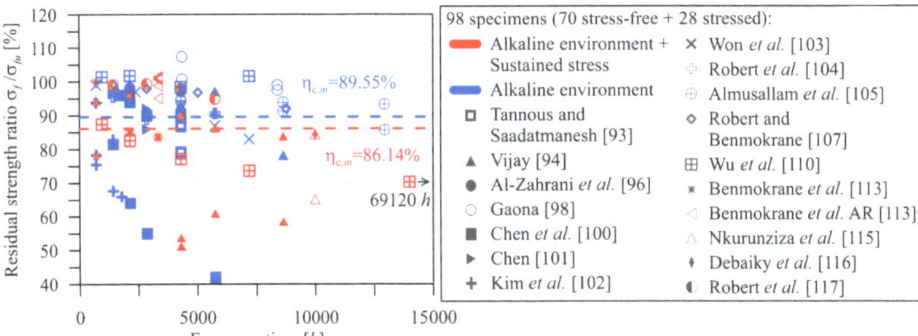

Fig. 14.15 Comparison between residual strength ratios of bars with and without sustained stress and exposed to alkaline environments at temperature 11 °C ≤ T ≤ 25 °C (results from [93, 94, 96, 98, 100–105, 107, 110, 113, 115–117]).

than those of smooth bars and also the quantity and quality of sand affected the results. Nevertheless, limited bond degradation was generally observed and statistical analysis of data available in literature, performed according to the model proposed by the *fib* bulletin 40 [11], provided residual bond strength ratios between 53 % and 85 % for a 50-year service life, depending on the member exposure environment (dry, moist, or saturated) [125]. However, further studies comprising specimens with bonded lengths higher than 5ϕ are needed to confirm these results, which would allow for fully establishing the FRP-concrete stress transfer mechanism and obtaining reliable estimations of the FRP-concrete bond durability.

14.5 Sustainability of FRP Reinforcement

14.5.1 Overview

This section presents a brief overview about the sustainability of FRP reinforcement, which, as for other construction materials, is presently a major issue, given the need to understand and reduce the footprint of the composites industry [126]. Here, sustainability is meant as environmental impact; in other words, the economic and social dimensions of sustainability are not considered.

When assessing the sustainability of a construction material, it is necessary to consider the different stages of its life cycle, which include the extraction and production of raw materials, manufacturing, distribution, execution, use, and end of life.

For these various life cycle stages, the following major concerns have been raised regarding FRP composites: (i) primary resources (raw materials) from which polymers are produced include crude oil, natural gas, chlorine, and nitrogen; (ii) the most commonly used polymeric resins and fibre reinforcements require high temperatures (i.e., energy) for their production (e.g., 1400 °C for glass and 1200–2400 °C for carbon) and, in some cases, petroleum by-products as precursors; (iii) FRP composites manufacturing requires energy and causes emissions/pollution [127, 128]; and (iv) the options at the end of the service life to manage FRP composites (waste) are limited [129, 130]. In fact, it is not possible to reprocess thermoset resins, presently used in the vast majority of FRP composites for structural applications, and this is a major concern in terms of sustainability, as the recycling options are very limited in practice. Consequently, the current prevalent approach for managing FRP composites at the end of their service life involves (i) converting them into granulates and disposing them in landfill (with minimal inherent value, an option already not allowed in several countries) or (ii) incinerating the waste with or without energy and material recovery, for instance in cement kilns [131].

On the other hand, compared to other construction materials, various potential advantages of FRP composites in terms of sustainability can be identified: (i) the high specific stiffness (for CFRP) and strength, which means that less material is needed for similar performance with respect to a traditional solution (thus minimizing the use of resources and waste production) and that transportation and construction costs (and energy) are lower, increasing the speed of construction and reducing environmental impacts; (ii) the application of FRP composites in rehabilitation projects (extending the service life of existing constructions and hence minimizing the generation of construction and demolition waste); (iii) the low thermal conductivity, which may provide energy savings in some building applications [132]; (iv) the improved durability (in various environments) when compared to alternative corrodible materials; and (v) the limited maintenance need during service life (resource savings). As mentioned in Section 14.4.3, there is a wealth of empirical evidence about the improved durability of FRP reinforcement. However, this is not yet fully established and, as discussed ahead, this makes it difficult to comprehensively assess the sustainability of concrete structures with FRP reinforcement considering the entire service life.

To make a comprehensive assessment of the environmental impact of FRP reinforcements, the above-mentioned concerns and potential advantages need to be duly balanced, using well-established and standardized tools, such as the life cycle assessment (LCA) methodology. The next section reviews the studies available in literature about the LCA of (i) FRP reinforcement compared to alternative metallic bars, and (ii) concrete structures reinforced with both types of bar.

14.5.2 LCA Studies

LCA is a standardised methodology (e.g., [133, 134]) that has been widely used for measuring and comparing the environmental impacts of different construction materials and structures. Ideally, LCA considers the whole life cycle of the material/structure to be studied, including the extraction of raw materials (cradle), the manufacturing process, the transportation to the construction site, the assembly and use (including maintenance), and the recycling and final disposal (grave) at the end of life. This broad analysis is typically called cradle-to-grave LCA. However, simplified LCA, known as cradle-to-gate, only includes the stages from the raw materials extraction until the product is ready to be used at the factory or at the site. It is often adopted when the necessary data for the remaining stages of the life cycle are not available and/or when the long-term behaviour of a certain material or structure is still not fully characterized/understood (as is the case for FRP reinforcements and FRP-RC structures).

The environmental impacts can be divided into several categories/indicators, based on the different types of environmental pollutant, which are selected according to specific environmental concerns. Due to the amount of reliable data available and because it is relatively easy to understand, the global warming potential (GWP, typically expressed in carbon dioxide equivalent) is one of the most accepted (and widely used) indicators.

As mentioned above, there is ample evidence of the improved durability of FRP reinforcements. However, the existing knowledge about their environmental performance is still limited owing to the reduced number of studies on this topic. In a recent investigation [135], a cradle-to-gate LCA was developed to compare the environmental impacts of conventional steel reinforcement with different types of FRP bar, including carbon, basalt, or glass fibres impregnated with the most common types of polymeric matrix (e.g., vinyl ester, epoxy, polyester). The environmental impact of the manufacturing stage of the FRP reinforcement (produced by pultrusion) was not considered in this analysis due to its reduced contribution when compared to the impacts associated to the extraction and processing of the constituent materials. In terms of GWP of the individual components of the various types of FRP reinforcement, epoxy presented the highest impact among the different matrices considered in this study, whereas the GWP of carbon fibres was more than 10 times higher than those of glass and basalt fibres, with the latter presenting the lowest impact. Regarding the comparison with conventional steel reinforcement, as the density and strength of FRPs are significantly different, the authors normalized the GWP indicators by the corresponding weights and tensile strengths. For both normalizations, BFRP and GFRP reinforcements presented significantly lower impacts than steel, while CFRP exhibited the highest normalized GWP value (cf. Fig. 14.16a). The worse performance of CFRP was also reported in [136]. Similar results were also reported in a recent study [137] in which BFRP bars presented the lowest environmental impact (in terms of GWP) among the various reinforcements considered: 74 % lower than conventional steel, 88 % lower than stainless steel, 49 % lower than galvanized steel, and 44 % lower than GFRP.

For what concerns FRP-RC structures, it is not possible to draw general conclusions about their environmental impacts directly from the (simplified) assessment of the impacts of the corresponding reinforcements, due to the inherent differences in their physical and mechanical properties. In most FRP-RC structural members, design is often governed by serviceability requirements. Consequently, compared to equivalent steel-RC members, this results in higher reinforcement ratios and often in thicker concrete members. Therefore, a valid comparison between the environmental performance of steel-RC and FRP-RC members shall be made for structural members comprehensively designed according to the same requirements/functions, considering their specific characteristics (e.g., geometry, loading) and all relevant strength and serviceability requirements. Moreover, besides the differences in physical and mechanical properties of both types of reinforcement, the different durability and maintenance requirements during the use-stage shall also be duly considered.

The number of studies published to date on the LCA of real-scale FRP-RC structures is still very limited, though different types of concrete structures have already been assessed: a pedestrian bridge [138], two vehicular bridges [138, 139], and the world's largest (at that date) FRP-RC flood mitigation channel [140].

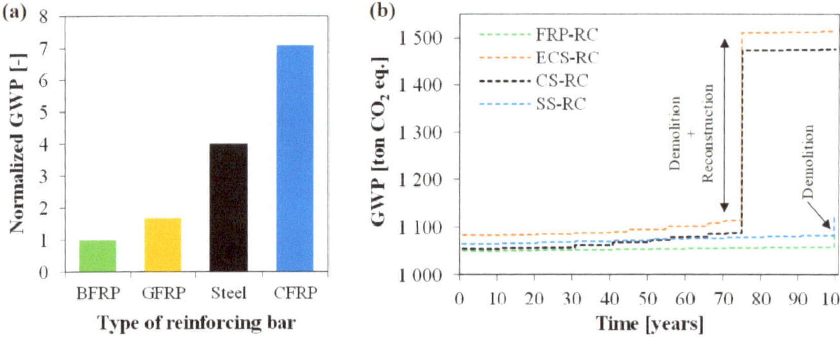

Fig. 14.16 (a) Normalized GWP (GWP/MPa/kg, with respect to BFRP) for different reinforcing bars (cradle-to-gate LCA, adapted from [138]); (b) GWP of four types of reinforcing bars for a vehicular RC bridge (cradle-to-grave LCA, adapted from [139]; CS=conventional steel; SS=stainless steel; FRP=combination of CFRP and GFRP; ECS=epoxy-coated steel).

In [138], a comparative cradle-to-gate LCA was performed for pedestrian and vehicular bridges. For each type of bridge, two separate design variants were considered, with conventional steel (as reference) and different types of CFRP reinforcement, passive or prestressed. For the vehicular bridge, a reduction of 18 % in the GWP was observed when both the passive and prestressed steel reinforcements were replaced with equivalent (passive and prestressed) CFRP reinforcements, whereas in the pedestrian bridge, a reduction of 28 % in the GWP was estimated when the passive steel reinforcement was replaced with prestressed CFRP. The authors concluded that, for bridges not subjected to heavy loads, the use of prestressed CFRP reinforcement may be an interesting environmental alternative to conventional (passive) steel reinforcing bars.

The study developed in [139], regarding a 57 metres long vehicular RC bridge, included a cradle-to-grave LCA of the following four alternative reinforcing bars: (i) conventional steel (CS), (ii) stainless steel (SS), (iii) FRP (combining both CFRP and GFRP), and (iv) epoxy-coated steel (ECS). The analysis was performed for a 100-year service life, assuming different maintenance models for the use stage of the various types of reinforcement and different strategies for their end-of-life stage (e.g., landfill disposal was assumed for FRP reinforcement, whereas CS and SS were considered as fully recyclable). Moreover, for the CS and ECS alternatives, a service life of 75 years was considered, followed by the demolition and reconstruction of a new bridge using the same design criteria and materials of the original one. Then, a second use stage of 25 years was considered. The analysis for the 100-year period (cf. Fig. 14.16b) showed that the FRP solution outperformed the CS and ECS alternatives in most of the environmental indicators (including GWP). The SS alternative, although presenting lower environmental impacts than the FRP solution in some indicators (e.g., ozone depletion), exhibited a slightly higher GWP, mainly due to the great amount of thermal energy needed to produce stainless steel.

The environmental features of a 21.3 kilometres long GFRP-RC flood mitigation channel built in Saudi Arabia were evaluated in [140] by means of a cradle-to-grave LCA, considering a 50-year period of analysis. The environmental impacts of an ECS-RC alternative (the original design) were also considered for comparison. Regarding the use stage, different maintenance schedules and activities were assumed for the two reinforcement solutions: for the FRP-RC option, only minor concrete repair was considered every 10 years, whereas for the ECS-RC alternative, significantly more extensive repairs (including both concrete and corroded steel) were assumed to be necessary. For the end-of-life stage, demolition, disposal, and recycling activities were included in the analysis. However, as in the previous study [139], GFRP rebars were considered as non-recyclable. The results of this study confirmed the lower environmental impacts of the GFRP-RC solution with respect to the ECS-RC alternative in all five impact indicators considered in the analysis (including the GWP). The lower environmental performance of the ECS-RC design was attributed to the higher number of reinforcing bars required in this specific design (resulting in higher impacts on both the production and construction phases), as well as the significantly higher impacts resulting from the more extensive maintenance actions required by the solution with (conventional) ECS reinforcement.

14.5.3 Future Trends

In the past few years, significant efforts and technological developments were made to improve the sustainability of FRP composites.

Part of these efforts include shifting the raw materials presently used in the manufacturing of FRP composites to renewable feedstock, namely by replacing (i) synthetic fibres with natural fibres [141, 142], and (ii) oil-based resins with bio-based resin systems [143]. While the former option seems to have major limitations for structural applications, due to the (limited) mechanical properties and much lower durability (due to the hydrophilicity) of natural fibres with respect to synthetic fibres, the latter option proved to be very promising, providing similar mechanical and thermomechanical properties to conventional composites [144].

Another major effort that is being pursued is the replacement of thermoset resins by thermoplastic resins, and this approach has already been successfully implemented in the production of FRP reinforcement [145]. This shift involves quite significant challenges in manufacturing. However, it offers many advantages, not only in terms of end-of-life options (thermoplastic resins are recyclable), but also during construction. Indeed, thermoplastic composites can be bent on site, which is a key advantage for FRP reinforcement used in concrete. This feature reduces time and cost of manufacturing and delivery, may allow for increasing the bar anchorage capacity [65], and it can also be used to design FRP-RC structures with improved fire resistance, following the principles outlined in [58].

14.6 Concluding Remarks

This chapter presented an overview of FRP bars for the reinforcement of concrete structures. This type of rebars is being increasingly considered for various applications where their inherent properties are an added value, namely the non-corrodibility, electromagnetic transparency, high tensile strength, and lightness. On the other hand, FRP bars present significant differences compared to conventional steel reinforcement, namely linear elastic behaviour up to failure and, in case of GFRP bars, a low elastic modulus. Consequently, the structural behaviour of FRP-RC members presents significant differences compared to their steel-RC counterparts that need to be duly considered in the design. This chapter presented a summary of these main behavioural differences, providing general concepts for the bending and shear design of FRP-RC members, also in view of the most recent design guidance. The final part of the chapter presented a brief review of three critical aspects concerning the behaviour of FRP-RC structures that are intimately concerned with the inherent properties of FRP bars: the fire behaviour, the seismic behaviour, and the long-term durability; for each of these three aspects, key findings from recent studies were highlighted and research needs were pointed out. Finally, the sustainability of using FRP reinforcement in RC structures was discussed and the limited information available in literature regarding this topic was reviewed. In spite of these critical aspects that still need to be investigated in further depth, and due to the increasingly competitive costs of FRP reinforcement when compared to other corrosion-resistant alternatives (*e.g.*, stainless steel or epoxy-coated mild steel), it is foreseen that the number of FRP-RC structures will continue to grow in the coming years.

14.7 References

[1]   American Society of Civil Engineering (ASCE), A comprehensive assessment of America's Infrastructure, Report Card for America's Infrastructure, 2021, Available at www.infrastructurereportcard.org.

[2]   GangaRao, H.V.S., Taly, N., Vijay, P.V., Reinforced Concrete Design with FRP Composites, CRC Press, 2006, p. 400. https://doi.org/10.1201/9781420020199

[3]   Bank, L.C., Composites for Construction. Structural Design with FRP Materials, Wiley. Hoboken, NJ, 2006, p. 560. https://doi.org/10.1002/9780470121429

[4]   ACI (American Concrete Institute), Guide for the Design and Construction of Structural Concrete Reinforced with FRP Bars, ACI 440.1R-15, Farmington Hills, Michigan, USA, 2015.

[5]   Faza, S.S., GangaRao, H.V.S., Glass FRP reinforcing bars for concrete. In "Fibre-Reinforced-Plastic (FRP) Reinforcement for Concrete Structures: Properties and Applications, Developments in Civil Engineering" (Ed. A. Nanni), Elsevier, Amsterdam, 1993, p. 167–188. https://doi.org/10.1016/b978-0-444-89689-6.50012-4

[6]   Souza., W.M., Behaviour of reinforced concrete columns with GFRP bars under different loading conditions. PhD Thesis in Civil Engineering. Instituto Superior Técnico, University of Lisbon, 2021 (in Portuguese).

[7]   Benmokrane, B., Brown, V.L., Mohamed, K., Nanni, A., Rossini, M., Shield, C., Creep-rupture limit for GFRP bars subjected to sustained loads. Journal of Composites for Construction, 2019, 23(6). https://doi.org/10.1061/(asce)cc.1943-5614.0000971

[8]   AASHTO LRFD Bridge Design Guide Specifications for GFRP–Reinforced Concrete. American Association of State Highway and Transportation Officials. 2018.

[9]   ACI (American Concrete Institute), Building Code Requirements for Structural Concrete Reinforced with Glass Fiber-Reinforced Polymer (GFRP) Bars, ACI CODE-440.11-22, Farmington Hills, Michigan, USA, 2022.

[10] Alarcão, J., Adherence between GFRP bars and concrete. Literature review and assessment of ACI 440 design equation, MSc Dissertation in Civil Engineering. Instituto Superior Técnico, University of Lisbon, 2014 (in Portuguese).

[11] FIB (International Federation for Structural Concrete), "Bulletin 40 – FRP Reinforcement for RC Structures", Lausanne, Switzerland, 2007. https:// doi. org/10.35789/fib.bull.0040

[12] Cosenza, E., Manfredi, G., Realfonzo, R., Bond characteristics and anchorage length of FRP rebars, Proceedings of the 2nd International Conference on Advanced Composite Materials in Bridges and Structures, Montreal, Canada, 1996.

[13] Cosenza E., Manfredi G., Realfonzo R., Analytical modelling of bond between FRP reinforcing bars and concrete, 2nd International RILEM Symposium (FRPRCS-2), Ghent, Belgium, 1995.

[14] Theriault, M., Benmokrame, B., Effects of FRP Reinforcement Ratio and Concrete Strength on Flexural Behaviour of Concrete Beams, Journal of Composites for Construction, 1998, 2(1), p. 7–16. https://doi.org/10.1061/(asce)1090-0268(1998)2:1(7)

[15] Nanni, A., Flexural Behavior and Design of RC Members Using FRP Reinforcement, Journal of Structural Engineering, 1993, 119(11), p. 3344–3359. https://doi. org/10.1061/(asce)0733-9445(1993)119:11(3344)

[16] Pilakoutas, K., Neocleous, K. Guadagnini, M. Design philosophy issues of fibres reinforced polymer reinforced concrete structures, Journal of Composites for Construction, 2002, 6(3), p. 154–161. https://doi.org/10.1061/(asce)1090-0268 (2002)6:3(154)

[17] Bischoff, P., Reevaluation of Deflection Prediction for Concrete Beams Reinforced with Steel and Fiber Reinforced Polymer Bars, Journal of Structural Engineering, 2005, 131(5), p. 752–767. https://doi.org/10.1061/(asce)0733-9445(2005)131:5(752)

[18] Charles E Bakis, Carlos E Ospina, Timothy E Bradberry, Brahim Benmokrane, Shawn P Gross, John P Newhook, Ganesh Thiagarajan, Evaluation of crack widths in concrete flexural members reinforced with FRP bars, 3rd International Conference on Composites in Civil Engineering – CICE 2006, International Institute for FRP in Construction (IIFC), 2006, p. 307–310.

[19] Rossini, M., Saqan, E., Nanni, A., Prediction of the creep rupture strength of GFRP bars, Construction and Building Materials, 2019, 227, 116620. https://doi. org/10.1016/j.conbuildmat.2019.08.001

[20] Benmokrane, B., Brown, V.L., Mohamed, K., Nanni, A., Rossini, M., Shield, C., Creep-rupture limit for GFRP bars subjected to sustained loads, Journal of Composites for Construction, 2019, 23(6), 06019001. https://doi.org/10.1061/(asce) cc.1943-5614.0000971

[21] Marí, A., Cladera, A., Oller, E., Bairán, J., Shear design of FRP reinforced concrete beams without transverse reinforcement, Composites: Part B, 2014, 57, p. 228–241. https://doi.org/10.1016/j.compositesb.2013.10.005

[22] Cavagnis, F., Ruiz, M. F., Muttoni, A., Shear failures in reinforced concrete members without transverse reinforcement: An analysis of the critical shear crack development on the basis of test results, Engineering Structures, 2015, 103, p. 157–173. https://doi.org/10.1016/j.engstruct.2015.09.015

[23] Mikani, H., Katoh, M., Takeuchi, H., Tamura, T., Flexural and Shear Behaviour of RC Beams Reinforced with Braided FRP Rods in Spiral Shape, Transactions of the Japan Concrete Institute, 1989, 11(1), p. 119–206.

[24] Kotsovos, M. D., Pavlovic, M. N., Ultimate Limit State Design of Concrete Structures – A New Approach, Thomas Telford, Ltd. London, UK, 1999. https://doi.org/10.1680/ulsdocs.26650

[25] Imjai, T., Guadagnini, M., Pilakoutas, K., Bend Strength of FRP Bars: Experimental a Investigation and Bond Modeling, Journal of Materials in Civil Engineering, 2017, 29(7), 04017024. https://doi.org/10.1061/(asce)mt.1943-5533.0001855

[26] Matthys, S., Taerwe, L., Concrete Slabs Reinforced with FRP grids. II: Punching Resistance, Journal of Composites for Construction, 2000, 4(3), p. 154–161. https://doi.org/10.1061/(asce)1090-0268(2000)4:3(154)

[27] Guadagnini, M., Pilakoutas, K., Waldron, P., Shear Resistance of FRP RC Beams: An Experimental Study, Journal of Composites for Construction, 2006, 10(6), p. 464–473. https://doi.org/10.1061/(asce)1090-0268(2006)10:6(464)

[28] Japan Society of Civil Engineers, JSCE, Recommendation for Design and Construction of Concrete Structures Using Continuous Fiber Reinforcing Materials, Concrete Engineering Series 23, Japan Society of Civil Engineers, 1997, Tokyo, Japan, p. 325.

[29] CNR-DT 203/2006, Guide for the Design and Construction of Concrete Structures Reinforced with Fiber-Reinforced Polymer Bar, National Research Council, 2007, Rome, Italy.

[30] CSA (Canadian Standards Association), Design and Construction of Building Structures with Fibre-Reinforced Polymers, CAN/CSA-S806-12, Canadian Standards Association, Rexdale, Ontario, Canada, 2017.

[31] Hajiloo, H., Green, M.F., Noël, M., Bénichou, N., Sultan, M., Fire tests on full-scale FRP reinforced concrete slabs, Composite Structures, 2017, 179, p. 705–719. https://doi.org/10.1016/j.compstruct.2017.07.060

[32] Rosa, I.C., Fire behaviour of concrete structures reinforced with GFRP bars, PhD Thesis in Civil Engineering, Instituto Superior Técnico, University of Lisbon, Portugal, 2022.

[33] AlAjarmeh, O., Manalo, A., Benmokrane, B., Schubel, P., Zeng, X., Ahmad, A., Hassanli, R., Sorbello, C.-D., Compression behavior of GFRP bars under elevated in-service temperatures, Construction and Building Materials, 2022, 314, p. 125675. https://doi.org/10.1016/j.conbuildmat.2021.125675

[34] Robert, M., Benmokrane, B., Behaviour of GFRP reinforcing bars subjected to extreme temperatures, Journal of Composites for Construction, 2010, 14 (4), p. 353–360. https://doi.org/10.1061/(asce)cc.1943-5614.0000092

[35] Mouritz, A.P., Gibson, A.G., Fire properties of polymer composite materials, Solid mechanics and its applications, Springer, 2006. p. 401. https://doi.org/10.1007/978-1-4020-5356-6

[36] Rosa, I.C., Firmo, J.P., Correia, J.R., Experimental study of the tensile behaviour of GFRP reinforcing bars at elevated temperatures, Construction and Building Materials, 2022, 324, 126676. https://doi.org/10.1016/j.conbuildmat.2022.126676

[37] Hajiloo, H., Green, M.F., Gales, J., Mechanical properties of GFRP reinforcing bars at high temperatures, Construction and Building Materials, 2018, 162, p. 142–154. https://doi.org/10.1016/j.conbuildmat.2017.12.025

[38] Wang, Y.C., Kodur, V., Variation of strength and stiffness of fibre reinforced polymer reinforcing bars with temperature, Cement and Concrete Composites, 2005, 27 (9–10), p. 864–874. https://doi.org/10.1016/j.cemconcomp.2005.03.012

[39] Rozsypalova, I., Prokes, J., Cairovic, D., Girgle, F., Danek, P., Stepanek, P., Research on the residual tensile strength of composite reinforcing bars exposed to elevated temperatures, IOP Conference Series: Materials Science and Engineering, 2021, 1205, 012011. https://doi.org/10.1088/1757-899x/1205/1/012011

[40] McIntyre, E., Fire performance of fibre reinforced polymer (FRP) bars in reinforced concrete: an experimental approach, PhD thesis in Civil Engineering, University of Edinburgh, United Kingdom, 2019.

[41] Najafabadi, E.P., Oskouei, A.V., Khaneghahi, M.H., Shoaei, P., Ozbakkaloglu, T., The tensile performance of FRP bars embedded in concrete under elevated temperatures, Construction and Building Materials, 2019, 211, p. 1138–1152. https://doi.org/10.1016/j.conbuildmat.2019.03.239

[42] Ashrafi, H., Bazli, M., Vatani Oskouei, A., Bazli, L., Effect of Sequential Exposure to UV Radiation and Water Vapor Condensation and Extreme Temperatures on the Mechanical Properties of GFRP Bars, Journal of Composites for Construction, 2018, 22(1), 04017047. https://doi.org/10.1061/(asce)cc.1943-5614.0000828

[43] Ashrafi, H., Bazli, M., Najafabadi, E.P., Vatani Oskouei, A., The effect of mechanical and thermal properties of FRP bars on their tensile performance under elevated temperatures, Construction and Building Materials, 2017, 157, p. 1001–1010. https://doi.org/10.1016/j.conbuildmat.2017.09.160

[44] McIntyre, E., Bilotta, A., Bisby, L. Nigro, E., Mechanical properties of fibre reinforced polymer reinforcement for concrete at high temperature, 8th International Conference on Structures in Fire, Shanghai, China, 2014.

[45] Wang, X., Zha, X., Experimental research on mechanical behavior of GFRP bars under high temperature, Applied Mechanics and Materials, 2011, 71–78, p. 3591–3594. https://doi.org/10.4028/www.scientific.net/amm.71-78.3591

[46] Weber, A., Fire-resistance tests on composite rebars”, 4th International Conference of FRP Composites in Civil Engineering – CICE 2008, Zurich, Switzerland, 2008.

[47] Abbasi, A., Hogg, P.J., Temperature and environmental effects on glass fibre rebar: modulus, strength and interfacial bond strength with concrete, Composites Part B: Engineering, 2005, 36 (5), p. 394–404. https://doi.org/10.1016/j.compositesb.2005.01.006

[48] Dimitrienko, Y.I., Thermomechanics of composites under high temperatures, 1999, Klewer Academic Publishers, London, UK, p. 352.

[49] Tanano, H., Masuda, Y., Kage, T., Fukuyama, H., Nishida, I., Hashimoto, T., Fire resistance of continuous fibre reinforced concrete, Non-metallic (FRP) Reinforcement for Concrete Structures, L. Taerwe (eds), E&FN Spon, London, United Kingdom, 1995, p. 9.

[50] Clarke, J., Alternative Materials for Reinforcement and Prestressing of Concrete, Blackie Academic and Professional (eds), Glasgow, United Kingdom, 1993, p. 218.

[51] Kumahara, S., Masuda, Y., Tanano, H., Shimizu, A., Tensile strength of continuous fibre bar under high temperature, Fibre-Reinforced-Plastic Reinforcement for Concrete Structures: An International Symposium, Edited by A. Nanni and C. W. Dolan, American Concrete Institute, 1993, Detroit, Michigan, p. 731–742.

[52] Zhou, C., Pan, J., Zhang, Z., Zhu, Y., Comparative study on the tensile mechanical behavior of GFRP bars under and after high temperature exposure, Case Studies in Construction Materials, June 2022, 16 , e00905. https://doi.org/10.1016/j.cscm.2022.e00905

[53] Khaneghahi, M.H., Najafabadi, E.P., Shoaei, P., Oskouei, A.V., Effect of intumescent paint coating on mechanical properties of FRP bars at elevated temperature, Polymer Testing, 2018, 71 (August), p. 72–86. https://doi.org/10.1016/j.polymertesting.2018.08.020

[54] Chaallal, O., Benmokrane, B., Physical and mechanical performance of an innovative glass-fiber-reinforced plastic rod for concrete and grouted anchorages, Canadian Journal of Civil Engineering, 1993, 20 (2), p. 254–268. https://doi.org/10.1139/l93-031

[55] Dimitrienko, Y.I., Thermomechanical behaviour of composite materials and structures under high temperatures: 1. Materials, Composites Part A: Applied Science and Manufacturing, 1997, 28 (5), p. 453–461. https://doi.org/10.1016/s1359-835x(96)00144-3

[56] Rosa, I.C., Firmo, J.P., Correia, J.R., Bisby, L.A., Fire behaviour of GRFP-reinforced concrete structural members: a state-of-the-art review, Journal of Composites for Construction, 2023, 27(5). https://doi.org/10.1061/jccof2.cceng-4268

[57] Rosa, I.C., Firmo, J.P., Correia, J.R., Barros, J.A.O., Bond behaviour of sand coated GFRP bars to concrete at elevated temperature – Definition of bond vs. slip relations, Composites Part B: Engineering, 2019, 160, p. 329–340. https://doi.org/10.1016/j.compositesb.2018.10.020

[58] Rosa, I.C., Arruda, M.R.T., Firmo, J.P., Correia, J.R., Bond behaviour of straight and bent GFRP bars at elevated temperature: pull-out tests and numerical simulations, Journal of Composites for Construction, 2022, 26 (3), 04022028. https://doi.org/10.1061/(asce)cc.1943-5614.0001213

[59] Rosa, I.C., Firmo, J.P., Correia, J.R., Mazzuca, P., Influence of elevated temperatures on the bond behaviour of ribbed GFRP bars in concrete, Cement and Concrete Composites, 2021, 122, p. 104119. https://doi.org/10.1016/j.cemconcomp.2021.104119

[60] Solyom, S., Di Benedetti, M., Guadagnini, M., Balázs, G.L., Effect of temperature on the bond behaviour of GFRP bars in concrete, Composites Part B: Engineering, 2020, 183, 107602. https://doi.org/10.1016/j.compositesb.2019.107602

[61] Hajiloo, H., Green, M.F., Bond strength of GFRP reinforcing bars at high temperatures with implications for performance in fire, Journal of Composites for Construction, 2018, 22 (6), 04018055. https://doi.org/10.1061/(asce)cc.1943-5614.0000897

[62] Katz, A., Berman, N., Bank, L.C., Effect of high temperature on bond strength of FRP rebars, Journal of Composites for Construction, 1999, 3 (2), p. 73–81. https://doi.org/10.1061/(asce)1090-0268(1999)3:2(73)

[63] Hajiloo, H., Green, M.F., Noël, M., GFRP-Reinforced Concrete Slabs: Fire Resistance and Design Efficiency, Journal of Composites for Construction, 2019, 23 (2), 04019009. https://doi.org/10.1061/(asce)cc.1943-5614.0000937

[64] Nigro, E., Bilotta, A., Cefarelli, G., Manfredi, G., Cosenza, E., Performance under fire situations of concrete members reinforced with FRP rods: bond models and design nomograms, Journal of Composites for Construction, 2012, 16 (4), p. 395–406. https://doi.org/10.1061/(asce)cc.1943-5614.0000279

[65] Rosa, I.C., Firmo, J.P., Correia, J.R., Fire behaviour of GFRP-reinforced concrete slab strips – effect of straight and 90° bent splices, Engineering Structures, 2022, 270, 114904. https://doi.org/10.1016/j.engstruct.2022.114904

[66] Rosa, I.C., Santos, P., Firmo, J.P., Correia, J.R., Fire behaviour of concrete slab strips reinforced with sand-coated GFRP bars, Composite Structures, 2020, 244, 112270. https://doi.org/10.1016/j.compstruct.2020.112270

[67] Abbasi, A., Hogg, P.J., Fire testing of concrete beams with fibre reinforced plastic rebar, Composites Part A: Applied Science and Manufacturing, 2006, 37 (8), p. 1142–1150. https://doi.org/10.1016/j.compositesa.2005.05.029

[68] Nigro, E., Cefarelli, G., Bilotta, A., Manfredi, G., Cosenza E., Fire resistance of concrete slabs reinforced with FRP bars. Part I: Experimental investigations on the mechanical behavior, Composites Part B: Engineering, 2011, 42, p. 1739–1750. https://doi.org/10.1016/j.compositesb.2011.02.025

[69] Nour, O., Salem, O., Mostafa, A., Experimental fire testing of large-scale glass fibre-reinforced polymer-reinforced concrete beams with mid-span straight-end bar lap splices, Fire and Materials, 2021, 46(2), p. 360–375. https://doi.org/10.1002/fam.2960

[70] Gurung, S., Salem, S., Effects of load level on the structural fire behaviour of GFRP-reinforced concrete beams with straight-end bar lap splices, 8th International Conference on Advanced Composite Materials in Bridges and Structures, Sherbrooke, Quebec, Canada, 2021.

[71] Grace, N.F., Soliman, A.K., Abdel-Sayed, G., Saleh, K.R., Behavior and Ductility of Simple and Continuous FRP Reinforced Beams. Journal of Composites for Construction, 2(4), 1998, p. 186–194. https://doi.org/10.1061/(asce)1090-0268(1998)2:4(186)

[72] Gravina, R.J., Smith, S.T., Flexural behaviour of indeterminate concrete beams reinforced with FRP bars. Engineering Structures, 2008, 30(9), p. 2370–2380. https://doi.org/10.1016/j.engstruct.2007.12.019

[73] Matos, B., Correia, J.R., Castro, L.M.S., França, P., Structural response of hyperstatic concrete beams reinforced with GFRP bars: Effect of increasing concrete confinement, Composite Structures, 2012, 94(3), p. 1200–1210. https://doi.org/10.1016/j.compstruct.2011.10.021

[74] Santos, P., Laranjeira, G., França, P.M., Correia J.R., Ductility and moment redistribution capacity of multi-span T-section concrete beams reinforced with GFRP bars, Construction and Building Materials, 2013, 49, p. 949–961. https://doi.org/10.1016/j.conbuildmat.2013.01.014

[75] Sharbatdar, M.K., Concrete columns and beams reinforced with FRP bars and grids under monotonic and reversed cyclic loading, PhD Thesis, University of Ottawa, 2003.

[76] Sharbatdar, M.K., Saarcioglu, M., Seismic design of FRP concrete structures, Asian Journal of Applied Sciences, 2(3), 2009, p. 211–222. https://doi.org/10.3923/ajaps.2009.211.222

[77] Ali M.A., El-Salakawy E., Seismic Performance of GFRP-Reinforced Concrete Rectangular Columns, Journal of Composites for Construction, 2016, 20(3), p. 04015074. https://doi.org/10.1061/(asce)cc.1943-5614.0000637

[78] Naqvi, S., El-Salakawy, E., Lap Splice in GFRP-RC Rectangular Columns Subjected to Cyclic-Reversed Loads, Journal of Composites for Construction, 2017, 21(4), p. 04016117(1–13). https://doi.org/10.1061/(asce)cc.1943-5614.0000777

[79] Said, A.M., Nehdi, M.L., Use of FRP for RC Frames in Seismic Zones: Part II. Performance of Steel-Free GFRP-Reinforced Beam-Column Joints, Applied Composite Materials, 2004. 11, p. 227–245. https://doi.org/10.1023/b:acma.0000035480.85721.b5

[80] Mady, M., El-Ragaby, A., El-Salakawy, E., Seismic Behavior of Beam-Column Joints Reinforced with GFRP Bars and Stirrups, Journal of Composites for Construction, 2011, 15(6), p. 875–886. https://doi.org/10.1061/(asce)cc.1943-5614.0000220

[81] Ghomi, S.K., El-Salakawy, E., Seismic Performance of GFRP-RC Exterior Beam–Column Joints with Lateral Beams, Journal of Composites for Construction, 2016, 20(1), p. 04015019(1–11). https://doi.org/10.1061/(asce)cc.1943-5614.0000582

[82] Nehdi, M.L., Said, A.M., Performance of RC frames with hybrid reinforcement under reversed cyclic loading, Materials and Structures, 2005, 38, p. 627–637. https://doi.org/10.1617/14221

[83] Nehdia, M., Alamb, M.S., Youssef, M.A., Development of corrosion-free concrete beam-column joint with adequate seismic energy dissipation, Engineering Structures, 2010, 32, p. 2518–2528. https://doi.org/10.1016/j.engstruct.2010.04.020

[84] ASTM (American Society for Testing and Materials), Standard test method for tensile properties of fiber reinforced polymer matrix composite bars, ASTM D7205, West Conshohocken, Pennsylvania, USA, 2006.

[85] ASTM (American Society for Testing and Materials), Standard test method for strength of Fiber Reinforced Polymer (FRP) bend bars in bend locations, ASTM D7914, West Conshohocken, Pennsylvania, USA, 2014.

[86] ACI (American Concrete Institute), Acceptance criteria for moment frames based on structural testing and commentary, ACI 374.1, Farmington Hills, USA, 2005.

[87] Broomfield, J., Corrosion of Steel in Concrete: Understanding, Investigation and Repair, Second Edition, 2nd edition, CRC Press, London, July 2003.

[88] Ahmed, A., Guo, S., Zhang, Z., Shi, C., Zhu, D., A review on durability of fiber reinforced polymer (FRP) bars reinforced seawater sea sand concrete. Construction and Building Materials, 2020, 256, 119484. https://doi.org/10.1016/j.conbuildmat.2020.119484

[89] Morales, C.N., Claure, G., Emparanza, A.R., et al.: Durability of GFRP reinforcing bars in seawater concrete. Construction and Building Materials, February 2021, 270, 121492. https://doi.org/10.1016/j.conbuildmat.2020.121492

[90] Emparanza, A.R., Kampmann, R., De Caso, F., Morales, C., Nanni, A., Durability assessment of GFRP rebars in marine environments. Construction and Building Materials, 2022, 329, 127028. https://doi.org/10.1016/j.conbuildmat.2022.127028

[91] Gonenc, O., Durability and Service Life Prediction of Concrete Reinforcing Materials. PhD Thesis, University of Wisconsin, Madison, 2003.

[92] D'Antino, T., Pisani, M.A.,Poggi, C., Effect of the environment on the performance of GFRP reinforcing bars. Composites Part B: Engineering, 2018, 141, p. 123–136. https://doi.org/10.1016/j.compositesb.2017.12.037

[93] Tannous, F.E., Saadatmanesh, H., Durability of AR glass fiber reinforced plastic bars. Journal of Composites for Construction, 1999, 3, p. 12–19. https://doi.org/10.1061/(asce)1090-0268(1999)3:1(12)

[94] Vijay, P.V., Aging and design of concrete members reinforced with GFRP bars, West Virginia University, Morgantown, West Virginia, 1999.

[95] Boris, T., Porter, M., Advancements in GFRP materials provide improved durability for reinforced concrete. Plastics in Building Construction, 2000, 24, p. 10.

[96] Al-Zahrani, M.M., Al-Dulaijan, S.U., Sharif, A., Maslehuddin, M., Durability performance of glass fiber reinforced plastic reinforcement in harsh environments. 6th Saudi engineering conference, 2002.

[97] Stone, D.K., Koenigsfeld, D., Myers, J., Nanni, A., Durability of GFRP Rods, Laminates, and Sandwich Panels Subjected to Various Environmental Conditioning. Proceeding of 2nd international conference on durability of fiber reinforced polymer (FRP) composites for construction, Montreal, Canada, 2002, p. 213–24.

[98] Gaona, F.A., Characterization of design parameters for fiber reinforced polymer composite reinforced concrete systems. PhD Thesis, Texas A&M University, 2003.

[99] Micelli, F., Nanni, A., Durability of FRP rods for concrete structures. Construction and Building Materials, September 2004, 18, p. 491–503. https://doi.org/10.1016/j.conbuildmat.2004.04.012

[100] Chen, Y., Davalos, J.F., Ray, I., Durability prediction for GFRP reinforcing bars using short-term data of accelerated aging tests. Journal of Composites for Construction, 2006, 10, p. 279–286. https://doi.org/10.1061/(asce)1090-0268(2006)10:4(279)

[101] Chen, Y., Accelerated ageing tests and long-term prediction models for durability of FRP bars in concrete, West Virginia University, Morgantown, West Virginia, 2007.

[102] Kim, H.-Y., Park, Y.-H., You, Y.-J., Moon, Y.-J., Short-term durability test for GFRP rods under various environmental conditions. Composite Structures, March 2008, 83, p. 37–47. https://doi.org/10.1016/j.compstruct.2007.03.005

[103] Won, J.-P., Lee, S.-J., Kim, Y.-J., Jang, C.-I., Lee, S.-W., The effect of exposure to alkaline solution and water on the strength–porosity relationship of GFRP rebar. Composites Part B: Engineering, July 2008, 39, p. 764–772. https://doi.org/10.1016/j.compositesb.2007.11.002

[104] Robert, M., Cousin, P., Benmokrane, B., Durability of GFRP reinforcing bars embedded in moist concrete. Journal of Composites for Construction, 2009, 13, p. 66–73. https://doi.org/10.1061/(asce)1090-0268(2009)13:2(66)

[105] Almusallam, T.H., Al-Salloum, Y.A., Alsayed, S.H., El-Gamal, S., Aqel, M., Tensile properties degradation of glass fiber-reinforced polymer bars embedded in concrete under severe laboratory and field environmental conditions. Journal of Composite Materials, 2013, 47, p. 393–407. https://doi.org/10.1177/0021998312440473

[106] Al-Salloum, Y.A., El-Gamal, S., Almusallam, T.H., Alsayed, S.H., Aqel, M., Effect of harsh environmental conditions on the tensile properties of GFRP bars.

Composites Part B: Engineering, 2013, 45, p. 835–844. https://doi.org/10.1016/j.compositesb.2012.05.004

[107] Robert, M., Benmokrane, B., Combined effects of saline solution and moist concrete on long-term durability of GFRP reinforcing bars. Construction and Building Materials, January 2013, 38, p. 274–284. https://doi.org/10.1016/j.conbuildmat.2012.08.021

[108] Sawpan, M.A., Mamun, A.A., Holdsworth, P.G., Long term durability of pultruded polymer composite rebar in concrete environment. Materials & Design, May 2014, 57, p. 616–624. https://doi.org/10.1016/j.matdes.2014.01.049

[109] Arczewska, P., Polak, M.A., Penlidis, A., Degradation of glass fiber reinforced polymer (GFRP) bars in concrete environment. Construction and Building Materials, July 2021, 293, 123451. https://doi.org/10.1016/j.conbuildmat.2021.123451

[110] Wu, W., He, X., Yang, W., Dai, L., Wang, Y. He, J., Long-time durability of GFRP bars in the alkaline concrete environment for eight years. Construction and Building Materials, January 2022, 314, 125573. https://doi.org/10.1016/j.conbuildmat.2021.125573

[111] EN1990: Eurocode: Basis of structural design. European Committee for Standardization: Brussels, Belgium, 2002.

[112] Davalos, J.F., Chen, Y., Ray, I., Long-term durability prediction models for GFRP bars in concrete environment. Journal of Composite Materials, August 2012, 46, p. 1899–1914. https://doi.org/10.1177/0021998311427777

[113] Benmokrane, B., Wang, P., Ton-That, T.M., Rahman, H., Durability of glass fiber-reinforced polymer reinforcing bars in concrete environment. Journal of Composites for Construction, 2002, 6, p. 143–153. https://doi.org/10.1061/(asce)1090-0268(2002)6:3(143)

[114] Masmoudi, R., Nkurunziza, G., Benmokrane, B., Cousin, P., Durability of glass FRP composite bars for concrete structure reinforcement under tensile sustained load in wet and alkaline environments. Proceedings, Annual Conference – Canadian Society for Civil Engineering, January 2003, p. 916–924.

[115] Nkurunziza, G., Benmokrane, B., Debaiky, A.S., Masmoudi, R., Effect of sustained load and environment on long-term tensile properties of glass fiber-reinforced polymer reinforcing bars. ACI Structural Journal, 2005,1 02, p. 615–621. https://doi.org/10.14359/14566

[116] Debaiky, A.S., Nkurunziza, G., Benmokrane, B., Cousin, P., Residual tensile properties of GFRP reinforcing bars after loading in severe environments. Journal of Composites for Construction, 2006, 10, p. 370–380. https://doi.org/10.1061/(asce)1090-0268(2006)10:5(370)

[117] Robert, M., Cousin, P., Benmokrane, B., Study of the performance of FRP reinforcing bars subjected to extreme conditions of application. Annales du bâtiment et des travaux publics, 1, 2011, p. 9–27.

[118] D'Antino, T., Pisani, M.A., Influence of sustained stress on the durability of glass FRP reinforcing bars. Construction and Building Materials, 2018, 187, p. 474–486. https://doi.org/10.1016/j.conbuildmat.2018.07.175

[119] D'Antino, T., Pisani, M.A., Long-term behavior of GFRP reinforcing bars. Composite Structures, 2019, 227, 111283. https://doi.org/10.1016/j.compstruct.2019.111283

[120] ASTM International: Standard Test Method for Bond Strength of Fiber-Reinforced Polymer Matrix Composite Bars to Concrete by Pullout Testing. ASTM D7913/D7913M-14, ASTM International, 2020.

[121] European Committee for Standardization: Fibre-reinforced polymer (FRP) reinforcement of concrete – Test methods – Part 1: FRP bars and grids. ISO 10406-1:2015, CEN, Brussels, Belgium, 2015.

[122] Achillides, Z., Pilakoutas, K., Bond Behavior of Fiber Reinforced Polymer Bars under Direct Pullout Conditions. Journal of Composites for Construction, April 2004, 8, p. 173–181. https://doi.org/10.1061/(asce)1090-0268(2004)8:2(173)

[123] Baena, M., Torres, L., Turon, A., Barris, C., Experimental study of bond behaviour between concrete and FRP bars using a pull-out test. Composites Part B: December 2009, Engineering, 40, p. 784–797.

[124] Solyom, S., Balázs, G.L., Bond of FRP bars with different surface characteristics. Construction and Building Materials, December 2020, 264, 119839. https://doi.org/10.1016/j.conbuildmat.2020.119839

[125] Nepomuceno, E., Sena-Cruz, J., Correia, L., D'Antino, T., Review on the bond behavior and durability of FRP bars to concrete. Construction and Building Materials, June 2021, 287, 123042. https://doi.org/10.1016/j.conbuildmat.2021.123042

[126] Lee, L.S., Jain, R. The role of FRP composites in a sustainable world. Clean Technology and Environmental Policy. 2009, 11, p. 247–249. https://doi.org/10.1007/s10098-009-0253-0

[127] Song, Y.S., Youn, J.R., Gutowski, T.G., Life cycle energy analysis of fiber-reinforced composites. Composites Part A: Applied Science and Manufacturing. 2009, 40, p. 1257–1265. https://doi.org/10.1016/j.compositesa.2009.05.020

[128] Belarbi, A., Dawood, M., Sustainability of fiber-reinforced polymers (FRPs) as a construction material. In Sustainability of Construction Materials (Chapter 20), Woodhead Publishing Series in Civil and Structural Engineering, 2016, p. 521–538. https://doi.org/10.1016/b978-0-08-100370-1.00020-2

[129] Pickering, S.J. (2006) Recycling technologies for thermoset composite materials—current status. Composites Part A: Applied Science and Manufacturing, 2006, 37, p. 1206–1215. https://doi.org/10.1016/j.compositesa.2005.05.030

[130] Conroy, A., Halliwell, S., Reynolds, T. Composite recycling in the construction industry. Composites Part A: Applied Science and Manufacturing. 2006, 37. p. 1216–1222. https://doi.org/10.1016/j.compositesa.2005.05.031

[131] EuCIA, Composites Recycling Made Easy. European Composites Industry Association. www.eucia.eu, 15/7/2023.

[132] Goulouti, K., De Castro, J., Vassilopoulos, A. P., Keller, T., Thermal performance evaluation of fiber-reinforced polymer thermal breaks for balcony connections. Energy and Buildings. 2014, 70, p. 365–371. https://doi.org/10.1016/j.enbuild.2013.11.070

[133] ISO 14040, 2006. Environmental Management d Life Cycle Assessment Principles and Framework. Swiss, Geneva. International Organization for Standardization. https://www.iso.org/standard/37456.html

[134] EN 15804, Sustainability of construction works— Environmental product declarations—core rules for the product category of construction products, CEN, 2013.

[135] Reichenbach S., Stoiber N., Kromoser B., Cradle-to-gate LCA for FRP reinforcement for concrete structures, fib Symposium 2023 – Building for the future: Durable, Sustainable, Resilient, Istanbul, Turkey, 5–7 June, 2023.

[136] Stoiber, N., Hammerl, M., Kromoser, B., Cradle-to-gate life cycle assessment of CFRP reinforcement for concrete structures: Calculation basis and exemplary application, Journal of Cleaner Production, January 2021, 280, Part 1, p. 124300. https://doi.org/10.1016/j.jclepro.2020.124300

[137] Pavlovic, A., Donchev, T., Petkova, D., Staletovic, N., Sustainability of alternative reinforcement for concrete structures: Life cycle assessment of basalt FRP bars, Construction and Building Materials, June 2022, 334, p. 127424. https://doi.org/10.1016/j.conbuildmat.2022.127424

[138] Khorgade, P., Rettinger, M., Burghartz, A., Schlaich, M., A comparative cradle-to-gate life cycle assessment of carbon fiber-reinforced polymer and steel-reinforced bridges. Structural Concrete, December 2022, 24(2), p. 1737–50. https://doi.org/10.1002/suco.202200229

[139] Cadenazzi, T., Dotelli, G., Rossini, M., Nolan, S., Nanni, A., Cost and environmental analyses of reinforcement alternatives for a concrete bridge, Structure and Infrastructure Engineering, 2020, 16(4), p. 787–802. https://doi.org/10.1080/15732479.2019.1662066

[140] Cadenazzi T., Keles B., Rahman M.K., Nanni A., Life-Cycle Cost and Life-Cycle Assessment of a Monumental Fiber-Reinforced Polymer Reinforced Concrete Structure, Journal of Construction Engineering and Management, July 2022, 148(9), 05022007. https://doi.org/10.1061/(asce)co.1943-7862.0002339

[141] Gurunathan, T., Mohanty, S., Nayak, S.K., A review of the recent developments in biocomposites based on natural fibres and their application perspectives. Composites Part A: Applied Science and Manufacturing, 2015, 77, p. 1–25. https://doi.org/10.1016/j.compositesa.2015.06.007

[142] Li, M., Pu, Y., Thomas, V.M., Yoo, C.C., Ozcan, S., Deng, Y., Nelson, K., Ragauskas, A.J., Recent advancements of plant-based natural fiber–reinforced composites and their applications, Composites Part B: Engineering, 2020, 200, 108254. https://doi.org/10.1016/j.compositesb.2020.108254

[143] Hofmann, M.A., Sustainable bio-based resins and fibre-polymer composites for civil engineering structural applications. PhD Thesis in Civil Engineering. University of Lisbon, 2022.

[144] Hofmann, M., Machado, M., Shahid, A., Dourado, F., Garrido, M., Bordado, J.C., Correia, J.R., Pultruded carbon fibre reinforced polymer strips produced with a novel bio-based thermoset polyester for structural strengthening. Composites Science and Technology, 2023, 234, 109936. https://doi.org/10.1016/j.compscitech.2023.109936

[145] D'Antino, T., Pisani, M.A., Tensile and compressive behavior of thermoset and thermoplastic GFRP bars, Construction and Building Materials, 2023, 366, 130104. https://doi.org/10.1016/j.conbuildmat.2022.130104

9 783857 482014